T0348374

PROGRESS IN

Molecular Biology and Translational Science

Volume 92

PROGRESS IN

Molecular Biology and Translational Science

Development of T Cell Immunity

edited by

Adrian Liston

VIB and University of Leuven
Leuven, Belgium

Volume 92

ELSEVIER

AMSTERDAM • BOSTON • HEIDELBERG • LONDON
NEW YORK • OXFORD • PARIS • SAN DIEGO
SAN FRANCISCO • SINGAPORE • SYDNEY • TOKYO
Academic Press is an imprint of Elsevier

Academic Press is an imprint of Elsevier
32 Jamestown Road, London, NW1 7BY, UK
Radarweg 29, PO Box 211, 1000 AE Amsterdam, The Netherlands
30 Corporate Drive, Suite 400, Burlington, MA 01803, USA
525 B Street, Suite 1900, San Diego, CA 92101-4495, USA

This book is printed on acid-free paper. ♾

Library of Congress Cataloging-in-Publication Data
A catalog record for this book is available from the Library of Congress

British Library Cataloguing in Publication Data
A catalogue record for this book is available from the British Library

ISBN: 978-0-12-381284-1
ISSN: 1877-1173

For information on all Academic Press publications
visit our website at elsevierdirect.com

Printed and bound by CPI Group (UK) Ltd, Croydon, CR0 4YY
Transferred to Digital Printing, 2013

Contents

Contributors

Numbers in parentheses indicate the pages on which the authors' contributions begin.

Graham Anderson, MRC Centre for Immune Regulation, Institute for Biomedical Research, Birmingham Medical School, Birmingham, United Kingdom (159)

Mark Coles, Centre for Immunology and Infection, Department of Biology and HYMS, University of York, York, United Kingdom (175, 177)

Brian G. Condie, Department of Genetics, University of Georgia, Athens, Georgia, USA (103)

Mark M. Davis, The Department of Microbiology and Immunology; and Howard Hughes Medical Institute, Stanford University School of Medicine, Stanford, California, USA (65)

Peter J.R. Ebert, The Department of Microbiology and Immunology, Stanford University School of Medicine, Stanford, California, USA (65)

Fabrina Gaspal, MRC Centre for Immune Regulation, Institute for Biomedical Research, Birmingham Medical School, Birmingham, United Kingdom (159)

Thu Hoang, Mathématiques Appliquées Paris-5, Université René Descartes, Paris, France (121)

Trang Hoang, Institute of Research in Immunology and Cancer, University of Montreal, Montréal; and Pharmacology, Biochemistry and Molecular Biology, Faculty of Medicine, University of Montréal, Québec, Canada (121)

Johannes B. Huppa, Howard Hughes Medical Institute, Stanford University School of Medicine, Stanford, California, USA (65)

Masanori Kasahara, Department of Pathology, Hokkaido University Graduate School of Medicine, Sapporo, Japan (7, 38, 61)

Jeong M. Kim, Genentech, 1 DNA Way, South San Francisco, California, USA (279)

Dimitris Kioussis, Division of Molecular Immunology, MRC National Institute for Medical Research, The Ridgeway, London, United Kingdom (175, 177)

Peter J.L. Lane, MRC Centre for Immune Regulation, Institute for Biomedical Research, Birmingham Medical School, Birmingham, United Kingdom (159)

Qi-Jing Li, The Department of Immunology, Duke University Medical Center, Durham, North Carolina, USA (65)

Michelle A. Linterman, Cambridge Institute for Medical Research and the Department of Medicine, Addenbrooke's Hospital, Cambridge, England, United Kingdom (207)

Adrian Liston, VIB and University of Leuven, Leuven, Belgium (1, 315)

Nancy R. Manley, Department of Genetics, University of Georgia, Athens, Georgia, USA (103)

Fiona M. McConnell, MRC Centre for Immune Regulation, Institute for Biomedical Research, Birmingham Medical School, Birmingham, United Kingdom (159)

Claude Perreault, Institute for Research in Immunology and Cancer, Université de Montréal, Montréal, Québec, Canada (37, 41, 62)

Paola Romagnoli, Tolerance and Autoimmunity Section, Centre de Physiopathologie de Toulouse Purpan, Institut National de la santé et de la Recherche Medicale (Inserm) U563, and University Paul Sabatier; and IFR150, Institut Fédératif de Recherche Bio-Médicale de Toulouse, Toulouse, France (251)

Manoj Saini, MRC Centre for Immune Regulation, Institute for Biomedical Research, Birmingham Medical School, Birmingham, United Kingdom (159)

Cédric S. Tremblay, Institute of Research in Immunology and Cancer, University of Montreal, Montréal, Québec, Canada (121)

Joost P.M. van Meerwijk, Tolerance and Autoimmunity Section, Centre de Physiopathologie de Toulouse Purpan, Institut National de la santé et de la Recherche Medicale (Inserm) U563, and University Paul Sabatier, Toulouse, France, and IFR150, Institut Fédératif de Recherche Bio-Médicale de Toulouse; and Institut Universitaire de France and Faculty of Life-Sciences (UFR-SVT), University Paul Sabatier, Toulouse, France (251)

Henrique Veiga-Fernandes, Immunobiology Unit, Instituto de Medicina Molecular, Faculdade de Medicina de Lisboa. Av. Prof. Egas Moniz. Edifício Egas Moniz, Lisboa, Portugal (175, 177)

Carola G. Vinuesa, Immunology Program, John Curtin School of Medical Research, Australian National University, Canberra, Australia (207)

David Withers, MRC Centre for Immune Regulation, Institute for Biomedical Research, Birmingham Medical School, Birmingham, United Kingdom (159)

The Development of T-Cell Immunity

ADRIAN LISTON

VIB and University of Leuven, Leuven, Belgium

The development of T-cell immunity covers a broad range of possible topics. In this volume, we have attempted to look at four of these topics in detail: the evolution of T-cell immunity, thymic requirements for T-cell immunity, T-cell immunity in the periphery, and prevention of T-cell-dependent autoimmunity. In each section, we have two to three chapters reviewing the latest developments in the field from different perspectives.

The evolution of T-cell immunity was once an area restricted only to speculation. Now with the rise of large-scale "omics" research, the various hypotheses for the origin of T-cell evolution are being rigorously tested. In Chapter 2, Kasahara outlines the genomic innovations that were required for the development of adaptive immunity, in particular T-cell immunity. T-cell immunity is evolutionarily ancient, tracing back to the common ancestor of jawed vertebrates, but adaptive immunity is still more ancient, with an analogous system of lymphocyte-like cells using evolutionarily unrelated and structurally different antigen–receptor systems. The appearance of two analogous systems within a relatively short time period suggests the requirement for a necessary precondition in the common ancestor, which Kasahara suggests may be the freeing up of genetic capacity via several rounds of whole genome duplication. Perhaps worthy of speculation is the idea that the critical importance of the innate immune system prevented excessive experimentation in immunity until redundant copies of the genome became available to allow the evolution of a second immune system layered over the first. In Chapter 3 by Perreault, advances in proteomics data are used to discuss the origin and function of self-peptide presentation on major histocompatibility complex (MHC) class I. Despite the initial assumption of random sampling, recent data indicates that peptide selection is nonrandom, with disproportionate representation from certain classes of proteins. It would be fascinating to know how much of this bias is dictated by biochemical necessity (e.g., easier to process during translation, hence orientation toward rapidly translated proteins) versus an evolutionarily selected bias to increase the efficiency of antiviral defense (i.e., rapidly translated proteins being targeted because they are enriched for viral antigens).

Progress in Molecular Biology
and Translational Science, Vol. 92
DOI: 10.1016/S1877-1173(10)92001-2

1877-1173/10 $35.00

In the third article of this section Davis and colleagues analyse the evolved complexity of the biochemical recognition between the T cell receptor (TCR) and its cognate antigen. Unlike the B cell receptor, the random rearrangement of TCR genes subsequently requires selection for affinity to the necessary ligand. This creates a tension between random generation of affinity, necessary interaction with self-ligand and yet highly specific and sensitive activation from foreign ligand. While largely unresolved, detailed biochemical analyses of well characterized examples of TCRs suggest potential evolutionary solutions to this conundrum.

T cells are unique in that they need a specialized organ, the thymus, for differentiation. Considered to be only a "lymphocyte graveyard" until the pivotal experiments by Jacques Miller in 1961, the microenvironmental conditions required for T-cell differentiation in the thymus have turned out to be remarkably complex. The two chapters of this section dissect the molecular control over the two sides of T-cell development—the thymocytes themselves and the essential thymic stromal support cells. In Chapter 5, Manley and Condie outline the transcription factor control over early thymic organogenesis, from initial fate specification to end-point differentiation. The authors make the telling point that at this stage a full catalog of the functional subsets of thymic stromal cells is unavailable, leaving a rich field of transcription factor control as yet unexplored. In Chapter 6, Tremblay, Hoang, and Hoang take the novel approach of using the molecular genetics of T-cell acute lymphoblastic leukemia to dissect the thymocyte signaling requirements for survival and differentiation. Together, these chapters show the complexity of the interplay between thymocyte and stroma.

While T-cell immunity is evolutionarily ancient, developing in the common ancestor of all jawed vertebrates, the full collaboration between T cells and B cells is a relatively modern innovation. It is only in eutherian mammals, 125 million years ago, that the sophisticated system of lymph node segregation of function evolved to allow effective T cell help and strong generation of B cell memory. Two chapters in this volume look at the development of the lymph nodes and other secondary lymphoid tissue, which are so important for effective immunity. In Chapter 8, Coles, Kioussis, and Veiga-Fernandes take us through a historical overview of research on secondary lymphoid tissue development, culminating in the conclusions from modern research techniques that have revealed the role of the lymphoid tissue inducer (LTi) cell in creating a structure to bring together CD T cells and B cells in a context to create high-affinity memory responses. In Chapter 7 on this topic, Lane and colleagues look at the role of LTi cells not only in the development of lymph nodes but also in thymic tolerance. They make a convincing argument that the evolution of a powerful CD4 helper T-cell response, complete with CD4 T-cell and B-cell memory and the stimulation of high-affinity antibody production by B cells, necessitated the coincident evolution of a more stringent mechanism of thymic negative selection. Linterman and Vinuesa follow this theme from the perspective of the key T-cell driver of antibody responses, the follicular T cell (TFH). In Chapter 9, the authors outline the differentiation of the TFH and the role it plays in enabling the germinal center reaction and affinity maturation in B cells. Like Lane and colleagues, Linterman and Vinuesa emphasize the importance of tolerance processes, with an increased risk of autoimmune pathology being the reciprocal cost for the capacity to generate high-affinity antibodies.

The evolution of a high-capacity effector response necessitates the evolution of an efficient suppressive mechanism to prevent catastrophic autoimmunity in the circumstance of the effector response being directed against self-targets. The recessive tolerance mechanisms of negative selection and anergy induction do not provide a fail-safe mechanism against those autoreactive T cells that evade tolerance induction; however, the evolution of dominant tolerance mechanisms ensures efficient suppression of undesirable reactions. In two chapters, one by Romagnoli and van Meerwijk and the other by Kim, the current status of research on the best understood form of dominant tolerance, that of Foxp3^{+} regulatory T cells, is outlined, from differentiation to peripheral function.

In addition to the review chapters outlined above, in this volume we have attempted to display the vibrancy of research in the field of the development of T-cell immunity through commentaries on the data and interpretations in the main reviews. With key breakthroughs occurring in each of the topics presented in this volume, several topics are still highly contentious, with the final models yet to be established.

Section I

The Evolution of T Cell Immunity

Genome Duplication and T Cell Immunity

Masanori Kasahara

Department of Pathology, Hokkaido University Graduate School of Medicine, Sapporo, Japan

The adaptive immune system (AIS) mediated by T cells and B cells arose ~ 450 million years ago in a common ancestor of jawed vertebrates. This system was so successful that, once established, it has been maintained in all classes of jawed vertebrates with only minor modifications. One event thought to have contributed to the emergence of this form of AIS is two rounds of whole-genome duplication. This event enabled jawed vertebrate ancestors to acquire many paralogous genes, known as ohnologs, with essential roles in T cell and B cell immunity. Ohnologs encode the key components of the antigen presentation machinery and signal transduction pathway for lymphocyte activation as well as numerous transcription factors important for lymphocyte development. Recently, it has been discovered that jawless vertebrates have developed an AIS employing antigen receptors unrelated to T/B cell receptors, but with marked overall similarities to the AIS of jawed vertebrates. Emerging evidence suggests that a common ancestor of all vertebrates was equipped with T-lymphoid and B-lymphoid lineages.

Progress in Molecular Biology
and Translational Science, Vol. 92
DOI: 10.1016/S1877-1173(10)92002-4

1877-1173/10 $35.00

I. Introduction

When and how T cell immunity emerged is an important issue in understanding the origin and evolution of the adaptive immune system (AIS). Thanks to the decades-long efforts of immunologists and the advances of genome projects, we now know that the key components of T cell immunity, such as T cell receptors (TCRs) and major histocompatibility complex (MHC) molecules, are present in all classes of jawed vertebrates (gnathostomes) ranging from mammals to the cartilaginous fish, but absent in jawless vertebrates (agnathans) and invertebrates[1–7] (Fig. 1).

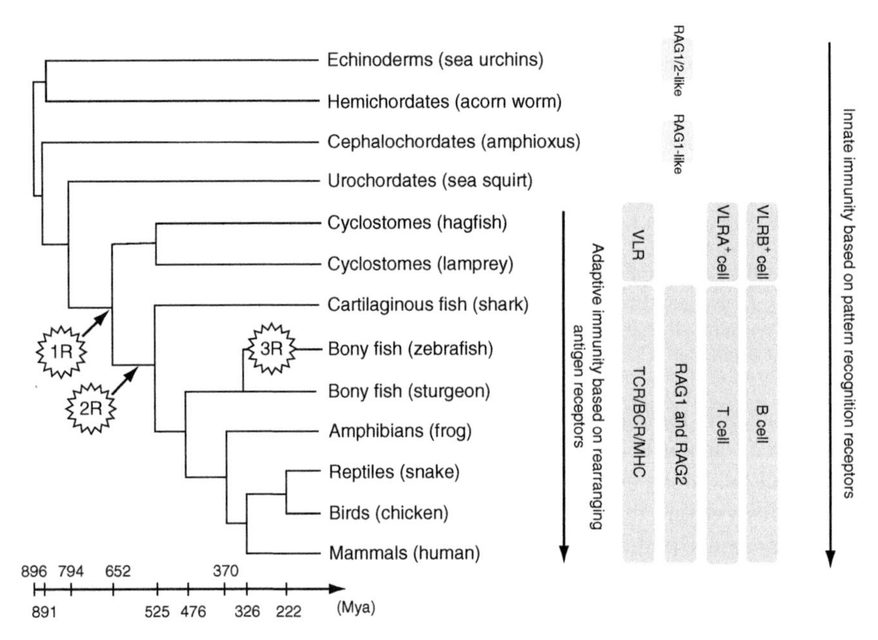

FIG. 1. Evolution of the AIS in deuterostomes. The figure shows schematically at which stage in phylogeny major immune molecules and cells emerged. RAG1-like genes are derived from a transposon; recently, they have been identified also in the genomes of sea urchins[8] and amphioxus.[9] "1R" and "2R" indicate the first and second rounds of WGD. The timing of WGD relative to the emergence of jawless vertebrates is controversial. For detailed discussions, see Section IV and Fig. 4. "3R" stands for a fish-specific WGD. Cephalochordates and urochordates are invertebrate chordates. Cyclostomes, represented by hagfish and lamprey, are jawless vertebrates. Cartilaginous fish, bony fish, amphibians, reptiles, birds, and mammals are jawed vertebrates. The divergence time of animals shown in Mya (million years ago) is based on Blair and Hedges.[12] Abbreviations: BCR, B cell receptor; MHC; major histocompatibility complex; RAG, recombination-activating gene; TCR, T cell receptor; VLR, variable lymphocyte receptor.

Jawless vertebrates represented by hagfish and lamprey are equipped with rearranging antigen receptors that are clonally expressed on lymphocyte-like cells.[13,14] However, their receptors, known as variable lymphocyte receptors (VLRs), generate diversity through somatic recombination of leucine-rich repeat (LRR) modules, and are hence structurally unrelated to TCRs or B cell receptors (BCRs).[15–19] In invertebrate chordates, such as urochordates (represented by sea squirts *Ciona intestinalis*) and cephalochordates (represented by amphioxus *Branchiostoma floridae*), draft genome sequence analysis has provided no evidence for the presence of the AIS.[9,20] Thus, accumulated evidence indicates that T cells as defined by the expression of TCRs are unique to jawed vertebrates and that authentic T cell immunity arose in a common ancestor of jawed vertebrates.

Less well understood is how T cell immunity, and more generally the AIS, emerged in evolution. In terms of molecular components, the cartilaginous fish have fully developed AISs essentially identical to those of mammals; they have not only TCRs of α/β and γ/δ types and BCRs,[2,21,22] but also MHC class I and class II molecules,[23–26] recombination-activating gene (RAG) recombinases,[27] and the components of the classical pathway of complement activation.[28] By sharp contrast, jawless vertebrates have none of these components, giving the impression that the TCR/BCR/MHC-based AIS emerged abruptly in a jawed vertebrate lineage.[3,29,30]

One event widely believed to have contributed to the emergence of the jawed vertebrate-type AIS is the acquisition of RAG recombinases that cut double-stranded DNA at the recombination signal sequence (RSS) and mediate V(D)J recombination in TCR/BCR loci.[31,32] Not only does the site-specific recombination process mediated by RAG share mechanistic similarities with the integration and excision process of transposable elements,[33] but also, RAG proteins can transpose an RSS-containing cleavage product to an unrelated target DNA *in vitro*.[34,35] Furthermore, the DNA-binding region of RAG1 shows sequence similarity to that of a *Transib* superfamily of DNA transposons.[36] Collectively, these observations have provided strong evidence that RAGs originated from transposons. The horizontal transfer of RAG transposons may have taken place multiple times or only once during deuterostome evolution.[8] However, the insertion of RAG transposons in an appropriate context ("appropriate" in the sense that the insertion disrupted an ancestral antigen receptor gene and eventually conferred upon it the ability to rearrange) seems to have taken place only in a common ancestor of jawed vertebrates. Exploitation of RAG transposons as V(D)J recombinases was most likely accidental, thus explaining why combinatorial antigen receptors such as TCRs and BCRs emerged abruptly in jawed vertebrates.

Another event assumed to have played a pivotal role in the emergence of the jawed vertebrate-type AIS is two rounds of whole-genome duplication (2R-WGD) that occurred early in vertebrate evolution.[3,37] The importance of this event in the evolution of T cell immunity was initially suggested by the observation that many of the genes encoded in the MHC, including those involved in antigen presentation, arose as a result of large-scale chromosomal duplication that presumably took place as part of WGD.[38,39] With the accumulation of genomic data from key vertebrate and invertebrate species, it is becoming increasingly clear that WGD was an important event that enabled the ancestor of jawed vertebrates to evolve highly sophisticated AISs.[7] Here I review the role of WGD in the emergence of the AIS, with particular emphasis on the evolution of T cell immunity. I then review the latest advances in our understanding of the immune system of jawless vertebrates. Surprisingly, the overall design of the agnathan AIS is similar to that of the gnathostome AIS, despite the fact that jawed and jawless vertebrates use completely different antigen receptors.

II. WGD: From a Hypothesis to the Fact

Exactly 40 years ago, Susumu Ohno proposed that the vertebrate genome underwent one or two rounds of WGD at the stage of fish or amphibians through a tetraploidization process.[40] This proposal was based mainly on the comparison of DNA content and karyotypes in various organisms, and the observation that tetraploid species occur naturally in fish and amphibians. Ohno argued that WGDs, which duplicate all genes in the genome simultaneously, were more effective than cumulative tandem duplications in bringing about major evolutionary changes because they would free an entire set of genes from purifying selection and allow it to coevolve, thus providing a unique opportunity to form novel genetic networks required for biologic innovations.

Although his proposal was quite influential from its inception, it was viewed with skepticism until the mid-1990s because of the paucity of experimental evidence. However, with the progress of genome projects, observations supporting Ohno's hypothesis, which became known as the 2R (two-round) hypothesis after some refinement,[41] accumulated exponentially. The major supporting evidence is twofold.[42–45] First, a gene, which occurs only in a single copy in invertebrate chordates such as urochordates and cephalochordates, often has multiple, typically up to four, copies (paralogous copies or paralogs) in jawed vertebrates, indicating that there were waves of gene duplication during the transition from invertebrates to jawed vertebrates. Second, such paralogs, often called ohnologs in honor of Ohno,[46] are not distributed randomly in the vertebrate genome, but tend to occur in clusters (called paralogons) on multiple, and typically four, separate chromosomes.[47] For example, the human genome contains four *HOX* clusters[48] (Fig. 2). Here, not only is the *HOX* gene cluster quadruplicated on four separate chromosomes, but also, many of

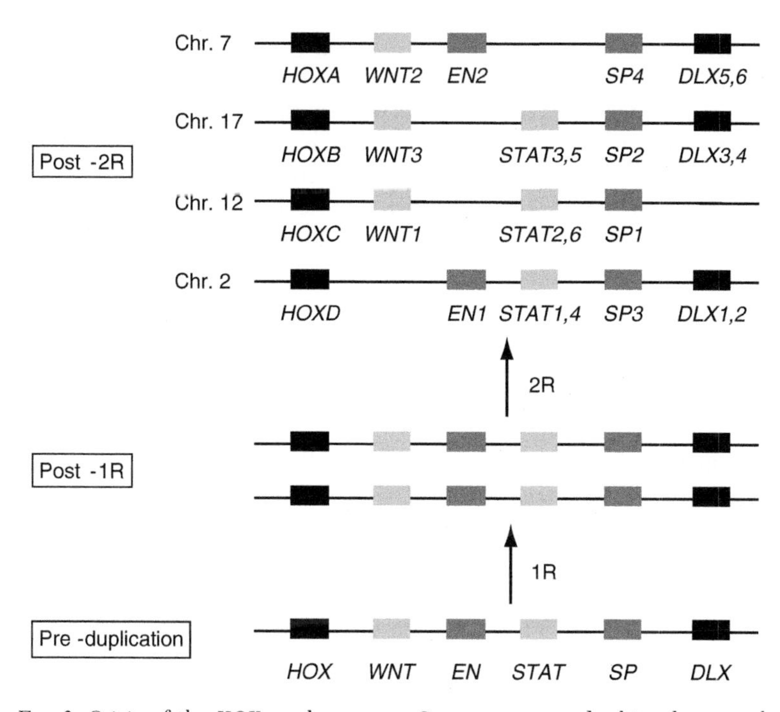

FIG. 2. Origin of the *HOX* paralogy group. Genes are arranged arbitrarily to emphasize corresponding paralogs. The upper panel shows four sets of paralogons constituting the human *HOX* paralogy group. Invertebrate chordates such as amphioxus have only a single *HOX* gene cluster.[49] "1R" and "2R" indicate the first and second rounds of WGD, respectively. Abbreviations: Chr, chromosome; DLX, distal-less homeobox; EN, engrailed homeobox: HOX, homeobox; SP, specificity protein transcription factors; STAT, signal transducer and activator of transcription; WNT, wingless-type MMTV integration site family member.

the genes adjacent to the *HOX* gene cluster are quadruplicated, triplicated, or duplicated on the same four sets of chromosomes, indicating that this unique arrangement of paralogs, known as genome paralogy, arose not as a result of individual gene duplications, but as a consequence of two rounds of large-scale block duplication.

A close inspection of the vertebrate genome indicates that genome paralogy is by no means an exceptional observation. Dehal and Boore[50] systematically identified ohnologs by comparing human and sea squirt genomes and examined their locations in the human genome; their analysis showed that ~25% of the human genome is covered by four sets of paralogons, indicating that genome paralogy is an essential feature of human genome architecture. More recently, comparison of the human and amphioxus genomes revealed widespread occurrence of quadruple conserved synteny, where four sets of human paralogons corresponded to one set of linked genes in amphioxus.[51] These observations provided incontrovertible evidence for the 2R hypothesis. It is now widely accepted that 2R-WGD took place in the vertebrate lineage after its separation from invertebrate chordates, but before the radiation of jawed vertebrates[41,45,52,53] (Fig. 1). Apart from the 2R-WGD discussed earlier, an ancestor of the majority of ray-finned fish experienced a lineage-specific WGD $\sim$320 million years ago.[11] This duplication is often called 3R (the third round of WGD).

III. Roles of Ohnologs in Adaptive Immunity

The function of the jawed vertebrate-type AIS depends on the participation of a large number of genes. Klein and Nikolaidis[4] have classified the genes deployed by the AIS into three categories. The first category includes genes that evolved long before the emergence of the AIS. Because these genes evolved for other biologic systems and were subsequently recruited to the AIS, they usually have functions not restricted to adaptive immune responses. For example, the proteasome, a proteolytic enzyme complex,[54] evolved as protein degradation machinery essential for cell survival and was later recruited as a supplier of peptides to MHC class I molecules.[55] Thus, most of the proteasome subunits are well conserved throughout eukaryotes, and their functions are not specialized for the AIS.[54]

The second category includes paralogs that emerged by duplication from preexisting genes and acquired functions involved in or specialized for adaptive immune responses. Many of these genes appear to be ohnologs generated by 2R-WGD.[56] For example, jawed vertebrates have a specialized type of proteasomes, called immunoproteasomes, that facilitates the production of peptides that serve as MHC class I ligands.[57,58] Instead of β1, β2, and β5 subunits found in regular proteasomes, immunoproteasomes contain three interferon (IFN)-γ-inducible subunits called β1i, β2i, and β5i.[55] These subunits alter the cleavage specificities of the proteasome so that peptides suitable for binding to MHC class I molecules are produced more efficiently. The genes coding for β1i, β2i, and β5i are related to those coding for β1, β2, and β5 subunits, respectively, and the former set of genes are ohnologs that arose by WGD from the latter set of evolutionarily more ancient genes with housekeeping functions.[55]

The third category includes a relatively small number of genes, such as those coding for MHC class I and class II molecules, TCRs, and BCRs, with functions dedicated to immune responses. These genes appear to have emerged by mechanisms other than simple duplication of preexisting genes; in the case of MHC class I and class II molecules, peptide-binding domains of unknown origin appear to have been grafted to the immunoglobulin (Ig)-like constant domains.[59,60] In the case of antigen receptors, an invasion by RAG transposons was instrumental in their emergence.[31]

Here, representative examples of ohnologs are discussed to highlight the importance of WGD in the emergence of the jawed vertebrate-type AIS.

A. Molecules of the MHC System

The MHC system is a cornerstone of T cell immunity because conventional α/β TCRs recognize antigen only in the form of peptides bound to MHC class I or class II molecules. Accumulated evidence indicates that many molecules involved in antigen presentation by class I and class II molecules are encoded by ohnologs[7,61] (Table I). Peptides presented by class I molecules are produced by proteasomes and transported to the endoplasmic reticulum by transporters associated with antigen processing (TAP), where they bind to nascent class I molecules with the help of tapasin.[62] Immunoproteasome subunits, β1i, β2i, and β5i, are encoded by ohnologs as discussed earlier, and so are the TAP and tapasin molecules.[7] Recently, a novel form of proteasomes, designated thymoproteasomes, has been identified in mice[63] and man.[64] Thymoproteasomes, expressed specifically in cortical thymic epithelial cells, are involved in positive selection of $CD8^+$ T cells.[65] β5t, a β-type subunit unique to thymoproteasomes, is also encoded by an ohnolog (Table I).

Peptides presented by MHC class II molecules are produced by endosomal/lysosomal proteases. Important among such proteases are cathepsins[66,67]; accumulated evidence indicates that cathepsins S, D, and L play particularly important roles in antigen presentation by class II molecules and that cathepsin L is involved in thymic selection of $CD4^+$ T cells and degradation of invariant chains.[68] As described below, the majority of cathepsin isoforms are encoded by ohnologs mapping to paralogons (Table I). Other examples of ohnologs directly related to the function of MHC molecules are *RXRB* (retinoid X receptor β) and *RFX5* (regulatory factor X, 5) genes, which regulate the expression of class I and class II molecules, respectively.[69,70]

The MHC is a prototypical region exhibiting genome paralogy.[61,71] Initially, the MHC paralogy group was defined as a set of paralogons located on human chromosomes 1, 6, 9, and 19.[37,39] Recent evidence indicates that the MHC paralogy group and the neurotrophin paralogy group[72] are partially overlapping and that they descended from a neighboring region on the same ancestral chromosome[41,73] (Fig. 3). It is remarkable that almost all of the ohnologs discussed earlier are located in the paralogons of the MHC/neurotrophin paralogy group.[7] This suggests that a preduplicated region that existed in the genome of our invertebrate chordate ancestor contained precursors of many genes coding for the components of the MHC system.[71,75]

TABLE I

REPRESENTATIVE HUMAN OHNOLOGS INVOLVED IN ANTIGEN PRESENTATION

Gene family	Genes	Location[a]	Gene products	Function	Other closely related ohnologs	Location[a]
Ohnologs involved in class I antigen presentation						
20S proteasome	*PSMB8*	6p21.3 (MHC)	β5i	Component of immunoproteasomes: production of MHC class I-binding peptides	*PSMB5*	14q11.2
β-subunits	*PSMB9*	6p21.3 (MHC)	β1i	Component of immunoproteasomes: production of MHC class I-binding peptides	*PSMB6*	17p13
	PSMB10	16q22.1	β2i	Component of immunoproteasomes: production of MHC class I-binding peptides	*PSMB7*	9q34.11–q34.12
	PSMB11	14q11.2	β5t	Component of thymoproteasomes: positive selection of $CD8^+$ T cells		
TAP	*TAP1*	6p21.3 (MHC)	TAP1	TAP1/TAP2 heterodimer transports peptides into the endoplasmic reticulum	*ABCB9 (TAPL)*	12q24
	TAP2	6p21.3 (MHC)	TAP2			
Tapasin	*TAPBP*	6p21.3 (MHC)	Tapasin	Promotes association of TAP and MHC class I molecules	*TAPBPL*	12p13.3
Retinoid X receptor	*RXRB*	6p21.3 (MHC)	RXRβ	Binds to the MHC class I promoter and regulates class I expression	*RXRA* *RXRG*	9q34.3 1q22–q23

Ohnologs involved in class II antigen presentation						
Cathepsins	*CTSL1*	9q21–q22	Cathepsin L1	$CD4^+$ T cell and NKT cell development	*CTSH*	15q24–q25
	CTSL2	9q22.2	Cathepsin L2	$CD4^+$ T cell and NKT cell development	*CTSK*	1q21
	CTSS	1q21	Cathepsin S	Removal of invariant chains in B cells and dendritic cells	*CTSG*	14q11.2
	CTSD	11p15.5	Cathepsin D	Production of MHC class II-binding peptides	*CTSC*	11q14.1–q14.3
					CTSF	11q13.1
					CTSW	11q13.1
Regulatory factor X	*RFX5*	1q21	RFX5	A component of RFX involved in MHC class II expression	*RFX1*	19p13.1
					RFX2	19p13.3–p13.2
					RFX3	9p24.2
					RFX4	12q24

[a]Chromosomal localization of human genes is based on the OMIM database (http://www.ncbi.nlm.nih.gov/omim) or Entrez gene (http://www.ncbi.nlm.nih.gov/sites/entrez?db=gene).

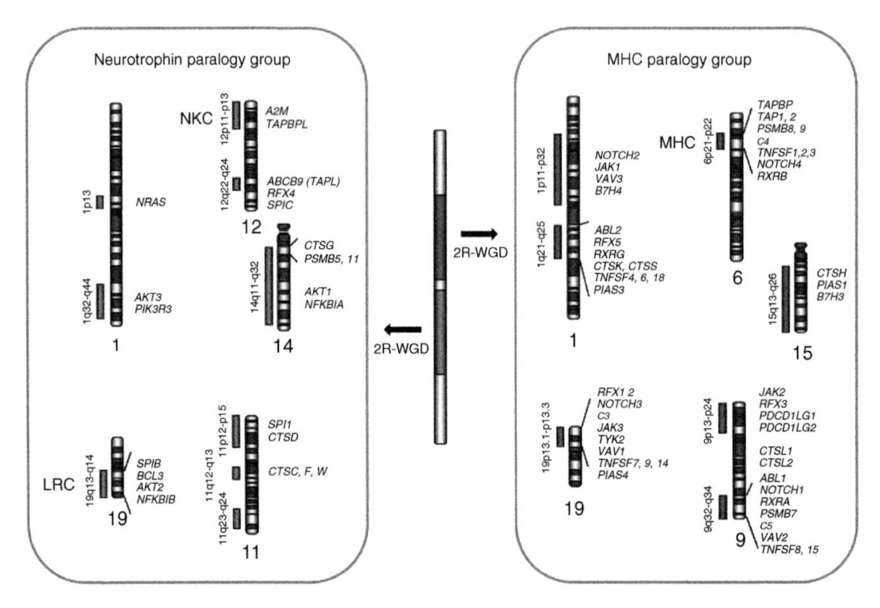

FIG. 3. The MHC/neurotrophin paralogy group in the human genome. The MHC paralogy group (right) is made up of four sets of paralogons located on human chromosomes 1, 6, 9, and 19. A number of smaller-sized MHC paralogons, which presumably originated from fragmentation and subsequent translocation of the major paralogons, have been identified. Among them, the paralogon located on 15q13-q26 appears to have been broken off from the paralogon on chromosome 6. Four sets of major neurotrophin paralogons (left) are located on chromosomes 1, 11, 12/14, and 19. The MHC- and neurotrophin-paralogy groups are partially overlapping and are thought to have descended from neighboring regions on a single ancestral chromosome. Hence, some paralogs are distributed across the two paralogy groups. In mammals, only two proteasome β-type subunit genes *PSMB8* and *PSMB9*, which encode β5i and β1i, respectively, are located in the MHC (Table II). However, β2i is also encoded in the MHC in the bony fish.[74] Abbreviations: A2M, α2-macroglobulin; ABL, Abelson murine leukemia viral oncogene homolog; AKT, V-AKT murine thymoma viral oncogene homolog; B7, B7 family; CTSC, CTSD, CTSF, CTSG, CTSH, CTSK, CTSL, CTSL2, CTSS, and CTSW, cathepsins C, D, F, G, H, K, L, L2, S, and W; JAK, Janus kinase; LRC, leukocyte receptor complex; MHC, major histocompatibility complex; NFKBIA, nuclear factor κ-B inhibitor; NKC, natural killer complex; NRAS, neuroblastoma RAS viral oncogene homolog; PD, programmed cell death 1 ligand; PIAS, protein inhibitors of activated STAT; PIK3R, phosphatidylinositol 3-kinase, regulatory subunit; PSMB, proteasome subunits, β-type; RFX, regulatory factor X; RXRA, RXRB, and RXRG, retinoid X receptors α, β, and γ; SPIC, SPIC transcription factor; SPI1, spleen focus forming virus proviral integration oncogene; TAP, transporter associated with antigen processing; TAPBP, TAP-binding protein (tapasin); TAPBPL, TAP-binding protein-like; ABCB9 (TAPL), transporter associated with antigen processing-like; TNFSF, tumor necrosis factor ligand superfamily; and 2R-WGD, two rounds of whole-genome duplication. This figure was modified from Flajnik and Kasahara.[7]

B. Signaling Molecules

Okada and Asai[76] systematically performed phylogenetic analysis of signaling molecules and came to the conclusion that 2R-WGD played a major role in the generation of over 100 ohnologs involved in lymphocyte signaling.

An example of ohnologs in this category is a family of genes coding for transcription factors known as signal transducers and activators of transcription (STAT) that mediate signal transduction in response to various cytokines[77] (Table II). STAT4 and STAT6 mediate transcriptional activation of target genes in response to IL-12 and IL-4, respectively. STAT4 deficiency causes a defect in T helper 1 (Th1) cell development,[78,79] whereas STAT6 deficiency impairs the development of T helper 2 (Th2) cells and IL4-dependent Ig class switching.[80,81] When cytokines are bound to cytokine receptors, Janus kinases (JAKs) are activated, and the activated JAKs phosphorylate STAT proteins, which then move to the nucleus and activate transcription of cytokine-responsive genes.[82] Four known members of the *JAK* family, *JAK1*, *JAK2*, *JAK3*, and *TYK2*, are ohnologs mapping to the MHC paralogy group (Table II, Fig. 3). The JAK/STAT pathway is negatively regulated by protein inhibitors of activated STAT (PIAS)[83] and suppressors of cytokine signaling (SOCS).[84] Four known members of the *PIAS* family are ohnologs, with three of them encoded in the MHC/neurotrophin paralogy group. It has also been suggested that four of the *SOCS* family members, *SOCS1*, *SOCS2*, *SOCS3*, and *CIS*, diverged by 2R-WGD.[85] These observations indicate that WGD was highly effective in creating a network of interacting molecules constituting the JAK/STAT pathway.

Other notable ohnologs involved in signal transduction include VAV family proteins[86] (Table II, Fig. 3), Abl tyrosine kinases[87] (Table II, Fig. 3) and TEC family kinases.[88]

C. Cytokines and Cytokine Receptors

Cytokine and cytokine receptor families are also known to have increased their family members by WGD (Table III). The best known example is a tumor necrosis factor (TNF) superfamily of cytokines with crucial roles in both adaptive and innate immunity. Most members of the TNF ligand superfamily, including CD40 ligand and 4-1BBL that function as co-stimulators for T cells,[89,90] are encoded by ohnologs mapping to the MHC paralogy group[61,91] (Fig. 3), indicating that 2R-WGD played a critical role in the diversification of TNF ligand genes.[92] It has been suggested that WGD was also involved in the diversification of TNF receptor superfamily genes.[93]

Most chemokine receptors and chemokines are encoded in the *HOX* paralogons.[94] Not only do chemokines recruit leukocytes including T and B cells to sites of infection, but they also regulate physiological migration of lymphocytes to and within various lymphoid tissues.[95] Detailed analysis of

TABLE II
REPRESENTATIVE HUMAN OHNOLOGS INVOLVED IN SIGNAL TRANSDUCTION

Gene family	Genes	Location[a]	Gene products	Function
Ohnologs encoded by the *HOX* paralogy group				
STAT transcription factors	*STAT1*	2q32.2	STAT1	Th1 cell development, cytokine signaling (IFN-α/β, IFN-γ)
	STAT2	12q13.2	STAT2	Cytokine signaling (IFN-α/β)
	STAT3	17q21.31	STAT3	Cell growth, suppression and induction of apoptosis, cytokine signaling (IL-6, IL-10)
	STAT4	2q32.2-q32.3	STAT4	Th1 cell development, cytokine signaling (IL-12)
	STAT5A	17q11.2	STAT5A	Cytokine signaling (IL-2, prolactin)
	STAT5B	17q11.2	STAT5B	Cytokine signaling (IL-2, IL-15, growth hormone)
	STAT6	12q13	STAT6	Th2 cell development, cytokine signaling (IL-4, IL-13)
Ohnologs encoded by the MHC/neurotrophin paralogy group				
Janus kinases	*JAK1*	1p31.3	JAK1	Response to IFNs, γc-dependent cytokines and gp130-dependent cytokines, involved in lymphopoiesis
	JAK2	9p24	JAK2	Response to erythropoietin, thrombopoietin, IL-3, GM-CSF and IFNγ
	JAK3	19p13.1	JAK3	γc-dependent lymphoid development
	TYK2	19p13.2	TYK2	Required for IL-12-induced T cell function
PIAS (Protein inhibitor of activated STAT)	*PIAS1*	15q22	PIAS1	Inhibitor of activated STAT1
	PIAS2[b]	18q21.1	PIAS2	Inhibitor of activated STAT2
	PIAS3	1q21	PIAS3	Inhibitor of activated STAT3
	PIAS4	19p13.3	PIAS4	Inhibitor of activated STAT4
VAV guanine nucleotide exchange factor	*VAV1*	19p13.3-p13.2	VAV1	T cell signaling
	VAV2	9q34.1	VAV2	BCR-induced proliferation, T cell-dependent antibody response
	VAV3	1p13.3	VAV3	B cell signaling

(*Continues*)

TABLE II (*Continued*)

Gene family	Genes	Location[a]	Gene products	Function
Abl (Abelson tyrosine kinases)	*ABL1*	9q34.1	ABL1 (ABL)	T cell signaling and T cell development
	ABL2	1q24-q25	ABL2 (ARG)	T cell signaling and T cell development

[a]Chromosomal localization of human genes is based on the OMIM database (http://www.ncbi.nlm.nih.gov/omim) or Entrez gene (http://www.ncbi.nlm.nih.gov/sites/entrez?db=gene)

[b]*PIAS2* seems to have been translocated secondarily; it is encoded outside the MHC/neurotrophin paralogy group.

chemokine and chemokine receptor genes indicated that they increased their copy number not only by tandem duplication but also by cluster duplication mediated by 2R-WGD.[96]

It has been shown recently that the transforming growth factor-β pathway also increased their complexity through 2R-WGD.[97]

D. Transcription Factors Involved in Lymphocyte Development

Lymphocyte development requires the participation of a number of transcription factors.[98] Many transcription factors critically involved in T cell development are encoded by ohnologs. For example, GATA3 is a transcription factor indispensable for Th2 cell development.[99] Phylogenetic analysis indicates that three *GATA* genes, *GATA1*, *GATA2*, and *GATA3*, map to *GATA* paralogons and diverged by 2R-WGD from a common ancestral *GATA1/2/3* gene.[100] Similarly, detailed analysis of the Ikaros-related family of zinc finger transcription factors showed that four of its members, *Ikaros*, *Aiolos*, *Helios*, and *Eos*, most likely diverged by 2R-WGD.[101–103] Mice lacking *Ikaros* ohnolog display multiple defects in T cell development.[104] Helios is also essential for T cell differentiation and homeostasis.[105]

Transcription factors of the IFN regulatory factor (IRF) family play essential roles in Th1-type immune responses, IFN-induced antiviral and antibacterial responses, and the development of natural killer (NK) cells.[106] Recent work has shown that they also increased their members by 2R-WGD.[107]

E. Co-stimulatory Molecules

Effective activation of naïve T cells requires a second signal known as the co-stimulatory signal. The best characterized co-stimulators for T cells are the B7 family of molecules expressed on antigen presenting cells.[108] Many

TABLE III

THE HUMAN TNF AND THE CHEMOKINE/CHEMOKINE RECEPTOR FAMILY

Gene family	Ohnologs located in the paralogons[a]	Location[b]	Gene products	Function	Other ohnologs	Location[b]
TNF superfamily ligands	*LTA (TNFSF1)*	6p21.3 (MHC)	LT α	Lymphoid organ, γ/δ T and NKT cell development	*TNFSF5*	Xq26
	TNF (TNFSF2)	6p21.3 (MHC)	TNF	Lymphoid organ development	*TNFSF10*	3q26
	LTB (TNFSF3)	6p21.3 (MHC)	LT β	Lymphoid organ, γ/δ T and NKT cell development	*TNFSF11*	13q14
	TNFSF4	1q25	OX40-L	Control of T cell function, and activation of B cells	*TNFSF12*	17p13
	TNFSF6	1q23	Fas-L	Apoptosis of Fas-expressing cells	*TNFSF13*	17p13.1
	TNFSF7	19p13	CD27-L, CD70	Control of T cell function	*TNFSF13B*	13q32–q34
	TNFSF8	9q33	CD30-L	Th1 response, B cell proliferation	*EDA*	Xq12–q13.1
	TNFSF9	19p13.3	4-1BB-L	Control of T cell function		
	TNFSF14	19p13.3	LIGHT	Control of T cell function		
	TNFSF15	9q33	TL1A	Control of T cell function		
	TNFSF18	1q23	GITRL	Modulation of T cell survival		
CC chemokines	*CCL1, 2, 3, 4, 5, 7, 8, 11, 13, 14, 15,16, 18, 23*	17q11-q12	CCL1, 2, 3, 4, 5, 7, 8, 11, 13, 14, 15, 16, 18, 23	Recruitment of leukocytes to sites of infection. Regulation of the traffic of leukocytes including T cells, B cells and dendritic cells. Development of nonlymphoid organs	*CCL19, 21, 27*	9p13

	CCL20	2q33-q37	CCL20	*CCL17, 22*	16p13
				CCL24, 26	7q11.23
				CCL25	19p13.2
				CCL28	5p12
CC chemokine receptors	*CCR1, 2, 3, 5, 8, 9*	3p21-p22	CCR1, 2, 3, 5, 8, 9	*CCR6*	6q27
	CCRL2	3p21-p22	CCRL2		
	CCR4	3p24	CCR4		
	CCR7, 10	17q12-q21.2	CCR7, 10		
CXC chemokine receptors	*CXCR1, CXCR2*	2q35	IL8RA, IL8RB	*CXCR3*	Xq13
	CXCR4	2q21	CXCR4	*CXCR5*	11q23.3
	CXCR6	3p21	CXCR6		
	CXCR7	2q37.3	CXCR7		
Other chemokine receptors	*CX3CR1, XCR1*	3p21.3-p21.1	CX3CR1, XCR1		

[a]For the TNF superfamily, copies mapping to the classical MHC paralogy group (chromosomes 1, 6, 9, and 19) are listed. For the chemokine/chemokine receptor family, copies mapping to the *HOX* paralogy group (chromosomes 2, 3, and 17) are listed. Some of the members listed here most likely arose by tandem duplication after WGD, and hence are not ohnologs in a strict sense.

[b]Chromosomal localization of human genes is based on the OMIM database (http://www.ncbi.nlm.nih.gov/omim) or Entrez gene (http://www.ncbi.nlm.nih.gov/sites/entrez?db=gene).

members of the B7 family, including B7-1, B7-2, ICOS, B7-H1, and B7-DC, are encoded by paralogous regions located on human chromosomes 1, 3, 11, 21, and 19, and are thought to have diverged by 2R-WGD.[109,110]

F. Complement

Genes coding for complement components C3, C4, and C5 are classic examples of ohnologs and are encoded in the MHC paralogons.[37] C4 and C5 are the components of the classical pathway of complement activation initiated by binding of antigen–antibody complexes to the C1 molecules. These components are present only in jawed vertebrates,[111] and are thought to have diverged from a C3-like precursor protein.[28,112]

IV. Controversies Surrounding the Timing of WGD

The genomes of invertebrate chordates show no evidence of WGD,[49,51,113,114] whereas those of jawed vertebrates exhibit clear evidence of 2R-WGD.[50,51] Thus, it is widely accepted that 2R-WGD took place in the vertebrate lineage after its divergence from invertebrate chordates. However, the exact timing of WGD in relation to the emergence of jawless vertebrates is still controversial. Three major possibilities have been suggested concerning the timing of WGD (Fig. 4): (i) the first round of WGD in a common ancestor of jawed and jawless vertebrates, and the second round in a common ancestor of jawed vertebrates (scenario A), (ii) both rounds before the separation of jawed and jawless vertebrates (scenario B); and (iii) both rounds in the jawed vertebrate lineage after its separation from the jawless vertebrate lineage (scenario C).

Initial analysis of 33 gene families by Escriva *et al.*[115] supported scenario A. This scenario is consistent with the observation that lampreys have single-chained hemoglobins whereas jawed vertebrate hemoglobins are composed of α and β chains.[116] It is also consistent with recent analysis of lamprey blood coagulation factors, using a draft genome assembly, which showed a simpler clotting system in jawless vertebrates.[117] Subsequent analysis of 358 gene families using sea lamprey expressed sequence tag sequences also favored scenario A, but it also raised the possibility that one or both rounds of WGD occurred nearly coincident with the lamprey lineage divergence.[51] More recently, Kuraku *et al.*[118] suggested, on the basis of the analysis of 55 gene families, that scenario B was most likely. However, this scenario appears inconsistent with the observation that lamprey genes can be assigned to one of the four human ohnologs in only 58% of gene families.[51] The absence of lamprey genes orthologous to one of the four human ohnologs may be

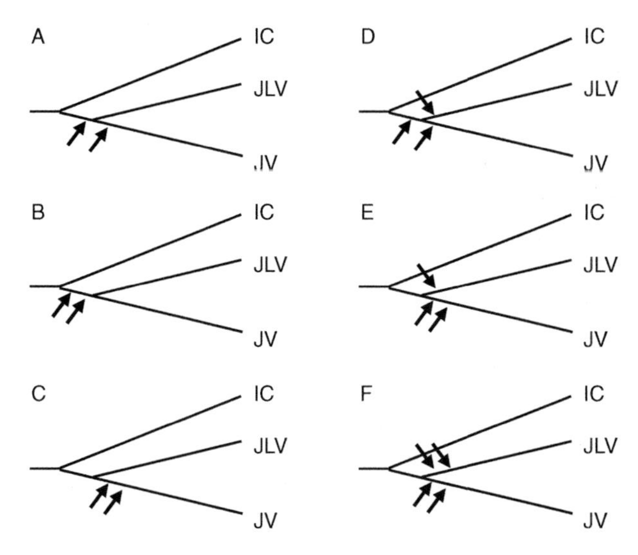

FIG. 4. Timing of WGDs in vertebrate evolution. Arrows indicate WGDs. Scenarios D–F assume independent WGD in a jawless vertebrate lineage. If scenario B is the case, jawed and jawless vertebrates should share corresponding ohnologs. In all other cases, they are in principle expected not to share strictly orthologous ohnologs. Two extant members of jawless vertebrates, hagfish and lamprey, are thought to have diverged 470–390 million years ago and are highly divergent from each other.[10] For simplicity, they are shown as a single group, but it is possible that their genome complexity is not the same as a result of independent lineage-specific WGDs. The third round of WGD, known to have occurred in some bony fish,[11] is not shown. Abbreviations: IC, invertebrate chordates; JLV, jawless vertebrates; and JV, jawed vertebrates.

accounted for by assuming extensive gene loss in the lamprey lineage; however, the presence of lamprey genes not orthologous to any of the four human ohnologs is difficult to explain.

Scenarios other than those discussed earlier are also possible (Fig. 4). For example, shortly after the separation of jawed and jawless vertebrate lineages, WGD may have taken place in both lineages independently, thus contributing to reduced gene orthology between the two lineages of vertebrates (scenarios D–F). Analysis of some gene families is consistent with these scenarios.[119,120] An ultimate resolution of the controversies surrounding the timing of WGD must await comprehensive analysis of the lamprey draft genome sequence.

So far as the genes involved in T cell and B cell immunity are concerned, there is ample evidence that jawless vertebrates lack many important ohnologs. For example, hagfish do not have authentic *GATA3* gene; their *GATA3*-like gene is equidistant from *GATA2* and *GATA3* and qualifies as a preduplicated form of *GATA2* and *GATA3*.[121] Likewise, hagfish do not have authentic *BTK* (gene coding for Bruton's tyrosine kinase), a member of the TEC family of tyrosine kinases required for B cell maturation; their *BTK*-like gene is equidistant from *BTK* and *BMX*.[121] Similarly, two *Ikaros*-like transcription factor genes of lampreys, *IKFL1* and *IKFL2*, are almost equidistant from *Ikaros*, *Helios*, *Eos*, and *Aiolos* of jawed vertebrates and not related to any specific members.[103] Also, lamprey *SPI*, a member of the Ets family of transcription factors, is not orthologous to any of the gnathostome genes *SPI1* (PU.1), *SPIB*, or *SPIC*,[122,123] which are encoded by the neurotrophin paralogy group (Fig. 3). These observations are more consistent with scenario A and some of the scenarios that assume independent WGD in the jawless vertebrate lineage.

It should be emphasized that the importance of 2R-WGD remains unchanged regardless of which scenario turns out to be the case, because it is clear that many ohnologs with essential roles in T cell and B cell immunity owe their existence to 2R-WGD. If scenario B is the case, 2R-WGD provided a basis required for the emergence of the jawed vertebrate-type AIS. On the other hand, if scenario A is the case as has been generally favored,[7] the second round of WGD likely played an important role as a trigger to the emergence of the jawed vertebrate-type AIS, along with the acquisition of RAG transposons. All the other scenarios are compatible with the idea that the second round of WGD served as a trigger to the emergence of the jawed vertebrate-type AIS.

V. The AIS of Jawless Vertebrates

Studies conducted in the 1960s and 1970s showed that lampreys were capable of producing specific agglutinins against particulate antigens and rejecting skin allografts with immunological memory,[124–130] suggesting that they are equipped with the AIS. In the following decades, extensive efforts were made to identify TCR, BCR, and MHC molecules in jawless vertebrates, but without any success.[121,131,132] This apparent paradox was resolved by the recent discovery that jawless vertebrates are equipped with a unique form of adaptive immunity that does not rely on TCR, BCR, or MHC molecules.[15,16]

A. Rearranging Antigen Receptors of Jawless Vertebrates

VLRs are antigen receptors of jawless vertebrates expressed clonally on lymphocyte-like cells; they generate diversity comparable to that of antigen receptors of jawed vertebrates by somatically rearranging LRR modules.[15,16] The germ line *VLR* gene has an incomplete structure incapable of encoding functional proteins (Fig. 5). In its vicinity are a large number of LLR-encoding modules with highly diverse sequences. During the development of lymphocyte-like cells, these modules are sequentially incorporated into the *VLR* gene

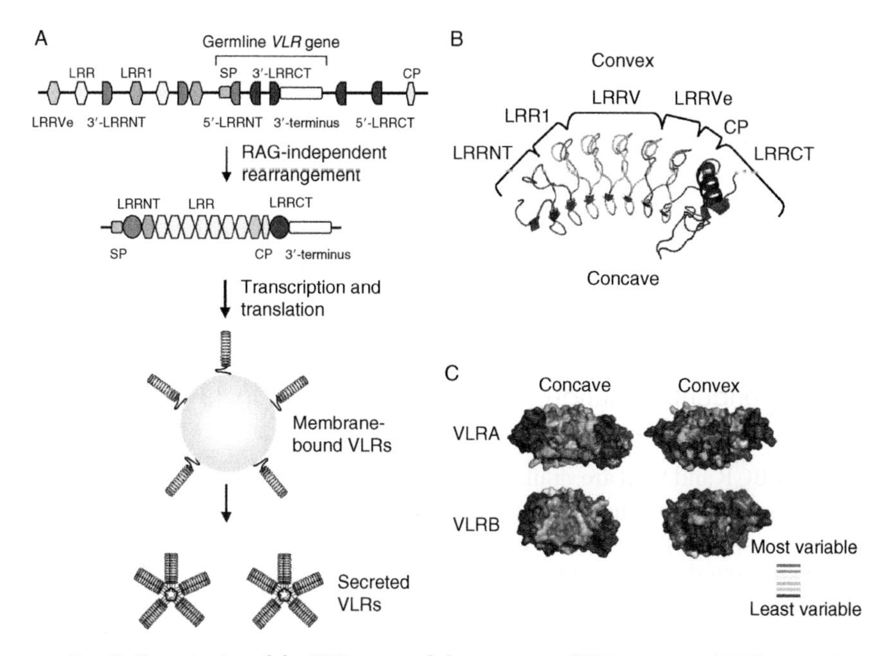

FIG. 5. Organization of the *VLR* gene and the structure of VLR proteins. (A) The germ line *VLR* gene has a structure incapable of encoding proteins. Modules coding for N-terminal caps (LRRNT), leucine-rich repeats (LRR), connecting peptides (CP) and C-terminal caps (LRRCT) occur in multiple copies adjacent to the germ line *VLR* gene. During the development of lymphoid cells, these modules are incorporated into the *VLR* gene. The rearranged *VLR* gene encodes a membrane-bound protein. The product of the *VLRB*, but not *VLRA*, gene is secreted and functions as antibodies. Whether membrane-bound VLRs occur as a monomer or multimer is not known. LRR1 and LRRVe denote LRR modules located at the N- and C-termini, respectively. The organization of the *VLR* locus shows considerable variation depending on loci and species. This figure, modified from Flajnik and Kasahara,[7] is intended to show salient features of *VLR* genes and does not accurately reproduce the organization of a specific *VLR* locus. (B) Crystal structure of hagfish VLRB molecules (PDB ID: 2o6S). The figure was generated using the PyMOL graphics tool (http://pymol.sourceforge.net/). (C) Sequence variability of hagfish VLR proteins. Variability is indicated by the color gradation from red to blue, where the most variable and the least variable patches are indicated in red and blue, respectively. Concave view (left); convex view (right). This figure was reproduced from Kim *et al.*[133] (See Color Insert.)

by a process called "copy choice,"[134,135] presumably assisted by cytidine deaminases of the AID-APOBEC family.[136] The rearranged, mature *VLR* gene encodes a glycosylphosphatidylinositol-anchored polypeptide composed of an N-terminal cap (LRRNT), multiple LRR modules, a connecting peptide (CP), a C-terminal cap (LRRCT), an invariant threonine/proline-rich stalk, and a C-terminal hydrophobic tail. Because the sequence of individual LRR modules is highly diverse, and the number of LRR modules incorporated into a rearranged gene shows considerable variation, a single *VLR* gene can generate combinatorial diversity comparable to that of BCR[137] (Fig. 5A). The crystal structures of hagfish VLR monomers indicate that they adopt a horseshoe-like solenoid structure characteristic of LRR family proteins (Fig. 5B), where seven-amino-acid LXXLXLX repeats of LRR modules form parallel β-strands in the concave surface (where L and X stands for leucine and any amino acid).[133] The majority of variable residues in VLR are located on the concave surface, suggesting that this surface is involved in antigen binding[133] (Fig. 5C). This suggestion was recently confirmed by *in vitro* mutagenesis experiments[138] and the crystal structure analysis of VLR–antigen complexes.[139,140]

B. Independent Evolution of Antigen Receptors in Jawed and Jawless Vertebrates

TCR/BCR and VLR are similar in that they both rely on combinatorial joining of gene segments to generate diversity. However, these receptors are evolutionarily unrelated and their structures are completely different. An analogous situation is found in NK cell receptors of mammals.[7] In primates, NK receptors interacting with classical MHC class I molecules are members of the killer cell Ig-like receptor (KIR) family. By contrast, the corresponding receptors in rodents are C-type lectin-like molecules known as Ly49.[141,142] Thus, primates and rodents use totally unrelated molecules as NK receptors. Available evidence indicates that a common ancestor of primates and rodents possessed precursor genes for both types of NK receptors and that the primate and rodent lineages adopted distinct gene families as their NK receptors.[143] The occurrence of two radically different antigen receptors in vertebrates appears to be accounted for in a similar manner.[7]

C. Two Types of Lymphoid Cells in Lamprey

Soon after the discovery of *VLR* in the sea lamprey,[13] it was shown that hagfish have two *VLR* genes.[14] These genes, named *VLRA* and *VLRB*, map to the same chromosome, but are distant from each other.[144] This suggested that they function as separate recombination units, with potentially distinct roles in host defense.

Subsequently, lampreys were shown to have two *VLR* genes orthologous to hagfish *VLRA* and *VLRB*.[136] Both genes are expressed exclusively on lymphocyte-like cells and seem to exhibit allelic exclusion.[145] In response to stimulation with antigen, VLRB$^+$ cells undergo clonal expansion and begin to secrete VLRs in a manner analogous to the secretion of Igs by B cells.[13,146] The secreted VLRB molecules, referred to as "VLRB antibodies," occur as pentamers or tetramers of dimers and have 8–10 antigen binding sites, thus resembling IgM in subunit organization[138] (Fig. 5A). Like IgM, VLRB binds antigens carrying repetitive epitopes with high avidity and specificity, and displays strong agglutinating activities,[146] accounting for the earlier observations that immunized lampreys produce specific agglutinins.[124–130] Thus, VLRB$^+$ cells resemble B cells in that they both expand clonally and secrete antibodies in response to antigen challenge (Fig. 6).

Specific agglutinins are exclusively or almost exclusively derived from the *VLRB* gene.[146] This suggested a role other than the production of antibodies for lamprey VLRA$^+$ cells. Remarkably, recent evidence indicates that VLRA$^+$ cells resemble T cells[145]; VLRA$^+$ cells not only undergo blastoid transformation in response to a T cell mitogen, but they also express such genes as *IL17*,

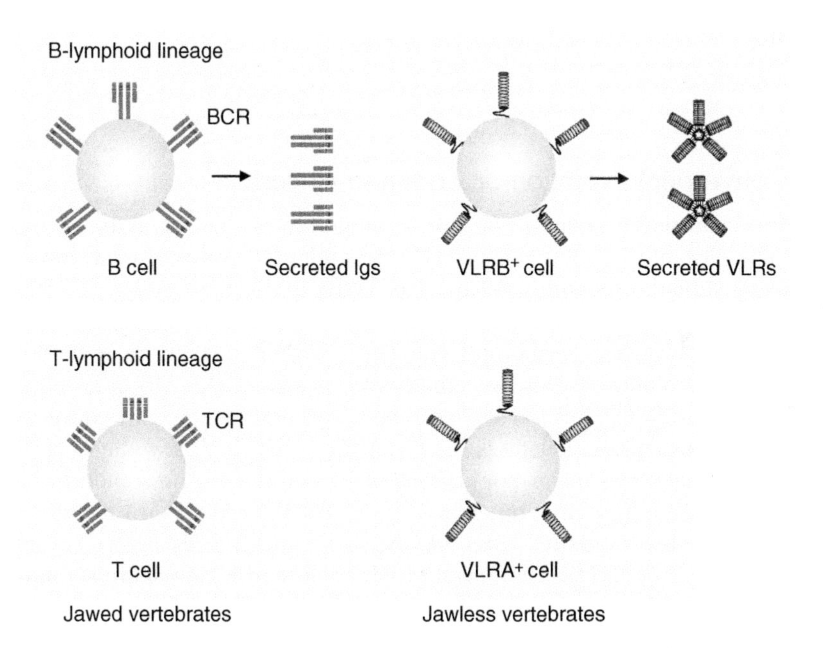

FIG. 6. Two lymphoid lineages in jawed and jawless vertebrates. VLR$^+$ cells of jawless vertebrates resemble lymphocytes of jawed vertebrates in that they clonally express specific antigen receptors and proliferate in response to antigen challenge. In both jawed and jawless vertebrates, two major populations of lymphoid cells have been identified. Like B cells, VLRB$^+$ cells secrete the antigen receptors as antibodies. By contrast, like T cells, VLRA$^+$ cells do not secrete the receptors. The gene expression profiles of VLRA$^+$ and VLRB$^+$ cells are remarkably similar to those of T cells and B cells, respectively.[145] Hence, it is likely that the two lineages of lymphoid cells emerged before the divergence of jawed and jawless vertebrates. After the divergence, jawed and jawless vertebrates appear to have adopted distinct molecules as their antigen receptors. If scenario A or D turns out to be the case (Fig. 4), it is possible that the first round of WGD was involved in the divergence of T-lymphoid and B-lymphoid lineages.

GATA2/3, and *NOTCH* whose jawed vertebrate counterparts are expressed in T cells and involved in their development and differentiation. Thus, lampreys have two major types of lymphocyte-like cells, with VLRA$^+$ and VLRB$^+$ cells likely involved in cellular and humoral arms of adaptive immunity, respectively[18,19,145] (Fig. 6).

More recently, a third VLR molecule, designated VLRC, was identified in the lamprey.[147] VLRC is expressed on a population of lymphocyte-like cells distinct from VLRA$^+$ or VLRB$^+$ cells. Because VLRC is more closely related in sequence to VLRA than to VLRB and is apparently not secreted, it was suggested that VLRC$^+$ cells might resemble T cells rather than B cells.[147] Interestingly, all classes of jawed vertebrates have two major lineages of T cells: $\alpha\beta$ and $\gamma\delta$ T cells.[148] It remains to be examined whether VLRA$^+$ and VLRC$^+$ cells are functionally specialized in a manner analogous to α/β and γ/δ T cells.

D. Convergent Evolution or Common Ancestry?

In evolutionary biology, convergent evolution is defined as the process whereby distantly related organisms independently evolve similar traits to adapt to similar necessities. VLRs and TCRs/BCRs both serve as antigen receptors, but are evolutionarily unrelated. Thus, the use of distinct receptors in jawed and jawless vertebrates can be regarded as a prime example of convergent evolution.[13] However, the overall design of the AIS in jawed and jawless vertebrates seems too similar to be accounted for solely by convergent evolution. Particularly striking is the observation that both jawed and jawless vertebrates have two major populations of lymphoid cells presumed to have similar specialized immune functions[145] (Fig. 6). To account for this, it seems more reasonable to assume that VLRA$^+$ cells and T cells evolved from a common ancestor and that, likewise, VLRB$^+$ cells and B cells shared common ancestry; most likely, a common ancestor of all vertebrates was equipped with two lineages of lymphoid cells.[7] Recent evidence indicates that, contrary to a commonly held belief, T cells and B cells do not share an immediate common ancestor, but differentiate from myeloid-T and myeloid-B progenitors, respectively.[149,150] If T cells and B cells are distantly related as suggested by these studies, it is not surprising if the two lineages of lymphoid cells diverged at an earlier stage in evolution than previously thought.[151]

In summary, authentic T cells and B cells, as defined by surface expression of TCRs and BCRs, are unique to jawed vertebrates (Fig. 1). However, jawless vertebrates have at least two populations of lymphoid cells that likely share common ancestry with T cells and B cells of jawed vertebrates.

VI. Concluding Remarks

The idea that the vertebrate genome underwent 2R-WGD close to the origin of vertebrates stirred hot debate for more than a decade.[41] This debate was finally settled in favor of the 2R hypothesis by systematic synteny comparison of human and amphioxus genomes.[51] As discussed earlier, many molecules that play essential roles in T cell and B cell immunity are encoded by ohnologs. Thus, the emergence of the AIS centered on T cells and B cells was critically dependent on WGD. A major unresolved issue at the moment is whether jawless vertebrates experienced 2R-WGD or not.[118] Despite this uncertainty, it is clear that jawless vertebrates lack many ohnologs indispensable for the function of the jawed vertebrate-type AIS.

It has been debated whether the AIS of jawed vertebrates emerged abruptly or gradually.[4,152] Evidence emerging from the studies of jawless vertebrates indicates that the two lineages of lymphoid cells that became T cells and B cells were already present in a common ancestor of all vertebrates.[7] Furthermore, the function of the gnathostome AIS is dependent on the participation of many genes that clearly evolved prior to the emergence of the AIS. In this sense, the AIS of jawed vertebrates emerged gradually, taking advantage of the resources already available in a common vertebrate ancestor.[4] On the other hand, it is also true that the two sudden accidents, the acquisition of RAG transposons[153] and the birth of a large number of ohnologs triggered by genome doubling,[30,37] transformed the nature of the immune system fundamentally, leading to the emergence of the jawed vertebrate-type AIS. In this regard, a novel form of immunity did emerge abruptly in a jawed vertebrate ancestor. Therefore, the evolutionary processes leading to the emergence of the jawed vertebrate-type AIS were gradual in some aspects, but abrupt in other aspects. This appears to be a more balanced view.

Acknowledgments

Experimental work described here has been supported by Grants-in-Aids from The Ministry of Education, Culture, Sports, Science and Technology of Japan, and by grants from The Japan Science and Technology Agency and The NOASTEC Foundation.

References

1. Flajnik MF, Kasahara M. Comparative genomics of the MHC: glimpses into the evolution of the adaptive immune system. *Immunity* 2001;**15**:351–62.
2. Flajnik MF. Comparative analyses of immunoglobulin genes: surprises and portents. *Nat Rev Immunol* 2002;**2**:688–98.

3. Kasahara M, Suzuki T, Du Pasquier L. On the origins of the adaptive immune system: novel insights from invertebrates and cold-blooded vertebrates. *Trends Immunol* 2004;**25**:105–11.
4. Klein J, Nikolaidis N. The descent of the antibody-based immune system by gradual evolution. *Proc Natl Acad Sci USA* 2005;**102**:169–74.
5. Litman GW, Cannon JP, Dishaw LJ. Reconstructing immune phylogeny: new perspectives. *Nat Rev Immunol* 2005;**5**:866–79.
6. Flajnik MF, Du Pasquier L. Evolution of the immune system. In: Paul WE, editor. *Fundamental immunology.* 6th ed. Philadelphia, NY: Lippincott Williams & Wilkins; 2008. p. 56–124.
7. Flajnik MF, Kasahara M. Origin and evolution of the adaptive immune system: genetic events and selective pressures. *Nat Rev Genet* 2010;**11**:47–59.
8. Fugmann SD, Messier C, Novack LA, Cameron RA, Rast JP. An ancient evolutionary origin of the *Rag1/2* gene locus. *Proc Natl Acad Sci USA* 2006;**103**:3728–33.
9. Holland LZ, Albalat R, Azumi K, Benito-Gutierrez E, Blow MJ, Bronner-Fraser M, et al. The amphioxus genome illuminates vertebrate origins and cephalochordate biology. *Genome Res* 2008;**18**:1100–11.
10. Kuraku S, Kuratani S. Time scale for cyclostome evolution inferred with a phylogenetic diagnosis of hagfish and lamprey cDNA sequences. *Zool Sci* 2006;**23**:1053–64.
11. Meyer A, Van de Peer Y. From 2R to 3R: evidence for a fish-specific genome duplication (FSGD). *BioEssays* 2005;**27**:937–45.
12. Blair JE, Hedges SB. Molecular phylogeny and divergence times of deuterostome animals. *Mol Biol Evol* 2005;**22**:2275–84.
13. Pancer Z, Amemiya CT, Ehrhardt GR, Ceitlin J, Gartland GL, Cooper MD. Somatic diversification of variable lymphocyte receptors in the agnathan sea lamprey. *Nature* 2004;**430**:174–80.
14. Pancer Z, Saha NR, Kasamatsu J, Suzuki T, Amemiya CT, Kasahara M, et al. Variable lymphocyte receptors in hagfish. *Proc Natl Acad Sci USA* 2005;**102**:9224–9.
15. Cooper MD, Alder MN. The evolution of adaptive immune systems. *Cell* 2006;**124**:815–22.
16. Pancer Z, Cooper MD. The evolution of adaptive immunity. *Annu Rev Immunol* 2006;**24**:497–518.
17. Kasahara M, Kasamatsu J, Sutoh Y. Two types of antigen receptor systems in vertebrates. *Zool Sci* 2008;**25**:969–75.
18. Boehm T. One problem, two solutions. *Nat Immunol* 2009;**10**:811–3.
19. Cooper MD, Herrin BR. How did our complex immune system evolve? *Nat Rev Immunol* 2010;**10**:2–3.
20. Azumi K, De Santis R, De Tomaso A, Rigoutsos I, Yoshizaki F, Pinto MR, et al. Genomic analysis of immunity in a urochordate and the emergence of the vertebrate immune system: waiting for Godot. *Immunogenetics* 2003;**55**:570–81.
21. Rast JP, Anderson MK, Strong SJ, Luer C, Litman RT, Litman GW. α, β, γ, and δ T cell antigen receptor genes arose early in vertebrate phylogeny. *Immunity* 1997;**6**:1–11.
22. Litman GW, Anderson MK, Rast JP. Evolution of antigen binding receptors. *Annu Rev Immunol* 1999;**17**:109–47.
23. Hashimoto K, Nakanishi T, Kurosawa Y. Identification of a shark sequence resembling the major histocompatibility complex class I α3 domain. *Proc Natl Acad Sci USA* 1992;**89**:2209–12.
24. Kasahara M, Vazquez M, Sato K, McKinney EC, Flajnik MF. Evolution of the major histocompatibility complex: isolation of class II A cDNA clones from the cartilaginous fish. *Proc Natl Acad Sci USA* 1992;**89**:6688–92.
25. Kasahara M, McKinney EC, Flajnik MF, Ishibashi T. The evolutionary origin of the major histocompatibility complex: polymorphism of class II α chain genes in the cartilaginous fish. *Eur J Immunol* 1993;**23**:2160–5.

26. Bartl S, Baish MA, Flajnik MF, Ohta Y. Identification of class I genes in cartilaginous fish, the most ancient group of vertebrates displaying an adaptive immune response. *J Immunol* 1997;**159**:6097–104.
27. Marchalonis JJ, Schluter SF, Bernstein RM, Shen S, Edmundson AB. Phylogenetic emergence and molecular evolution of the immunoglobulin family. *Adv Immunol* 1998;**70**:417–506.
28. Nonaka M, Kimura A. Genomic view of the evolution of the complement system. *Immunogenetics* 2006;**58**:701–13.
29. Bernstein RM, Schluter SF, Bernstein H, Marchalonis JJ. Primordial emergence of the recombination activating gene 1 (RAG1): sequence of the complete shark gene indicates homology to microbial integrases. *Proc Natl Acad Sci USA* 1996;**93**:9454–9.
30. Abi Rached L, McDermott MF, Potarotti P. The MHC big bang. *Immunol Rev* 1999;**167**:33–44.
31. Thompson CB. New insights into V(D)J recombination and its role in the evolution of the immune system. *Immunity* 1995;**3**:531–9.
32. Fugmann SD, Lee AI, Shockett PE, Villey IJ, Schatz DG. The RAG proteins and V(D)J recombination: complexes, ends, and transposition. *Annu Rev Immunol* 2000;**18**:495–527.
33. van Gent DC, Mizuuchi K, Gellert M. Similarities between initiation of V(D)J recombination and retroviral integration. *Science* 1996;**271**:1592–4.
34. Agrawal A, Eastman QM, Schatz DG. Transposition mediated by RAG1 and RAG2 and its implications for the evolution of the immune system. *Nature* 1998;**394**:744–51.
35. Hiom K, Melek M, Gellert M. DNA transposition by the RAG1 and RAG2 proteins: a possible source of oncogenic translocations. *Cell* 1998;**94**:463–70.
36. Kapitonov VV, Jurka J. RAG1 core and V(D)J recombination signal sequences were derived from Transib transposons. *PLoS Biol* 2005;**3**:e181.
37. Kasahara M, Nakaya J, Satta Y, Takahata N. Chromosomal duplication and the emergence of the adaptive immune system. *Trends Genet* 1997;**13**:90–2.
38. Kasahara M, Hayashi M, Tanaka K, Inoko H, Sugaya K, Ikemura T, et al. Chromosomal localization of the proteasome Z subunit gene reveals an ancient chromosomal duplication involving the major histocompatibility complex. *Proc Natl Acad Sci USA* 1996;**93**:9096–101.
39. Katsanis N, Fitzgibbon J, Fischer EMC. Paralogy mapping: Identification of a region in the human MHC triplicated onto human chromosomes 1 and 9 allows the prediction and isolation of novel *PBX* and *NOTCH* loci. *Genomics* 1996;**35**:101–8.
40. Ohno S. *Evolution by gene duplication.* New York: Springer-Verlag; 1970.
41. Kasahara M. The 2R hypothesis: an update. *Curr Opin Immunol* 2007;**19**:547–52.
42. Furlong RF, Holland PW. Were vertebrates octoploid? *Philos Trans R Soc Lond B Biol Sci* 2002;**357**:531–44.
43. Kasahara M. Polyploid origin of the human genome. In: Cooper DN, editor. *Nature encyclopedia of the human genome*. London: Nature Publishing Group; 2003. p. 614–8.
44. Hufton AL, Panopoulou G. Polyploidy and genome restructuring: a variety of outcomes. *Curr Opin Genet Dev* 2009;**19**:600–6.
45. Van de Peer Y, Maere S, Meyer A. The evolutionary significance of ancient genome duplications. *Nat Rev Genet* 2009;**10**:725–32.
46. Wolfe K. Robustness—it's not where you think it is. *Nat Genet* 2000;**25**:3–4.
47. Lundin LG. Evolution of the vertebrate genome as reflected in paralogous chromosomal regions in man and the house mouse. *Genomics* 1993;**16**:1–19.
48. Garcia-Fernandez J. The genesis and evolution of homeobox gene clusters. *Nat Rev Genet* 2005;**6**:881–92.
49. Garcia-Fernandez J, Holland PWH. Archetypal organization of the amphioxus *Hox* gene cluster. *Nature* 1994;**370**:563–6.

50. Dehal P, Boore JL. Two rounds of whole genome duplication in the ancestral vertebrate. *PLoS Biol* 2005;**3**:e314.
51. Putnam NH, Butts T, Ferrier DE, Furlong RF, Hellsten U, Kawashima T, et al. The amphioxus genome and the evolution of the chordate karyotype. *Nature* 2008;**453**:1064–71.
52. Zhang G, Cohn MJ. Genome duplication and the origin of the vertebrate skeleton. *Curr Opin Genet Dev* 2008;**18**:387–93.
53. Van de Peer Y, Maere S, Meyer A. 2R or not 2R is not the question anymore. *Nat Rev Genet* 2010;**11**:166.
54. Tanaka K. The proteasome: overview of structure and functions. *Proc Jpn Acad Ser B Phys Biol Sci* 2009;**85**:12–36.
55. Tanaka K, Kasahara M. The MHC class I ligand-generating system: roles of immunoproteasomes and the interferon-γ-inducible proteasome activator PA28. *Immunol Rev* 1998;**163**:161–76.
56. Kasahara M. What do the paralogous regions in the genome tell us about the origin of the adaptive immune system? *Immunol Rev* 1998;**166**:159–75.
57. Kandil E, Namikawa C, Nonaka M, Greenberg AS, Flajnik MF, Ishibashi T, et al. Isolation of low molecular mass polypeptide complementary DNA clones from primitive vertebrates: implications for the origin of MHC class I-restricted antigen presentation. *J Immunol* 1996;**156**:4245–53.
58. Ohta Y, McKinney EC, Criscitiello MF, Flajnik MF. Proteasome, transporter associated with antigen processing, and class I genes in the nurse shark *Ginglymostoma cirratum*: evidence for a stable class I region and MHC haplotype lineages. *J Immunol* 2002;**168**:771–81.
59. Flajnik MF, Canel C, Kramer J, Kasahara M. Which came first, MHC class I or class II? *Immunogenetics* 1991;**33**:295–300.
60. Klein J, O'hUigin C. Composite origin of major histocompatibility complex genes. *Curr Opin Genet Dev* 1993;**3**:923–30.
61. Kasahara M. The chromosomal duplication model of the major histocompatibility complex. *Immunol Rev* 1999;**167**:17–32.
62. Rock KL, Goldberg AL. Degradation of cell proteins and the generation of MHC class I-presented peptides. *Annu Rev Immunol* 1999;**17**:739–79.
63. Murata S, Sasaki K, Kishimoto T, Niwa S, Hayashi H, Takahama Y, et al. Regulation of CD8+ T cell development by thymus-specific proteasomes. *Science* 2007;**316**:1349–53.
64. Tomaru U, Ishizu A, Murata S, Miyatake Y, Suzuki S, Takahashi S, et al. Exclusive expression of proteasome subunit β5t in the human thymic cortex. *Blood* 2009;**113**:5186–91.
65. Nitta T, Murata S, Sasaki K, Fujii H, Ripen AM, Ishimaru N, et al. Thymoproteasome shapes immunocompetent repertoire of CD8+ T cells. *Immunity* 2010;**32**:29–40.
66. Honey K, Rudensky AY. Lysosomal cysteine proteases regulate antigen presentation. *Nat Rev Immunol* 2003;**3**:472–82.
67. Zavasnik-Bergant T, Turk B. Cysteine cathepsins in the immune response. *Tissue Antigens* 2006;**67**:349–55.
68. Nakagawa T, Roth W, Wong P, Nelson A, Farr A, Deussing J, et al. Cathepsin L: critical role in Ii degradation and CD4 T cell selection in the thymus. *Science* 1998;**280**:450–3.
69. Ting JP, Baldwin AS. Regulation of MHC gene expression. *Curr Opin Immunol* 1993;**5**:8–16.
70. Reith W, Mach B. The bare lymphocyte syndrome and the regulation of MHC expression. *Annu Rev Immunol* 2001;**19**:331–73.
71. Kasahara M. Genome dynamics of the major histocompatibility complex: insights from genome paralogy. *Immunogenetics* 1999;**49**:134–45.
72. Hallbook F, Wilson K, Thorndyke M, Olinski RP. Formation and evolution of the chordate neurotrophin and Trk receptor genes. *Brain Behav Evol* 2006;**68**:133–44.

73. Olinski RP, Lundin LG, Hallbook F. Conserved synteny between the Ciona genome and human paralogons identifies large duplication events in the molecular evolution of the insulin-relaxin gene family. *Mol Biol Evol* 2006;**23**:10–22.
74. Clark MS, Pontarotti P, Gilles A, Kelly A, Elgar G. Identification and characterization of a β proteasome subunit cluster in the Japanese pufferfish (*Fugu rubripes*). *J Immunol* 2000;**165**:4446–52.
75. Kasahara M. Genome paralogy: a new perspective on the organization and origin of the major histocompatibility complex. *Curr Top Microbiol Immunol* 2000;**248**:53–66.
76. Okada K, Asai K. Expansion of signaling genes for adaptive immune system evolution in early vertebrates. *BMC Genomics* 2008;**9**:218.
77. Shuai K, Liu B. Regulation of JAK-STAT signalling in the immune system. *Nat Rev Immunol* 2003;**3**:900–11.
78. Kaplan MH, Sun YL, Hoey T, Grusby MJ. Impaired IL-12 responses and enhanced development of Th2 cells in Stat4-deficient mice. *Nature* 1996;**382**:174–7.
79. Thierfelder WE, van Deursen JM, Yamamoto K, Tripp RA, Sarawar SR, Carson RT, et al. Requirement for Stat4 in interleukin-12-mediated responses of natural killer and T cells. *Nature* 1996;**382**:171–4.
80. Kaplan MH, Schindler U, Smiley ST, Grusby MJ. Stat6 is required for mediating responses to IL-4 and for development of Th2 cells. *Immunity* 1996;**4**:313–9.
81. Shimoda K, van Deursen J, Sangster MY, Sarawar SR, Carson RT, Tripp RA, et al. Lack of IL-4-induced Th2 response and IgE class switching in mice with disrupted Stat6 gene. *Nature* 1996;**380**:630–3.
82. Leonard WJ, O'Shea JJ. Jaks and STATs: biological implications. *Annu Rev Immunol* 1998;**16**:293–322.
83. Shuai K, Liu B. Regulation of gene-activation pathways by PIAS proteins in the immune system. *Nat Rev Immunol* 2005;**5**:593–605.
84. Croker BA, Kiu H, Nicholson SE. SOCS regulation of the JAK/STAT signalling pathway. *Semin Cell Dev Biol* 2008;**19**:414–22.
85. Jin HJ, Shao JZ, Xiang LX, Wang H, Sun LL. Global identification and comparative analysis of SOCS genes in fish: insights into the molecular evolution of SOCS family. *Mol Immunol* 2008;**45**:1258–68.
86. Tybulewicz VL. Vav-family proteins in T-cell signalling. *Curr Opin Immunol* 2005;**17**:267–74.
87. Gu JJ, Ryu JR, Pendergast AM. Abl tyrosine kinases in T-cell signaling. *Immunol Rev* 2009;**228**:170–83.
88. Berg LJ, Finkelstein LD, Lucas JA, Schwartzberg PL. Tec family kinases in T lymphocyte development and function. *Annu Rev Immunol* 2005;**23**:549–600.
89. Watts TH. TNF/TNFR family members in costimulation of T cell responses. *Annu Rev Immunol* 2005;**23**:23–68.
90. Croft M. The role of TNF superfamily members in T-cell function and diseases. *Nat Rev Immunol* 2009;**9**:271–85.
91. Ware CF. The TNF superfamily. *Cytokine Growth Factor Rev* 2003;**14**:181–4.
92. Glenney GW, Wiens GD. Early diversification of the TNF superfamily in teleosts: genomic characterization and expression analysis. *J Immunol* 2007;**178**:7955–73.
93. Collette Y, Gilles A, Pontarotti P, Olive D. A co-evolution perspective of the TNFSF and TNFRSF families in the immune system. *Trends Immunol* 2003;**24**:387–94.
94. Zlotnik A, Yoshie O, Nomiyama H. The chemokine and chemokine receptor superfamilies and their molecular evolution. *Genome Biol* 2006;**7**:243.
95. Rossi D, Zlotnik A. The biology of chemokines and their receptors. *Annu Rev Immunol* 2000;**18**:217–42.

96. DeVries ME, Kelvin AA, Xu L, Ran L, Robinson J, Kelvin DJ. Defining the origins and evolution of the chemokine/chemokine receptor system. *J Immunol* 2006;**176**:401–15.
97. Huminiecki L, Goldovsky L, Freilich S, Moustakas A, Ouzounis C, Heldin CH. Emergence, development and diversification of the TGF-beta signalling pathway within the animal kingdom. *BMC Evol Biol* 2009;**9**:28.
98. Rothenberg EV, Pant R. Origins of lymphocyte developmental programs: transcription factor evidence. *Semin Immunol* 2004;**16**:227–38.
99. Ho IC, Tai TS, Pai SY. GATA3 and the T-cell lineage: essential functions before and after T-helper-2-cell differentiation. *Nat Rev Immunol* 2009;**9**:125–35.
100. Gillis WQ, St John J, Bowerman B, Schneider SQ. Whole genome duplications and expansion of the vertebrate GATA transcription factor gene family. *BMC Evol Biol* 2009;**9**:207.
101. Mayer WE, O'hUigin C, Tichy H, Terzic J, Saraga-Babic M. Identification of two Ikaros-like transcription factors in lamprey. *Scand J Immunol* 2002;**55**:162–70.
102. Cupit PM, Hansen JD, McCarty AS, White G, Chioda M, Spada F, et al. Ikaros family members from the agnathan *Myxine glutinosa* and the urochordate *Oikopleura dioica*: emergence of an essential transcription factor for adaptive immunity. *J Immunol* 2003;**171**:6006–13.
103. John LB, Yoong S, Ward AC. Evolution of the Ikaros gene family: implications for the origins of adaptive immunity. *J Immunol* 2009;**182**:4792–9.
104. Wang JH, Nichogiannopoulou A, Wu L, Sun L, Sharpe AH, Bigby M, et al. Selective defects in the development of the fetal and adult lymphoid system in mice with an Ikaros null mutation. *Immunity* 1996;**5**:537–49.
105. Hahm K, Cobb BS, McCarty AS, Brown KE, Klug CA, Lee R, et al. Helios, a T cell-restricted Ikaros family member that quantitatively associates with Ikaros at centromeric heterochromatin. *Genes Dev* 1998;**12**:782–96.
106. Takaoka A, Tamura T, Taniguchi T. Interferon regulatory factor family of transcription factors and regulation of oncogenesis. *Cancer Sci* 2008;**99**:467–78.
107. Nehyba J, Hrdlickova R, Bose HR. Dynamic evolution of immune system regulators: the history of the interferon regulatory factor family. *Mol Biol Evol* 2009;**26**:2539–50.
108. Greenwald RJ, Freeman GJ, Sharpe AH. The B7 family revisited. *Annu Rev Immunol* 2005;**23**:515–48.
109. Du Pasquier L, Zucchetti I, De Santis R. Immunoglobulin superfamily receptors in protochordates: before RAG time. *Immunol Rev* 2004;**198**:233–48.
110. Hansen JD, Pasquier LD, Lefranc M-P, Lopez V, Benmansour A, Boudinot P. The B7 family of immunoregulatory receptors: a comparative and evolutionary perspective. *Mol Immunol* 2009;**46**:457–72.
111. Kimura A, Ikeo K, Nonaka M. Evolutionary origin of the vertebrate blood complement and coagulation systems inferred from liver EST analysis of lamprey. *Dev Comp Immunol* 2009;**33**:77–87.
112. Fujita T. Evolution of the lectin-complement pathway and its role in innate immunity. *Nat Rev Immunol* 2002;**2**:346–53.
113. Abi-Rached L, Gilles A, Shiina T, Pontarotti P, Inoko H. Evidence of *en bloc* duplication in vertebrate genomes. *Nat Genet* 2002;**31**:100–5.
114. Dehal P, Satou Y, Campbell RK, Chapman J, Degnan B, De Tomaso A, et al. The draft genome of *Ciona intestinalis*: insights into chordate and vertebrate origins. *Science* 2002;**298**:2157–67.
115. Escriva H, Manzon L, Youson J, Laudet V. Analysis of lamprey and hagfish genes reveals a complex history of gene duplications during early vertebrate evolution. *Mol Biol Evol* 2002;**19**:1440–50.
116. Honig GR, Adams III JG. *Human hemoglobin genetics*. Wien: Springer-Verlag; 1986.

117. Doolittle RF, Jiang Y, Nand J. Genomic evidence for a simpler clotting scheme in jawless vertebrates. *J Mol Evol* 2008;**66**:185–96.
118. Kuraku S, Meyer A, Kuratani S. Timing of genome duplications relative to the origin of the vertebrates: did cyclostomes diverge before or after? *Mol Biol Evol* 2009;**26**:47–59.
119. Huang X, Hui MN, Liu Y, Yuen DS, Zhang Y, Chan WY, et al. Discovery of a novel prolactin in non-mammalian vertebrates: evolutionary perspectives and its involvement in teleost retina development. *PLoS ONE* 2009;**4**:e6163.
120. Tank EM, Dekker RG, Beauchamp K, Wilson KA, Boehmke AE, Langeland JA. Patterns and consequences of vertebrate *Emx* gene duplications. *Evol Dev* 2009;**11**:343–53.
121. Suzuki T, Shin-I T, Kohara Y, Kasahara M. Transcriptome analysis of hagfish leukocytes: a framework for understanding the immune system of jawless fishes. *Dev Comp Immunol* 2004;**28**:993–1003.
122. Shintani S, Terzic J, Sato A, Saraga-Babic M, O'hUigin C, Tichy H, et al. Do lampreys have lymphocytes? The Spi evidence. *Proc Natl Acad Sci USA* 2000;**97**:7417–22.
123. Anderson MK, Sun X, Miracle AL, Litman GW, Rothenberg EV. Evolution of hematopoiesis: three members of the PU.1 transcription factor family in a cartilaginous fish, *Raja eglanteria*. *Proc Natl Acad Sci USA* 2001;**98**:553–8.
124. Finstad J, Good RA. The evolution of the immune response. III. Immunologic responses in the lamprey. *J Exp Med* 1964;**120**:1151–68.
125. Marchalonis JJ, Edelman GM. Phylogenetic origins of antibody structure. III. Antibodies in the primary immune response of the sea lamprey, *Petromyzon marinus*. *J Exp Med* 1968;**127**:891–914.
126. Hildemann WH. Transplantation immunity in fishes: Agnatha, Chondrichthyes and Osteichthyes. *Transplant Proc* 1970;**2**:253–9.
127. Linthicum DS, Hildemann WH. Immunologic responses of Pacific hagfish. III. Serum antibodies to cellular antigens. *J Immunol* 1970;**105**:912–8.
128. Litman GW, Finstad FJ, Howell J, Pollara BW, Good RA. The evolution of the immune response. III. Structural studies of the lamprey immunoglobulin. *J Immunol* 1970;**105**:1278–85.
129. Pollara B, Litman GW, Finstad J, Howell J, Good RA. The evolution of the immune response. VII. Antibody to human "O" cells and properties of the immunoglobulin in lamprey. *J Immunol* 1970;**105**:738–45.
130. Fujii T, Nakagawa H, Murakawa S. Immunity in lamprey. II. Antigen-binding responses to sheep erythrocytes and hapten in the ammocoete. *Dev Comp Immunol* 1979;**3**:609–20.
131. Mayer WE, Uinuk-Ool T, Tichy H, Gartland LA, Klein J, Cooper MD. Isolation and characterization of lymphocyte-like cells from a lamprey. *Proc Natl Acad Sci USA* 2002;**99**:14350–5.
132. Uinuk-Ool T, Mayer WE, Sato A, Dongak R, Cooper MD, Klein J. Lamprey lymphocyte-like cells express homologs of genes involved in immunologically relevant activities of mammalian lymphocytes. *Proc Natl Acad Sci USA* 2002;**99**:14356–61.
133. Kim HM, Oh SC, Lim KJ, Kasamatsu J, Heo JY, Park BS, et al. Structural diversity of the hagfish variable lymphocyte receptors. *J Biol Chem* 2007;**282**:6726–32.
134. Nagawa F, Kishishita N, Shimizu K, Hirose S, Miyoshi M, Nezu J, et al. Antigen-receptor genes of the agnathan lamprey are assembled by a process involving copy choice. *Nat Immunol* 2007;**8**:206–13.
135. Kishishita N, Matsuno T, Takahashi Y, Takaba H, Nishizumi H, Nagawa F. Regulation of antigen-receptor gene assembly in hagfish. *EMBO Rep* 2010;**11**:126–32.
136. Rogozin IB, Iyer LM, Liang L, Glazko GV, Liston VG, Pavlov YI, et al. Evolution and diversification of lamprey antigen receptors: evidence for involvement of an AID-APOBEC family cytosine deaminase. *Nat Immunol* 2007;**8**:647–56.

137. Alder MN, Rogozin IB, Iyer LM, Glazko GV, Cooper MD, Pancer Z. Diversity and function of adaptive immune receptors in a jawless vertebrate. *Science* 2005;**310**:1970–3.
138. Herrin BR, Alder MN, Roux KH, Sina C, Ehrhardt GR, Boydston JA, et al. Structure and specificity of lamprey monoclonal antibodies. *Proc Natl Acad Sci USA* 2008;**105**:2040–5.
139. Han BW, Herrin BR, Cooper MD, Wilson IA. Antigen recognition by variable lymphocyte receptors. *Science* 2008;**321**:1834–7.
140. Velikovsky CA, Deng L, Tasumi S, Iyer LM, Kerzic MC, Aravind L, et al. Structure of a lamprey variable lymphocyte receptor in complex with a protein antigen. *Nat Struct Mol Biol* 2009;**16**:725–30.
141. Gumperz JE, Parham P. The enigma of the natural killer cell. *Nature* 1995;**378**:245–8.
142. Trowsdale J, Parham P. Defense strategies and immunity-related genes. *Eur J Immunol* 2004;**34**:7–17.
143. Barten R, Torkar M, Haude A, Trowsdale J, Wilson MJ. Divergent and convergent evolution of NK-cell receptors. *Trends Immunol* 2001;**22**:52–7.
144. Kasamatsu J, Suzuki T, Ishijima J, Matsuda Y, Kasahara M. Two variable lymphocyte receptor genes of the inshore hagfish are located far apart on the same chromosome. *Immunogenetics* 2007;**59**:329–31.
145. Guo P, Hirano M, Herrin BR, Li J, Yu C, Sadlonova A, et al. Dual nature of the adaptive immune system in lampreys. *Nature* 2009;**459**:796–802.
146. Alder MN, Herrin BR, Sadlonova A, Stockard CR, Grizzle WE, Gartland LA, et al. Antibody responses of variable lymphocyte receptors in the lamprey. *Nat Immunol* 2008;**9**:319–27.
147. Kasamatsu J, Sutoh Y, Fugo K, Otsuka N, Iwabuchi K, Kasahara M. Identification of a third variable lymphocyte receptor in the lamprey. *Proc Natl Acad Sci USA* 2010;**107**: [in press].
148. Hayday AC. γδ cells: a right time and a right place for a conserved third way of protection. *Annu Rev Immunol* 2000;**18**:975–1026.
149. Bell JJ, Bhandoola A. The earliest thymic progenitors for T cells possess myeloid lineage potential. *Nature* 2008;**452**:764–7.
150. Wada H, Masuda K, Satoh R, Kakugawa K, Ikawa T, Katsura Y, et al. Adult T-cell progenitors retain myeloid potential. *Nature* 2008;**452**:768–72.
151. Kawamoto H, Katsura Y. A new paradigm for hematopoietic cell lineages: revision of the classical concept of the myeloid-lymphoid dichotomy. *Trends Immunol* 2009;**30**:193–200.
152. Klein J, Sato A, Mayer WE. Jaws and AIS. In: Kasahara M, editor. *Major histocompatibility complex: evolution, structure, and function*. Tokyo, Berlin, Heidelberg: Springer-Verlag; 2000. p. 3–26.
153. Marchalonis JJ, Kaveri S, Lacroix-Desmazes S, Kazatchkine MD. Natural recognition repertoire and the evolutionary emergence of the combinatorial immune system. *FASEB J* 2002;**16**:842–8.

Questions Arising from "Genome Duplication and T Cell Immunity"

Claude Perreault

Institute for Research in Immunology and Cancer, Université de Montréal, Montréal, Québec, Canada

(1) Rag1 and Rag2 genes are coexpressed during development and in the adult tissues of the purple sea urchin.[1] Is it possible to speculate on the role that Rag1 and Rag2 may play in animals that do not have an adaptive immune system?

(2) Analyses performed mainly in humans have led to the concept of an "extended MHC" (7.6 Mb) compared with the classical MHC (3.6 Mb).[2] The extended MHC would include large gene clusters such as tRNA and histone genes. From a phylogenetic perspective, is the extended MHC concept a solid and instructive paradigm?

References

1. Fugmann SD, Messier C, Novack LA, Cameron RA, Rast JP. An ancient evolutionary origin of the Rag1/2 gene locus. *Proc Natl Acad Sci USA* 2006;**103**:3728–33.
2. Horton R, et al. Gene map of the extended human MHC. *Nat Rev Genet* 2004;**5**:889–99.

Progress in Molecular Biology
and Translational Science, Vol. 92
DOI: 10.1016/S1877-1173(10)92012-7

Response to Questions

Masanori Kasahara

Department of Pathology, Hokkaido University Graduate School of Medicine, Sapporo, Japan

I. Response to Question #1

Unlike jawed vertebrate RAG1 and RAG2, coordinate expression of which is restricted to lymphocytes, sea urchin RAG1 and RAG2 are expressed in a variety of adult tissues. Among them are coelomocytes, the putative immune cells of the echinoderm.[1] Thus, it is possible that sea urchin RAG1/2 plays a role in immunity. However, to my knowledge, there is no experimental evidence supporting this speculation.

II. Response to Question #2

The idea of an "extended MHC" was proposed in humans because the regions surrounding the classical MHC contain MHC-related genes (most notably, the tapasin gene on the centromeric side and the *HFE* class I gene on the telomeric side), show synteny to the mouse MHC, and display linkage disequilibrium. Genes in the extended class II and class I regions are reasonably well conserved even in the frog, excluding class I genes themselves.[2] Thus, in terms of synteny conservation, the concept of an "extended MHC" is not limited to mammalian species. However, in the bony fish, the MHC does not occur as a single, contiguous region; here, class I, II, and III regions are not linked.[3] The fragmentation of the MHC in the bony fish is believed to be a consequence of secondary genomic rearrangements because class I and II genes are linked in the cartilaginous fish.[4] Genome paralogy observed in the MHC encompasses the "extended MHC."

Progress in Molecular Biology
and Translational Science, Vol. 92
DOI: 10.1016/S1877-1173(10)92013-9

1877-1173/10 $35.00

References

1. Fugmann SD, Messier C, Novack LA, Cameron RA, Rast JP. An ancient evolutionary origin of the *Rag1/2* gene locus. *Proc Natl Acad Sci USA* 2006;**103**:3728–33.
2. Ohta Y, Goetz W, Hossain MZ, Nonaka M, Flajnik MF. Ancestral organization of the MHC revealed in the amphibian *Xenopus*. *J Immunol* 2006;**176**:3674–85.
3. Bingulac-Popovic J, Figueroa F, Sato A, Talbot WS, Johnson SL, Gates M, et al. Mapping of *Mhc* class I and class II regions to different linkage groups in the zebrafish, *Danio rerio*. *Immunogenetics* 1997;**46**:129–34.
4. Ohta Y, Okamura K, McKinney EC, Bartl S, Hashimoto K, Flajnik MF. Primitive synteny of vertebrate major histocompatibility complex class I and class II genes. *Proc Natl Acad Sci USA* 2000;**97**:4712–7.

The Origin and Role of MHC Class I-Associated Self-Peptides

Claude Perreault

Institute for Research in Immunology and Cancer, Université de Montréal, Montréal, Québec, Canada

Under steady-state conditions, cell surface major histocompatibility complex (MHC) I molecules are associated with self-peptides collectively referred to as the self-MHC I immunopeptidome (SMII). The SMII regulates all key events that occur during the lifetime of CD8 T cells in the thymus and in the periphery. The SMII derives mainly from rapidly degraded proteins and contains a tissue-specific signature. Peptide-source proteins derive from all cell compartments but are enriched in RNA- and DNA-binding proteins, cyclins and cyclin-dependent kinases, ribosomal constituents, and chaperones. Cell stress, infection, and transformation can modify the repertoire of peptides in the SMII. Constitutive MHC I presentation of self-peptides is fraught with the risk of autoimmunity, and there is the need for complex self-tolerance mechanisms. However, self-peptide/MHC I complexes are essential for the development of "classic adaptive" TCR$\alpha\beta$ CD8 T cells and directly contribute to CD8 T-cell responses against pathogens and transformed cells.

Progress in Molecular Biology
and Translational Science, Vol. 92
DOI: 10.1016/S1877-1173(10)92003-6

I. Key Features of the Adaptive Immune System

The survival of vertebrates depends on the flawless functioning of their adaptive immune system. Whereas innate immune cells rely on invariant germline-encoded receptors for sensing pathogens, adaptive immune cells (B and T lymphocytes) require productive somatic gene rearrangement of their antigen receptor.[1] The dominant paradigm is that, as a complement to the innate immune system, the adaptive immune system confers two decisive advantages in resistance to pathogens: (i) greater specificity and versatility in immune defenses, and (ii) immune memory.[2,3] Still, it must be acknowledged that innate immune cells are necessary and sufficient to eradicate most pathogens, and that natural killer (NK) cells, for example, are endowed with a rudimentary form of memory.[4,5] Adaptive immune memory is particularly important during the perinatal period.[6] During this period of physiological immunoincompetence, protective maternal immunological memory (that allows transfer of neutralizing antibodies) is essential for survival of the fetus, the newborn, and the infant: that is, the survival of the species.[7] Recently, a second role has been proposed to explain conservation of the adaptive immune system in vertebrates: its ability to temper innate immune response.[8,9] Indeed, in the absence of T cells, mice are hyper-responsive to toll-like receptor (TLR) agonists such as lipopolysacharide (LPS) and poly(I:C) and develop exaggerated and potentially lethal inflammatory responses following viral infection.[8,9]

Though vertebrates can no longer live without an adaptive immune system, Hedrick, in a brilliant essay, casted some doubts on the dogma stating that emergence of the adaptive immune system conferred a long-term evolutionary advantage to vertebrates.[10] Since hosts and pathogens coevolve, there is never a "solution" to infectious agents, and more resistant hosts ultimately become colonized by more virulent pathogens. Accordingly, vertebrates pay a hefty price for possessing an adaptive immune system because, compared with invertebrates, they are colonized with more virulent pathogens and have a unique susceptibility to allergies and autoimmune diseases.[10] Nevertheless, the evolution around 500 million years ago of two very different somatically assembled "anticipatory receptors" of comparable diversity in jawless and jawed vertebrates, while conserving similar compartmentalization of T/B lymphocyte differentiation, strongly supports the survival value of adaptive immunity.[11]

As eloquently pointed out by C.A. Janeway and M.M. Davis, the adaptive immune system of gnathostomes is self-referential as it functions on internal cues to prepare it to react to the vast range of external cues that it encounters over a lifetime.[12,13] While that is true for B cells, CD4 T cells, and CD8 T cells, I will focus here on CD8 T cells because the self-ligands that they interact with have been more extensively characterized. Under steady-state conditions, that is in the absence of infection, cell surface MHC I molecules are associated solely with self-peptides. These peptides, collectively referred to as the self-MHC I immunopeptidome (SMII), regulate all key events that occur during the lifetime of CD8 T cells in the thymus and in the periphery. I will therefore examine two related questions in the present essay: what are the origin and the role of the SMII?

II. Genesis of the SMII

A. The SMII Is Shaped by Protein Translation and Degradation

Despite the tremendous importance of the SMII, we know very little about its genesis and molecular composition. Nevertheless, it is clear that the content of the SMII is ultimately shaped by two processes: protein translation and degradation.[14-17] Thus, generation of MHC I-associated peptides ceases abruptly in the presence of cycloheximide (that blocks protein synthesis) or proteasome inhibitors.[15,18,19] However, the SMII is not a random sample of the proteome: many abundant proteins do not generate MHC I-associated peptides, while some low-abundance proteins have a major contribution to the SMII.[20-22] A likely explanation for this discrepancy is that the SMII preferentially derives from rapidly degraded polypeptides (RDPs; $t_{1/2}$ of $\sim$10 min) relative to slowly degraded proteins (SDPs; $t_{1/2}$ of $\sim$1000 min).[14,15,17,21,23-28] Thus, in a sense, MHC I preferentially sample what is being translated, as opposed to what has been translated.[15] Two structural features are typically required for proteins to be degraded: covalently attached ubiquitin polypeptides and an unstructured region in the targeted protein.[29] These two structural features can act in *trans* when separated onto different proteins in a multi-subunit complex.[30] Ubiquitination allows protein binding to the proteasome 19S regulatory particle via a diverse set of ubiquitin receptors.[31] The unstructured region serves as the initiation site for degradation and is hydrolyzed first, after which the rest of the protein is digested sequentially.[29,30] In addition, evidence suggests that proteins with unstructured regions can undergo NQO1-dependent but ubiquitin-independent degradation by the 20S proteasome.[32-34] One reason why RDPs are the main source of MHC I-associated peptides is that many cytosolic proteins proceed through a "fragile period" during which they are sensitive to ubiquitin-dependent proteasomal degradation induced by superoxide radicals or increased temperatures.[35] The fragile period lasts for 30–60 min after synthesis, presumably until proteins have assumed their final conformations in stable multimeric complexes.[35] Degradation of a substantial fraction of newly synthesized proteins is probably important to maintain the proper steady-state levels of proteins and to preclude the interference of excess protein monomers with the functioning of multiprotein complexes.[36]

Many RDPs represent defective ribosomal products (DRiPs), that is, polypeptides that fail to achieve native structure owing to inevitable imperfections in transcription, translation, posttranslational modifications, or protein folding.[21,23,28,37] That is the case for STT3B, whose high DRiP rate contributes to the generation of STT3B-derived MHC I-associated peptides.[21] Approximately 25% of cell proteins are targeted to the endoplasmic reticulum (ER) for secretion or cell surface display and are eventually degraded by extracytosolic proteases. Using DRiPs (generated before protein exportation into the secretory pathway) as a source of peptides allows those proteins to be represented in the SMII.[38] Nevertheless, on a per-minute basis, cells synthesize $\sim 3 \times 10^6$ proteins (with an average length of 450 residues), of which $\sim 30\%$ are DRiPs, but only need to fill $\sim 10^2$ class I molecules with short peptides ($\sim$ 9mers).[23,27] Thus, even among RDPs/DRiPs, only a small proportion of the polypeptide pool is effectively presented by the MHC I.

Genesis of the SMII is initiated by proteasomal degradation of proteins in the cytosol and nucleus.[39-41] Only a few MHC I peptides (MIPs) can be produced in the absence of proteasomes.[19,42] The proteasome digests proteins mainly in fragments ranging in size from 4 to 20 amino acids.[43,44] Thus, many peptides are too small for presentation by MHC I molecules and are recycled into amino acids that can be used for synthesis of new proteins. Another fraction is appropriate or too long for MHC class I molecules. These, too, are substrates for various cytosolic peptidases, such as tripeptidyl aminopeptidase II (TPPII) and thimet oligopeptidase (TOP), from which they can be protected by chaperones like T-complex protein 1 (TCP-1) and heat-shock protein 90 (HSP90).[45-50] In the end, most proteasomal degradation products are degraded to amino acids by cytosolic peptidases, but a few ($< 0.1\%$) can bind the transporter for antigen processing (TAP) and are then translocated into the ER.[14] In the ER, peptides longer than 10–11 amino acids can be further trimmed by the ER aminopeptidase ERAP1/ERAAP to generate peptides of optimal length (8–11 amino acids), which can then associate with MHC I proteins with the help of a peptide loading complex whose main components are calnexin, calreticulin, ERp57, and tapasin.[51-55] Successful peptide binding releases the MHC I molecules from the peptide loading complex for delivery to the cell surface through the secretory pathway.

B. The Elusive Role of Immunoproteasomes

While all eukaryotic cells express constitutive proteasomes (CPs), gnathostomes also express another form of proteasome, the immunoproteasomes (IPs). In IPs, the three catalytic β-subunits expressed in CPs are replaced by three

IFN-γ-inducible homologs (immunosubunits): low molecular weight protein (LMP)-2 (or β1i) for β1, multicatalytic endopeptidase complex-like (MECL)-1 (or β2i) for β2, and LMP7 (or β5i) for β5. In gnathostomes, most cells express only CPs under steady-state conditions and harbor IPs when exposed to IFN-γ.[56] In contrast, DCs constitutively express both CPs and IPs. IPs represents half of the proteasome population in immature DCs, and LPS-triggered DC maturation slightly increases the IP:CP ratio.[57] IPs are closely linked to the adaptive immune system, being present in all gnathostomes but absent in invertebrates. Phylogenetic analyses revealed that proteasome immunosubunits evolved faster than their constitutive counterparts.[58,59] This finding points to a functional differentiation between IPs and CPs. However, the ultimate role of IPs, that is, their ecologically relevant and evolutionarily selected function, remains elusive.

It has been assumed that the key role of IPs may hinge on their impact on the repertoire of peptides associated with MHC I molecules. Studies of selected epitopes revealed that some MHC I-associated peptides can be generated only by CPs, some only by IPs, and others by both types of proteasomes.[60-66] *In vitro* proteasome digestion experiments suggest that, compared to CPs, IPs have greater efflux and cleavage rates and generate more N-extended versions of MHC I epitopes.[44,67] In addition, immunosubunits alter proteasome structure and cleavage site preferences.[65,68] Nevertheless, the aforementioned studies cannot predict the overall impact of IPs on the MIP repertoire *in vivo*, mainly for three reasons. First, *in vitro* proteasomal digestion may not reproduce *in vivo* conditions, where most MIPs derive from rapidly degraded proteins that translocate into the ER a few seconds after cleavage by proteasomes.[14,69] Second, MIP presentation is orchestrated by several steps downstream of proteasomal digestion so that only a small fraction of peptides generated by proteasomes are presented by MHC I molecules.[46,70,71] Finally, previous studies did not take into account potential differences in transcription regulation by CPs and IPs. While CPs clearly regulate transcriptional activation,[72-75] the potential impact of IPs on transcription remains unexplored. Conceivably, IPs and CPs might differentially regulate MHC I presentation of a given peptide not only by affecting degradation of the peptide's source protein but also by modulating transcription of the gene encoding that peptide. The overall impact of IPs on the SMII has yet to be studied by high-throughput mass spectrometry (MS)-based methods.

C. The SMII Repertoire is Exceedingly Complex

Proteomic analysis of the SMII is a daunting task since it encompasses thousands of peptides present in low copy numbers per cell.[76,77] Each MHC molecule recognizes peptides through a broadly defined consensus motif of amino acids serving as anchors to the appropriate binding pockets on the MHC molecules. Such motifs were first established by pool Edman sequencing of unfractionated peptide mixtures eluted from MHC molecules.[78] Direct biochemical characterization of specific MHC I-associated peptides has typically involved immunoaffinity purification of MHC molecules after cell lysis, fractionation of the peptides by chromatography, and sequencing. Refinements in MS methods pioneered by Engelhard and Hunt et al. represented a major progress in MHC-peptide analysis and sequencing that led to characterization of several MHC I-associated peptides.[76,78-80] More recently, several MS-based methods have been developed for high-throughput analyses and sequencing of the SMII.[81-88] These studies have shown that MHC I proteins present peptides derived from all cell compartments.[81,85] Statistical analyses of peptide-source proteins revealed a preference for RNA- and DNA-binding proteins, proteins related to cyclins and cyclin-dependent kinases, ribosomal constituents, and chaperones.[81,85] Thus, proteins that regulate transcription, translation, and cell cycle progression are overrepresented in the SMII.

With rare exception, discrete MHC I molecules present peptides derived from different sets of source proteins.[85] A corollary is that expression of multiple MHC I allelic products in a given subject (a consequence of gene duplication and diversification) favors representation of largely nonoverlapping sets of source proteins in the SMII. By integrating global profiling of the mouse protein-encoding transcriptome[89] with the MIP repertoire of thymocytes, we reported that the thymocytes' SMII is enriched in peptides derived from highly abundant transcripts.[85] In addition, we found that level of overlap between the SMII of primary C57BL/6 mouse thymocytes and dendritic cells was only 40% (unpublished). These data demonstrate that, at least for the cell types examined, the SMII contains a tissue-specific signature. Notably, the SMII contains peptides bearing posttranslational modifications: glycosylation, phosphorylation, and protein splicing.[90-95] Furthermore, MHC I-associated peptides can originate from cryptic translation products. Cryptic translation refers to polypeptides that are synthesized in the cells by unconventional translational mechanisms. These include peptides encoded by introns, intron/exon junctions, 5′ and 3′ untranslated regions, and alternate translational reading frames.[96,97] The SMII therefore contains a very diversified peptide repertoire that contains a tissue-specific signature and is shaped by poorly defined mechanisms.

D. Infection, Transformation, and Cell Stress can Modify the SMII

In view of the fact that viral infection and neoplastic transformation trigger ER stress, it can be assumed that a large proportion of the cells that must be recognized by CD8 T lymphocytes are stressed cells.[98-101] ER stress regulates the two processes that shape the SMII, protein translation and degradation.

Different stress stimuli (DNA damage, hypoxia, misfolded proteins, reactive oxygen species, starvation) ultimately lead to accumulation of unfolded proteins in the ER and thereby elicit the unfolded protein response (UPR).[98,99,101-105] The UPR has pervasive effects on cell protein economy: (i) increased transcription of genes involved in protein folding and maturation in the ER, (ii) decreased initiation of protein translation, and (iii) increased degradation of ER proteins and of mRNAs for ER proteins.[106-111]

As ER stress affects protein translation and degradation in many ways, it can potentially induce pleiotropic changes in the SMII. In line with this, MS-based studies of peptides presented by HLA-B*0702 in HIV-infected and -uninfected cells yielded quite spectacular results. Most peptides overexpressed by HIV-infected cells did not derive from HIV proteins but from host cellular proteins that may all relate to the stress response.[112] Furthermore, HSPs, whose upregulation is a hallmark of cell stress, stand out as a predominant source of MHC I-associated peptides.[113] In several models, a variety of stimuli (pharmacological agents, mutant proteins, glucose starvation, and saturated fatty acid) were found to decrease expression of MHC I-peptide complexes.[114,115] Evidence suggests that decreased MHC I presentation during ER stress was due, at least in part, to global attenuation of protein synthesis.[115] While ~20–25% of peptides are differentially expressed in the SMII of normal versus neoplastic cells,[82,85,116] we ignore to what extent such discrepancies might be due to ongoing ER stress in cancer cells.

What might be the impact of ER stress on immune recognition of infected and neoplastic cells? Paradoxically, at least in some cases, it could facilitate recognition of virus-infected cells. By inducing phosphorylation of eIF2α, ER stress hampers canonical cap-dependent translation initiation which regulates synthesis of 95–98% of cellular mRNAs.[117] However, some viruses can use internal ribosomal entry sites in their 5′ noncoding region to initiate cap-independent translation.[98,117] Thus, by preferentially repressing presentation of self-peptides, the UPR could facilitate recognition of viral peptides (the needle in the haystack[118]). The potential impact of the UPR on recognition of neoplastic cells is not inherently obvious. On the one hand, by repressing production of MHC I-peptide complexes, the UPR may hinder presentation of tumor antigens to CD8 T cells and facilitate immune evasion. Indeed, generation of optimal CD8 T-cell responses is promoted by high epitope density on antigen presenting cells.[119,120] However, an elegant study by Schwab et al. has shown that, upon induction of eIF2α phosphorylation by ER stress, cells can generate MHC I-associated peptides derived from cryptic cap-independent translational reading frames.[121] Expression of such cryptic peptides by neoplastic cells might trigger recognition of stressed cells by CD8 T lymphocytes. The fragmentary data summarized in this section indicate that ER stress can impinge in many ways on the SMII. However, high-throughput sequencing of MHC I-associated peptides will be necessary to comprehensively evaluate how ER stress molds the peptide repertoire (in terms of both abundance and diversity) and to gain further insights into the global impact of the UPR on recognition of stressed cells by CD8 T lymphocytes.

III. The Role of Constitutive Self-Peptide Presentation

Constitutive self-peptide presentation incurs a number of costs.[118] First, it has necessitated the evolution of elaborate central and peripheral self-tolerance mechanisms which, despite their sophistication, fail to prevent the emergence of autoimmunity in 3–5% of humans.[122-126] Maintenance of T-cell tolerance is a byzantine process because T-cell receptors are expressed following somatic gene rearrangements and are extremely diverse. Second, at face value, the SMII appears to create a problem of signal-to-noise discrimination: at the time of infection, T cells that scan MHC I-associated peptides must specifically react to a few pathogen-derived peptides admixed with $\sim 10^5$ self-peptides. Then, what is the ecologically relevant and evolutionarily selected function (the "ultimate role" according to Casanova and Abel[127]) of the SMII?

A. T-Cell Development and Homeostasis

Development of "conventional" or "classic" TCRαβ T cells is finely regulated by MHC-associated peptides. That is true for positive selection, negative selection, and peripheral survival of naive T cells.[124,128,129] However, development of several T-cell subsets is independent of MHC-associated peptides. These "unconventional" or "innate" T cells include γδT cells, CD1-specific NK T cells, CD8αα T cells in the intraepithelial compartment, and some MHC Ib-restricted TCRαβ T cells.[130-133] Therefore, are SMII-dependent and -independent T cells similar? The answer is clearly no. Innate T cells exhibit a surface phenotype similar to activated or memory conventional T cells, and generate almost immediate effector responses.[131] Furthermore, in contrast to classic T cells, many, if not all innate T cells do not require thymic epithelium for positive selection and are exquisitely IL-15-dependent.[134]

While innate T cells can undoubtedly contribute to eradication of pathogens, can they substitute for classic T cells? This question was addressed in oncostatin M-transgenic (OMTg) mice. In OMTg mice, T cells develop extrathymically, in the lymph nodes, and share all attributes of innate T cells.[135-137] To evaluate their protective value against pathogens, we assessed *in vivo* responses of OMTg innate T cells against two model infections: lymphocytic choriomeningitis virus (LCMV) and vesicular stomatitis virus (VSV). OMTg innate CD4 T cells were found to be functionally deficient. They did not expand properly, they did not provide adequate help to B cells in mice infected with VSV or LCMV, and they contained an increased proportion (40%) of Foxp3$^+$ Treg cells.[138,139] The functionality of OMTg innate CD8 T cells was superior to that of their CD4 counterpart but they could not substitute for classic CD8 T cells. Thus, following LCMV infection, GP33-41 tetramer$^+$ OMTg innate CD8 T cells initiated proliferation and IFN-γ secretion swiftly but were unduly susceptible to apoptosis and exhaustion.[138,139] As a result, OMTg innate CD8 T cells specific for the GP33-41 epitope accumulated to lower levels than conventional T cells and provided only transient control of LCMV infection.[138] We must therefore conclude that, in contrast to innate T cells, generation of classic TCR$\alpha\beta$ T cells is contingent on molding by the SMII.

B. Response to Pathogens

The prime function of the adaptive immune system is protection against infectious agents, and its functioning has been molded by coevolution with pathogens.[6,10] It is therefore logical to postulate that a putative key selective advantage conferred by constitutive expression of the SMII would probably involve CD8 T-cell responses against bacteria and viruses. Infection by intracellular pathogens is commonly depicted as a race between “invaders” and CD8 T cells where time is of the essence. Thus, since viral infections are characterized by rates of replication exceeding 10^{12} virions per day, a short delay in T-cell expansion can make the difference between life and death.[140,141] Experimental studies suggest two nonmutually exclusive ways by which the SMII may help the adaptive immune system to win the race with pathogens: the first is indirect; the second, direct. The first strategy hinges on enhancement of CD8 T-cell response to pathogen-derived MHC I-associated peptides. Since only a few T cells bear a TCR reactive to a given pathogen, these T cells must proliferate very rapidly. An uninfected mouse has only about 100–200 CD8 T cells per epitope that must expand ≥ 1000-fold to control infection.[142-145] Remarkably, the MHC I molecules that do not present pathogen peptides substantially enhance the ability of CD8 T cells to detect foreign antigens, an effect measurable by upregulation of activation or maturation markers and by increased effector function.[146-148] In other words, cooperation between self-peptide/MHC I complexes and foreign peptide/MHC I complexes augment CD8 T-cell responses against pathogens. Synergism between the self-MHC-peptide complexes and foreign peptide/MHC I complexes seems to be initiated by interactions between MHC I (on APCs) and CD8 (on T cells) that recruit Lck to the immunological synapse and thereby enhance TCR signal transduction.[146-149] Since MHC Ia molecules (classical or “modern” MHC I molecules) are not exported at the cell surface unless they are loaded with a peptide,[150] the SMII is essential for MHC Ia expression and is therefore required for enhancement of T-cell responses to foreign epitopes. However, as interactions between CD8 and MHC I are not peptide-dependent, the devil’s advocate might wonder whether exploitation of CD8 interactions with peptide-less MHC Ib molecules (e.g., HFE) would have provided a more economical alternative by not requiring generation of the SMII.

The SMII may also be a direct target of antipathogen T-cell responses. As mentioned above, the ER stress induced by infection can enhance MHC I presentation of self-peptides derived from stress proteins.[112] Several stress peptides were found to be immunogenic for their hosts and to elicit CD8 T-cell responses.[100,113] Specific examples include peptides derived from vinculin in HIV-infected subjects,[151] from HSP90 and IFI-6-16 following measles infection,[152] and from HSP70 in patients with epithelial cancers.[153,154] These data support the exciting but largely unexplored possibility that one function of the SMII might be to present peptides from stress proteins to CD8 T cells.[100,113] In theory, CD8 T cells primed against stress peptides would have the ability to recognize rapidly cells stressed by a variety of pathogens or even by neoplastic transformation.

C. Mate Selection

A provocative concept is that MHC-associated peptides serve as signals of individuality in the context of sexual selection.[155] MHC-peptide complexes are shed from the cell surface and are found in body fluids. Evidence suggests that olfactory assessment of the SMII by sensory neurons influences mate choice in mice, fishes, birds, and humans.[156-160] However, several studies argue against the notion that the MHC genotype influences the urinary volatile composition or mating preferences.[161-163] For the time being, that fascinating and complex issue remains unsettled. Nonetheless, the potential evolutionary impact of the SMII on sexual selection could be considerable.

D. Cancer Immunosurveillance

The realms of infection and neoplasia overlap substantially. About 16% of malignancies worldwide can be attributed to infection with hepatitis B and C viruses, the human papillomaviruses, Epstein-Barr virus (EBV), human T-cell lymphotrophic virus I, human immunodeficiency virus (HIV), the bacterium *Helicobacter pylori*, schistosomes, or liver flukes.[164] The adaptive immune system undoubtedly contributes to the eradication of the aforementioned pathogens and therefore, indirectly decreases cancer incidence. A more contentious issue is whether T cells mediate immunosurveillance against spontaneous nonmicrobial cancers (~84% of cancers).[165,166] The main argument developed in this section is that, although CD8 T cells interact in many ways with cancer cells, immunosurveillance against spontaneous nonmicrobial cancers is not the raison d'être of constitutive self-peptide presentation.

What is the impact of T-cell deficiency on the rate of spontaneous tumors and cancer deaths? In the seminal study of Shankaran, neither wild-type nor $RAG2^{-/-}$ mice showed outward signs of disease during the observation period

(15–21 months), but RAG2$^{-/-}$ mice presented a greater frequency of tumors at necropsy.[167] These data suggest that lifelong deficiency of adaptive immunity increases cancer incidence in senescent laboratory mice. However, in our view, they also suggest that cancer immunosurveillance is not a fundamental role of the adaptive immune system (and therefore of the SMII). Indeed, RAG2$^{-/-}$ mice remained clinically fit until the term of the study when they reached an age (15–21 months) that outbred mice usually do not attain in the wild.[167] Ailments restricted to such aged animals are unlikely to have any evolutionary relevance. In humans, substantial evidence suggests that, during tumor progression, neoplastic cells have to deal with CD8 T cells and may adopt two strategies: decrease their immunogenicity (immunoediting) or induce tolerance.[167-171] Furthermore, lack of T-cell infiltrates in breast, colon, and lung malignant tumors correlates with a poor prognosis.[172-174] It therefore seems sound to assume that T cells can mitigate the growth of some tumor types. However, it is important to note that the median age at diagnosis for breast, colon, and lung cancers is 61–71 years of age. Cancer immunosurveillance in this context is certainly medically relevant[170] but probably of no evolutionary significance because natural selection is not influenced by traits that affect subjects in that age group.

By no means do we imply that T cells cannot be used to treat cancer. On the contrary, we believe that vaccines and adoptive T-cell immunotherapies offer exciting possibilities for cancer treatment and perhaps cancer prevention.[175-183] However, current evidence does not support the idea that immunosurveillance against nonmicrobial malignancies has been a driving force that might justify constitutive self-peptide presentation. We wish to emphasize that further studies on ecological immunology could shed new light on this question. MHC-peptide presentation and adaptive immunity are found only in jawed vertebrates. If they had been positively selected during evolution for cancer immunosurveillance, one might expect to find high cancer rates in long-lived invertebrates from the phyla of arthropods, mollusks, and echinoderms.[10] To the best of our knowledge, this question has yet to be addressed.[184] In addition, further insights might emerge from studies on cancer incidence in agnathans (lamprey and hagfish) who possess an adaptive immune system but no MHC-like genes.[185]

Acknowledgments

I am indebted to my students (Étienne Caron, Diana Granados, Danièle de Verteuil, and Krystel Vincent) and colleagues (Pierre Thibault, Sébastien Lemieux, Jonathan Bramson, Brad Nelson, and Réjean Lapointe) for thoughtful discussion and suggestions. The experimental work from my laboratory was supported by the Canadian Cancer Society Research Institute (grant # 019475), the Terry Fox New Frontiers Program in Cancer Immunotherapy (grant# 018005) and the Canada Research Chair program.

References

1. Lanier LL, Sun JC. Do the terms innate and adaptive immunity create conceptual barriers? *Nat Rev Immunol* 2009;**9**:302–3.
2. Flajnik MF, Du Pasquier L. Evolution of the immune system. In: Paul WE, editor. *Fundamental immunology*. Philadelphia: Lippincott Williams & Wilkins; 2008. p. 56–124.
3. Parham P. Principles of adaptive immunity. In: Parham P, editor. *The immune system*. New York: Garland Science; 2009. p. 71–92.
4. Kurtz J. Specific memory within innate immune systems. *Trends Immunol* 2005;**26**:186–92.
5. O'leary JG, Goodarzi M, Drayton DL, von Andrian UH. T cell- and B cell-independent adaptive immunity mediated by natural killer cells. *Nat Immunol* 2006;**7**:507–16.
6. Zinkernagel RM. Uncertainties—discrepancies in immunology. *Immunol Rev* 2002;**185**:103–25.
7. Zinkernagel RM. On natural and artificial vaccinations. *Annu Rev Immunol* 2003;**21**:515–46.
8. Kim KD, Zhao J, Auh S, Yang X, Du P, Tang H, et al. Adaptive immune cells temper initial innate responses. *Nat Med* 2007;**13**:1248–52.
9. Zhao J, Kim KD, Yang X, Auh S, Fu YX, Tang H. Hyper innate responses in neonates lead to increased morbidity and mortality after infection. *Proc Natl Acad Sci USA* 2008;**105**:7528–33.
10. Hedrick SM. The acquired immune system; a vantage from beneath. *Immunity* 2004;**21**:607–15.
11. Guo P, Hirano M, Herrin BR, Li J, Yu C, Sadlonova A, et al. Dual nature of the adaptive immune system in lampreys. *Nature* 2009;**459**:796–801.
12. Janeway Jr. CA. How the immune system works to protect the host from infection: a personal view. *Proc Natl Acad Sci USA* 2001;**98**:7461–8.
13. Davis MM, Krogsgaard M, Huse M, Huppa J, Lillemeier BF, Li QJ. T cells as a self-referential, sensory organ. *Annu Rev Immunol* 2007;**25**:681–95.
14. Yewdell JW, Reits E, Neefjes J. Making sense of mass destruction: quantitating MHC class I antigen presentation. *Nat Rev Immunol* 2003;**3**:952–61.
15. Qian S-B, Reits E, Neefjes J, Deslich JM, Bennink JR, Yewdell JW. Tight linkage between translation and MHC class I peptide ligand generation implies specialized antigen processing for defective ribosomal products. *J Immunol* 2006;**177**:227–33.
16. Rock KL, York IA, Saric T, Goldberg AL. Protein degradation and the generation of MHC class I-presented peptides. *Adv Immunol* 2002;**80**:1–70.
17. Reits EA, Vos JC, Grommé M, Neefjes J. The major substrates for TAP *in vivo* are derived from newly synthesized proteins. *Nature* 2000;**404**:774–8.
18. Michalek MT, Grant EP, Gramm C, Goldberg AL, Rock KL. A role for the ubiquitin-dependent proteolytic pathway in MHC class I-restricted antigen presentation. *Nature* 1993;**363**:552–4.
19. Wherry EJ, Golovina TN, Morrison SE, Sinnathamby G, McElhaugh MJ, Shockey DC, et al. Re-evaluating the generation of a "proteasome-independent" MHC class I-restricted CD8 T cell epitope. *J Immunol* 2006;**176**:2249–61.
20. Frumento G, Corradi A, Ferrara GB, Rubartelli A. Activation-related differences in HLA class I-bound peptides: presentation of an IL-1 receptor antagonist-derived peptide by activated, but not resting, $CD4^+$ T lymphocytes. *J Immunol* 1997;**159**:5993–9.
21. Caron É, Charbonneau R, Huppé G, Brochu S, Perreault C. The structure and location of SIMP/STT3B account for its prominent imprint on the MHC I immunopeptidome. *Int Immunol* 2005;**17**:1583–96.
22. Milner E, Barnea E, Beer I, Admon A. The turnover kinetics of MHC peptides of human cancer cells. *Mol Cell Proteomics* 2006;**5**:357–65.

23. Yewdell JW, Nicchitta CV. The DRiP hypothesis decennial: support, controversy, refinement and extension. *Trends Immunol* 2006;**27**:368–73.
24. Poole B, Wibo M. Protein degradation in cultured cells. The effect of fresh medium, fluoride, and iodoacetate on the digestion of cellular protein of rat fibroblasts. *J Biol Chem* 1973;**248**:6221–6.
25. Wheatley DN, Giddings MR, Inglis MS. Kinetics of degradation of "short-" and "long-lived" proteins in cultured mammalian cells. *Cell Biol Int Rep* 1980;**4**:1081–90.
26. Schubert U, Anton LC, Gibbs J, Norbury CC, Yewdell JW, Bennink JR. Rapid degradation of a large fraction of newly synthesized proteins by proteasomes. *Nature* 2000;**404**:770–4.
27. Princiotta MF, Finzi D, Qian SB, Gibbs J, Schuchmann S, Buttgereit F, et al. Quantitating protein synthesis, degradation, and endogenous antigen processing. *Immunity* 2003;**18**:343–54.
28. Qian SB, Princiotta MF, Bennink JR, Yewdell JW. Characterization of rapidly degraded polypeptides in mammalian cells reveals a novel layer of nascent protein quality control. *J Biol Chem* 2006;**281**:392–400.
29. Prakash S, Tian L, Ratliff KS, Lehotzky RE, Matouschek A. An unstructured initiation site is required for efficient proteasome-mediated degradation. *Nat Struct Mol Biol* 2004;**11**:830–7.
30. Prakash S, Inobe T, Hatch AJ, Matouschek A. Substrate selection by the proteasome during degradation of protein complexes. *Nat Chem Biol* 2009;**5**:29–36.
31. Finley D. Recognition and processing of ubiquitin-protein conjugates by the proteasome. *Annu Rev Biochem* 2009;**78**:477–513.
32. Asher G, Lotem J, Sachs L, Kahana C, Shaul Y. Mdm-2 and ubiquitin-independent p53 proteasomal degradation regulated by NQO1. *Proc Natl Acad Sci USA* 2002;**99**:13125–30.
33. Asher G, Tsvetkov P, Kahana C, Shaul Y. A mechanism of ubiquitin-independent proteasomal degradation of the tumor suppressors p53 and p73. *Genes Dev* 2005;**19**:316–21.
34. Tsvetkov P, Asher G, Paz A, Reuven N, Sussman JL, Silman I, et al. Operational definition of intrinsically unstructured protein sequences based on susceptibility to the 20S proteasome. *Proteins* 2008;**70**:1357–66.
35. Medicherla B, Goldberg AL. Heat shock and oxygen radicals stimulate ubiquitin-dependent degradation mainly of newly synthesized proteins. *J Cell Biol* 2008;**182**:663–73.
36. Asher G, Reuven N, Shaul Y. 20S proteasomes and protein degradation "by default" *Bioessays* 2006;**28**:844–9.
37. Ostankovitch M, Robila V, Engelhard VH. Regulated folding of tyrosinase in the endoplasmic reticulum demonstrates that misfolded full-length proteins are efficient substrates for class I processing and presentation. *J Immunol* 2005;**174**:2544–51.
38. Yewdell JW, Hickman HD. New lane in the information highway: alternative reading frame peptides elicit T cells with potent antiretrovirus activity. *J Exp Med* 2007;**204**:2501–4.
39. Rock KL, Gramm C, Rothstein L, Clark K, Stein R, Dick L, et al. Inhibitors of the proteasome block the degradation of most cell proteins and the generation of peptides presented on MHC class I molecules. *Cell* 1994;**78**:761–71.
40. Anton LC, Schubert U, Bacik I, Princiotta MF, Wearsch PA, Gibbs J, et al. Intracellular localization of proteasomal degradation of a viral antigen. *J Cell Biol* 1999;**146**:113–24.
41. Wojcik C, DeMartino GN. Intracellular localization of proteasomes. *Int J Biochem Cell Biol* 2003;**35**:579–89.
42. Kessler B, Hong X, Petrovic J, Borodovsky A, Dantuma NP, Bogyo M, et al. Pathways accessory to proteasomal proteolysis are less efficient in major histocompatibility complex class I antigen production. *J Biol Chem* 2003;**278**:10013–21.
43. Kisselev AF, Akopian TN, Woo KM, Goldberg AL. The sizes of peptides generated from protein by mammalian 26 and 20 S proteasomes. Implications for understanding the degradative mechanism and antigen presentation. *J Biol Chem* 1999;**274**:3363–71.

44. Cascio P, Hilton C, Kisselev AF, Rock KL, Goldberg AL. 26S proteasomes and immunoproteasomes produce mainly N-extended versions of an antigenic peptide. *EMBO J* 2001;**20**:2357–66.
45. Kunisawa J, Shastri N. The group II chaperonin TRiC protects proteolytic intermediates from degradation in the MHC class I antigen processing pathway. *Mol Cell* 2003;**12**:565–76.
46. York IA, Mo AXY, Lemerise K, Zeng W, Shen Y, Abraham CR, et al. The cytosolic endopeptidase, thimet oligopeptidase, destroys antigenic peptides and limits the extent of MHC class I antigen presentation. *Immunity* 2003;**18**:429–40.
47. Kloetzel PM. Generation of major histocompatibility complex class I antigens: functional interplay between proteasomes and TPPII. *Nat Immunol* 2004;**5**:661–9.
48. Reits E, Neijssen J, Herberts C, Benckhuijsen W, Janssen L, Drijfhout JW, et al. A major role for TPPII in trimming proteasomal degradation products for MHC class I antigen presentation. *Immunity* 2004;**20**:495–506.
49. Kunisawa J, Shastri N. Hsp90α chaperones large C-terminally extended proteolytic intermediates in the MHC class I antigen processing pathway. *Immunity* 2006;**24**:523–34.
50. Callahan MK, Garg M, Srivastava PK. Heat-shock protein 90 associates with N-terminal extended peptides and is required for direct and indirect antigen presentation. *Proc Natl Acad Sci USA* 2008;**105**:1662–7.
51. Zhang Y, Williams DB. Assembly of MHC class I molecules within the endoplasmic reticulum. *Immunol Res* 2006;**35**:151–62.
52. York IA, Brehm MA, Zendzian S, Towne CF, Rock KL. Endoplasmic reticulum aminopeptidase 1 (ERAP1) trims MHC class I-presented peptides in vivo and plays an important role in immunodominance. *Proc Natl Acad Sci USA* 2006;**103**:9202–7.
53. Hammer GE, Gonzalez F, Champsaur M, Cado D, Shastri N. The aminopeptidase ERAAP shapes the peptide repertoire displayed by major histocompatibility complex class I molecules. *Nat Immunol* 2006;**7**:103–12.
54. Dong G, Wearsch PA, Peaper DR, Cresswell P, Reinisch KM. Insights into MHC class I peptide loading from the structure of the tapasin-ERp57 thiol oxidoreductase heterodimer. *Immunity* 2009;**30**:21–32.
55. Purcell AW, Elliott T. Molecular machinations of the MHC-I peptide loading complex. *Curr Opin Immunol* 2008;**20**:75–81.
56. Heink S, Ludwig D, Kloetzel PM, Kruger E. IFN-γ-induced immune adaptation of the proteasome system is an accelerated and transient response. *Proc Natl Acad Sci USA* 2005;**102**:9241–6.
57. Macagno A, Kuehn L, de Giuli R, Groettrup M. Pronounced up-regulation of the PA28α/β proteasome regulator but little increase in the steady-state content of immunoproteasome during dendritic cell maturation. *Eur J Immunol* 2001;**31**:3271–80.
58. Hughes AL. Evolution of the proteasome components. *Immunogenetics* 1997;**46**:82–92.
59. Kesmir C, van Noort V, de Boer RJ, Hogeweg P. Bioinformatic analysis of functional differences between the immunoproteasome and the constitutive proteasome. *Immunogenetics* 2003;**55**:437–49.
60. Sewell AK, Price DA, Teisserenc H, Booth BLJ, Gileadi U, Flavin FM, et al. IFN-γ exposes a cryptic cytotoxic T lymphocyte epitope in HIV-1 reverse transcriptase. *J Immunol* 1999;**162**:7075–9.
61. Morel S, Lévy F, Burlet-Schiltz O, Brasseur F, Probst-Kepper M, Peitrequin AL, et al. Processing of some antigens by the standard proteasome but not by the immunoproteasome results in poor presentation by dendritic cells. *Immunity* 2000;**12**:107–17.
62. Chen W, Norbury CC, Cho Y, Yewdell JW, Bennink JR. Immunoproteasomes shape immunodominance hierarchies of antiviral $CD8^+$ T cells at the levels of T cell repertoire and presentation of viral antigens. *J Exp Med* 2001;**193**:1319–26.

63. Chapiro J, Claverol S, Piette F, Ma W, Stroobant V, Guillaume B, et al. Destructive cleavage of antigenic peptides either by the immunoproteasome or by the standard proteasome results in differential antigen presentation. *J Immunol* 2006;**176**:1053–61.
64. Deol P, Zaiss DMW, Monaco JJ, Sijts AJAM. Rates of processing determine the immunogenicity of immunoproteasome-generated epitopes. *J Immunol* 2007;**178**:7557–62.
65. Kloetzel PM. Antigen processing by the proteasome. *Nat Rev Mol Cell Biol* 2001;**2**:179–87.
66. Dannull J, Lesher DT, Holzknecht R, Qi W, Hanna G, Seigler H, et al. Immunoproteasome down-modulation enhances the ability of dendritic cells to stimulate anti-tumor immunity. *Blood* 2007;**110**:4341–50.
67. Mishto M, Luciani F, Holzhutter HG, Bellavista E, Santoro A, Textoris-Taube K, et al. Modeling the in vitro 20S proteasome activity: the effect of PA28-αβ and of the sequence and length of polypeptides on the degradation kinetics. *J Mol Biol* 2008;**377**:1607–17.
68. Toes RE, Nussbaum AK, Degermann S, Schirle M, Emmerich NP, Kraft M, et al. Discrete cleavage motifs of constitutive and immunoproteasomes revealed by quantitative analysis of cleavage products. *J Exp Med* 2001;**194**:1–12.
69. Eisenlohr LC, Huang L, Golovina TN. Rethinking peptide supply to MHC class I molecules. *Nat Rev Immunol* 2007;**7**:403–10.
70. Hammer GE, Kanaseki T, Shastri N. The final touches make perfect the peptide-MHC class I repertoire. *Immunity* 2007;**26**:397–406.
71. Wearsch PA, Cresswell P. Selective loading of high-affinity peptides onto major histocompatibility complex class I molecules by the tapasin-ERp57 heterodimer. *Nat Immunol* 2007;**8**:873–81.
72. Lipford JR, Deshaies RJ. Diverse roles for ubiquitin-dependent proteolysis in transcriptional activation. *Nat Cell Biol* 2003;**5**:845–50.
73. Lipford JR, Smith GT, Chi Y, Deshaies RJ. A putative stimulatory role for activator turnover in gene expression. *Nature* 2005;**438**:113–6.
74. Collins GA, Tansey WP. The proteasome: a utility tool for transcription? *Curr Opin Genet Dev* 2006;**16**:197–202.
75. Bhaumik SR, Malik S. Diverse regulatory mechanisms of eukaryotic transcriptional activation by the proteasome complex. *Crit Rev Biochem Mol Biol* 2008;**43**:419–33.
76. Hunt DF, Henderson RA, Shabanowitz J, Sakaguchi K, Michel H, Sevilir N, et al. Characterization of peptides bound to the class I MHC molecule HLA-A2.1 by mass spectrometry. *Science* 1992;**255**:1261–3.
77. Rammensee HG, Falk K, Rotzschke O. Peptides naturally presented by MHC class I molecules. *Annu Rev Immunol* 1993;**11**:213–44.
78. Falk K, Rotzschke O, Stevanovic S, Jung G, Rammensee HG. Allele-specific motifs revealed by sequencing of self-peptides eluted from MHC molecules. *Nature* 1991;**351**:290–6.
79. Huczko EL, Bodnar WM, Benjamin D, Sakaguchi K, Zhu NZ, Shabanowitz J, et al. Characteristics of endogenous peptides eluted from the class I MHC molecule HLA-B7 determined by mass spectrometry and computer modeling. *J Immunol* 1993;**151**:2572–87.
80. den Haan JM, Sherman NE, Blokland E, Huczko E, Koning F, Drijfhout JW, et al. Identification of a graft versus host disease-associated human minor histocompatibility antigen. *Science* 1995;**268**:1476–80.
81. Hickman HD, Luis AD, Buchli R, Few SR, Sathiamurthy M, VanGundy RS, et al. Toward a definition of self: proteomic evaluation of the class I peptide repertoire. *J Immunol* 2004;**172**:2944–52.
82. Lemmel C, Weik S, Eberle U, Dengjel J, Kratt T, Becker HD, et al. Differential quantitative analysis of MHC ligands by mass spectrometry using stable isotope labeling. *Nat Biotechnol* 2004;**22**:450–4.

83. Weinzierl AO, Rudolf D, Hillen N, Tenzer S, van Endert P, Schild H, et al. Features of TAP-independent MHC class I ligands revealed by quantitative mass spectrometry. *Eur J Immunol* 2008;**38**:1503–10.
84. Weinzierl AO, Lemmel C, Schoor O, Muller M, Kruger T, Wernet D, et al. Distorted relation between mRNA copy number and corresponding major histocompatibility complex ligand density on the cell surface. *Mol Cell Proteomics* 2007;**6**:102–13.
85. Fortier MH, Caron E, Hardy MP, Voisin G, Lemieux S, Perreault C, et al. The MHC class I peptide repertoire is molded by the transcriptome. *J Exp Med* 2008;**205**:595–610.
86. Escobar H, Crockett DK, Reyes-Vargas E, Baena A, Rockwood AL, Jensen PE, et al. Large scale mass spectrometric profiling of peptides eluted from HLA molecules reveals N-terminal-extended peptide motifs. *J Immunol* 2008;**181**:4874–82.
87. Buchsbaum S, Barnea E, Dassau L, Beer I, Milner E, Admon A. Large-scale analysis of HLA peptides presented by HLA-Cw4. *Immunogenetics* 2003;**55**:172–6.
88. Fissolo N, Haag S, de Graaf KL, Drews O, Stevanovic S, Rammensee HG, et al. Naturally presented peptides on MHC I and II molecules eluted from central nervous system of multiple sclerosis patients. *Mol Cell Proteomics* 2009;**8**:2090–101.
89. Su AI, Wiltshire T, Batalov S, Lapp H, Ching KA, Block D, et al. A gene atlas of the mouse and human protein-encoding transcriptomes. *Proc Natl Acad Sci USA* 2004;**101**:6062–7.
90. Haurum JS, Bjerring Hoier I, Arsequell G, Neisig A, Valencia G, Zeuthen J, et al. Presentation of cytosolic glycosylated peptides by human class I major histocompatibility complex molecules in vivo. *J Exp Med* 1999;**190**:145–50.
91. Hanada K, Yewdell JW, Yang JC. Immune recognition of a human renal cancer antigen through post-translational protein splicing. *Nature* 2004;**427**:252–6.
92. Vigneron N, Stroobant V, Chapiro J, Ooms A, Degiovanni G, Morel S, et al. An antigenic peptide produced by peptide splicing in the proteasome. *Science* 2004;**304**:587–90.
93. Warren EH, Vigneron NJ, Gavin MA, Coulie PG, Stroobant V, Dalet A, et al. An antigen produced by splicing of noncontiguous peptides in the reverse order. *Science* 2006;**313**:1444–7.
94. Zarling AL, Polefrone JM, Evans AM, Mikesh LM, Shabanowitz J, Lewis ST, et al. Identification of class I MHC-associated phosphopeptides as targets for cancer immunotherapy. *Proc Natl Acad Sci USA* 2006;**103**:14889–94.
95. Meyer VS, Drews O, Gunder M, Hennenlotter J, Rammensee HG, Stevanovic S. Identification of natural MHC class II presented phosphopeptides and tumor-derived MHC class I phospholigands. *J Proteome Res* 2009;**8**:3666–74.
96. Shastri N, Schwab S, Serwold T. Producing nature's gene-chips: the generation of peptides for display by MHC class I molecules. *Annu Rev Immunol* 2002;**20**:463–93.
97. Shastri N, Cardinaud S, Schwab SR, Serwold T, Kunisawa J. All the peptides that fit: the beginning, the middle, and the end of the MHC class I antigen-processing pathway. *Immunol Rev* 2005;**207**:31–41.
98. Tardif KD, Mori K, Siddiqui A. Hepatitis C virus subgenomic replicons induce endoplasmic reticulum stress activating an intracellular signaling pathway. *J Virol* 2002;**76**:7453–9.
99. Shin BK, Wang H, Yim AM, Le Naour F, Brichory F, Jang JH, et al. Global profiling of the cell surface proteome of cancer cells uncovers an abundance of proteins with chaperone function. *J Biol Chem* 2003;**278**:7607–16.
100. Gleimer M, Parham P. Stress management: MHC class I and class I-like molecules as reporters of cellular stress. *Immunity* 2003;**19**:469–77.
101. Marciniak SJ, Ron D. Endoplasmic reticulum stress signaling in disease. *Physiol Rev* 2006;**86**:1133–49.

102. Dimcheff DE, Faasse MA, McAtee FJ, Portis JL. Endoplasmic reticulum (ER) stress induced by a neurovirulent mouse retrovirus is associated with prolonged BiP binding and retention of a viral protein in the ER. *J Biol Chem* 2004;**279**:33782–90.
103. Yu CY, Hsu YW, Liao CL, Lin YL. Flavivirus infection activates the XBP1 pathway of the unfolded protein response to cope with endoplasmic reticulum stress. *J Virol* 2006;**80**:11868–80.
104. Bi M, Naczki C, Koritzinsky M, Fels D, Blais J, Hu N, et al. ER stress-regulated translation increases tolerance to extreme hypoxia and promotes tumor growth. *EMBO J* 2005;**24**:3470–81.
105. Smith JA, Schmechel SC, Raghavan A, Abelson M, Reilly C, Katze MG, et al. Reovirus induces and benefits from an integrated cellular stress response. *J Virol* 2006;**80**:2019–33.
106. Patil C, Walter P. Intracellular signaling from the endoplasmic reticulum to the nucleus: the unfolded protein response in yeast and mammals. *Curr Opin Cell Biol* 2001;**13**:349–55.
107. Harding HP, Calfon M, Urano F, Novoa I, Ron D. Transcriptional and translational control in the mammalian unfolded protein response. *Annu Rev Cell Dev Biol* 2002;**18**:575–99.
108. Rutkowski DT, Kaufman RJ. A trip to the ER: coping with stress. *Trends Cell Biol* 2004;**14**:20–8.
109. Hollien J, Weissman JS. Decay of endoplasmic reticulum-localized mRNAs during the unfolded protein response. *Science* 2006;**313**:104–7.
110. Oyadomari S, Yun C, Fisher EA, Kreglinger N, Kreibich G, Oyadomari M, et al. Cotranslocational degradation protects the stressed endoplasmic reticulum from protein overload. *Cell* 2006;**126**:727–39.
111. Pearse BR, Hebert DN. Cotranslocational degradation: utilitarianism in the ER stress response. *Mol Cell* 2006;**23**:773–5.
112. Hickman HD, Luis AD, Bardet W, Buchli R, Battson CL, Shearer MH, et al. Cutting edge: class I presentation of host peptides following HIV infection. *J Immunol* 2003;**171**:22–6.
113. Hickman-Miller HD, Hildebrand WH. The immune response under stress: the role of HSP-derived peptides. *Trends Immunol* 2004;**25**:427–33.
114. de Almeida SF, Carvalho IF, Cardoso CS, Cordeiro JV, Azevedo JE, Neefjes J, et al. HFE cross-talks with the MHC class I antigen presentation pathway. *Blood* 2005;**106**:971–7.
115. Granados DP, Tanguay PL, Hardy MP, Caron E, De Verteuil D, Meloche S, et al. ER stress affects processing of MHC class I-associated peptides. *BMC Immunol* 2009;**10**:10.
116. Stickel JS, Weinzierl AO, Hillen N, Drews O, Schuler MM, Hennenlotter J, et al. HLA ligand profiles of primary renal cell carcinoma maintained in metastases. *Cancer Immunol Immunother* 2009;**58**:1407–17.
117. Holcik M, Sonenberg N. Translational control in stress and apoptosis. *Nat Rev Mol Cell Biol* 2005;**6**:318–27.
118. Yewdell JW. Plumbing the sources of endogenous MHC class I peptide ligands. *Curr Opin Immunol* 2007;**19**:79–86.
119. Wherry EJ, Puorro KA, Porgador A, Eisenlohr LC. The induction of virus-specific CTL as a function of increasing epitope expression: responses rise steadily until excessively high levels of epitope are attained. *J Immunol* 1999;**163**:3735–45.
120. Henrickson SE, Mempel TR, Mazo IB, Liu B, Artyomov MN, Zheng H, et al. T cell sensing of antigen dose governs interactive behavior with dendritic cells and sets a threshold for T cell activation. *Nat Immunol* 2008;**9**:282–91.
121. Schwab SR, Shugart JA, Horng T, Malarkannan S, Shastri N. Unanticipated antigens: translation initiation at CUG with leucine. *PLoS Biol* 2004;**2**:e366.
122. Liston A, Lesage S, Wilson J, Peltonen L, Goodnow CC. Aire regulates negative selection of organ-specific T cells. *Nat Immunol* 2003;**4**:350–4.

123. Cooper GS, Stroehla BC. The epidemiology of autoimmune diseases. *Autoimmun Rev* 2003;**2**:119–25.
124. Hogquist KA, Baldwin TA, Jameson SC. Central tolerance: learning self-control in the thymus. *Nat Rev Immunol* 2005;**5**:772–82.
125. Levings MK, Allan S, d'Hennezel E, Piccirillo CA. Functional dynamics of naturally occurring regulatory T cells in health and autoimmunity. *Adv Immunol* 2006;**92**:119–55.
126. Liston A, Nutsch KM, Farr AG, Lund JM, Rasmussen JP, Koni PA, et al. Differentiation of regulatory Foxp3$^+$ T cells in the thymic cortex. *Proc Natl Acad Sci USA* 2008;**105**:11903–8.
127. Casanova JL, Abel L. The human model: a genetic dissection of immunity to infection in natural conditions. *Nat Rev Immunol* 2004;**4**:55–66.
128. Goldrath AW, Bevan MJ. Selecting and maintaining a diverse T-cell repertoire. *Nature* 1999;**402**:255–62.
129. Santori FR, Kieper WC, Brown SM, Ly Y, Neubert TA, Johnson KL, et al. Rare, structurally homologous self-peptides promote thymocyte positive selection. *Immunity* 2002;**17**:131–42.
130. Urdahl KB, Sun JC, Bevan MJ. Positive selection of MHC class Ib restricted CD8$^+$ T cells on hematopoietic cells. *Nat Immunol* 2002;**3**:772–9.
131. Berg LJ. Signalling through TEC kinases regulates conventional versus innate CD8$^+$ T-cell development. *Nat Rev Immunol* 2007;**7**:479–85.
132. Bendelac A, Savage PB, Teyton L. The biology of NKT cells. *Annu Rev Immunol* 2007;**25**:297–336.
133. Hayday AC. γδ T cells and the lymphoid stress-surveillance response. *Immunity* 2009;**31**:184–96.
134. Prince AL, Yin CC, Enos ME, Felices M, Berg LJ. The Tec kinases Itk and Rlk regulate conventional versus innate T-cell development. *Immunol Rev* 2009;**228**:115–31.
135. Terra R, Labrecque N, Perreault C. Thymic and extrathymic T cell development pathways follow different rules. *J Immunol* 2002;**169**:684–92.
136. Blais ME, Louis I, Perreault C. T-cell development: an extrathymic perspective. *Immunol Rev* 2006;**209**:103–14.
137. Heinonen KM, Perreault C. Development and functional properties of thymic and extrathymic T lymphocytes. *Crit Rev Immunol* 2008;**28**:441–66.
138. Blais ME, Gérard G, Martinic MM, Roy-Proulx G, Zinkernagel RM, Perreault C. Do thymically and strictly extrathymically developing T cells generate similar immune responses? *Blood* 2004;**103**:3102–10.
139. Blais ME, Brochu S, Giroux M, Bélanger MP, Dulude G, Sékaly RP, et al. Why T cells of thymic versus extrathymic origin are functionally different. *J Immunol* 2008;**180**:2299–312.
140. Ehl S, Klenerman P, Aichele P, Hengartner H, Zinkernagel RM. A functional and kinetic comparison of antiviral effector and memory cytotoxic T lymphocyte populations *in vivo* and *in vitro*. *Eur J Immunol* 1997;**27**:3404–13.
141. Neumann AU, Lam NP, Dahari H, Gretch DR, Wiley TE, Layden TJ, et al. Hepatitis C viral dynamics in vivo and the antiviral efficacy of interferon-α therapy. *Science* 1998;**282**:103–7.
142. Flynn KJ, Belz GT, Altman JD, Ahmed R, Woodland DL, Doherty PC. Virus-specific CD8$^+$ T cells in primary and secondary influenza pneumonia. *Immunity* 1998;**8**:683–91.
143. Casrouge A, Beaudoing E, Dalle S, Pannetier C, Kanellopoulos J, Kourilsky P. Size estimate of the αβ TCR repertoire of naive mouse splenocytes. *J Immunol* 2000;**164**:5782–7.
144. Blattman JN, Antia R, Sourdive DJD, Wang X, Kaech SM, Murali-Krishna K, et al. Estimating the precursor frequency of naive antigen-specific CD8 T cells. *J Exp Med* 2002;**195**:657–64.
145. Roy-Proulx G, Baron C, Perreault C. CD8 T-cell ability to exert immunodomination correlates with T-cell receptor:epitope association rate. *Biol Blood Marrow Transplant* 2005;**11**:260–71.

146. Yachi PP, Ampudia J, Gascoigne NR, Zal T. Nonstimulatory peptides contribute to antigen-induced CD8-T cell receptor interaction at the immunological synapse. *Nat Immunol* 2005;**6**:785–92.
147. Anikeeva N, Lebedeva T, Clapp AR, Goldman ER, Dustin ML, Mattoussi H, et al. Quantum dot/peptide-MHC biosensors reveal strong CD8-dependent cooperation between self and viral antigens that augment the T cell response. *Proc Natl Acad Sci USA* 2006;**103**:16846–51.
148. Yachi PP, Lotz C, Ampudia J, Gascoigne NR. T cell activation enhancement by endogenous pMHC acts for both weak and strong agonists but varies with differentiation state. *J Exp Med* 2007;**204**:2747–57.
149. Gascoigne NRJ. Do T cells need endogenous peptides for activation? *Nat Rev Immunol* 2008;**8**:895–900.
150. Van Kaer L, Ashton-Rickardt PG, Ploegh HL, Tonegawa S. TAP1 mutant mice are deficient in antigen presentation, surface class I molecules, and $CD4^{-}8^{+}$ T cells. *Cell* 1992;**71**:1205–14.
151. di Marzo VF, Arnott D, Barnaba V, Loftus DJ, Sakaguchi K, Thompson CB, et al. Autoreactive cytotoxic T lymphocytes in human immunodeficiency virus type 1-infected subjects. *J Exp Med* 1996;**183**:2509–16.
152. Herberts CA, van Gaans-van den Brink J, van der Heeft E, van Wijk M, Hoekman J, Jaye A, et al. Autoreactivity against induced or upregulated abundant self-peptides in HLA-A*0201 following measles virus infection. *Hum Immunol* 2003;**64**:44–55.
153. Azuma K, Shichijo S, Takedatsu H, Komatsu N, Sawamizu H, Itoh K. Heat shock cognate protein 70 encodes antigenic epitopes recognised by HLA-B4601-restricted cytotoxic T lymphocytes from cancer patients. *Br J Cancer* 2003;**89**:1079–85.
154. Faure O, Graff-Dubois S, Bretaudeau L, Derre L, Gross DA, Alves PM, et al. Inducible Hsp70 as target of anticancer immunotherapy: identification of HLA-A*0201-restricted epitopes. *Int J Cancer* 2004;**108**:863–70.
155. Boehm T. Co-evolution of a primordial peptide-presentation system and cellular immunity. *Nat Rev Immunol* 2006;**6**:79–84.
156. Jacob S, McClintock MK, Zelano B, Ober C. Paternally inherited HLA alleles are associated with women's choice of male odor. *Nat Genet* 2002;**30**:175–9.
157. Leinders-Zufall T, Brennan P, Widmayer P, PC S, Maul-Pavicic A, Jager M, et al. MHC class I peptides as chemosensory signals in the vomeronasal organ. *Science* 2004;**306**:1033–7.
158. Milinski M, Griffiths S, Wegner KM, Reusch TB, Haas-Assenbaum A, Boehm T. Mate choice decisions of stickleback females predictably modified by MHC peptide ligands. *Proc Natl Acad Sci USA* 2005;**102**:4414–8.
159. Bonneaud C, Chastel O, Federici P, Westerdahl H, Sorci G. Complex *Mhc*-based mate choice in a wild passerine. *Proc Biol Sci* 2006;**273**:1111–6.
160. Garver-Apgar CE, Gangestad SW, Thornhill R, Miller RD, Olp JJ. Major histocompatibility complex alleles, sexual responsivity, and unfaithfulness in romantic couples. *Psychol Sci* 2006;**17**:830–5.
161. Hedrick PW, Black FL. HLA and mate selection: no evidence in South Amerindians. *Am J Hum Genet* 1997;**61**:505–11.
162. Westerdahl H. No evidence of an MHC-based female mating preference in great reed warblers. *Mol Ecol* 2004;**13**:2465–70.
163. Rock F, Hadeler KP, Rammensee HG, Overath P. Quantitative analysis of mouse urine volatiles: in search of MHC-dependent differences. *PLoS ONE* 2007;**2**:e429.
164. Pisani P, Parkin DM, Munoz N, Ferlay J. Cancer and infection: estimates of the attributable fraction in 1990. *Cancer Epidemiol Biomarkers Prev* 1997;**6**:387–400.
165. Bui JD, Schreiber RD. Cancer immunosurveillance, immunoediting and inflammation: independent or interdependent processes? *Curr Opin Immunol* 2007;**19**:203–8.

166. Willimsky G, Blankenstein T. The adaptive immune response to sporadic cancer. *Immunol Rev* 2007;**220**:102–12.
167. Shankaran V, Ikeda H, Bruce AT, White JM, Swanson PE, Old LJ, et al. IFNγ and lymphocytes prevent primary tumour development and shape tumour immunogenicity. *Nature* 2001;**410**:1107–11.
168. Koebel CM, Vermi W, Swann JB, Zerafa N, Rodig SJ, Old LJ, et al. Adaptive immunity maintains occult cancer in an equilibrium state. *Nature* 2007;**450**:903–7.
169. Willimsky G, Blankenstein T. Sporadic immunogenic tumours avoid destruction by inducing T-cell tolerance. *Nature* 2005;**437**:141–6.
170. Zitvogel L, Tesniere A, Kroemer G. Cancer despite immunosurveillance: immunoselection and immunosubversion. *Nat Rev Immunol* 2006;**6**:715–27.
171. Willimsky G, Czeh M, Loddenkemper C, Gellermann J, Schmidt K, Wust P, et al. Immunogenicity of premalignant lesions is the primary cause of general cytotoxic T lymphocyte unresponsiveness. *J Exp Med* 2008;**205**:1687–700.
172. Galon J, Costes A, Sanchez-Cabo F, Kirilovsky A, Mlecnik B, Lagorce-Pages C, et al. Type, density, and location of immune cells within human colorectal tumors predict clinical outcome. *Science* 2006;**313**:1960–4.
173. Hiraoka K, Miyamoto M, Cho Y, Suzuoki M, Oshikiri T, Nakakubo Y, et al. Concurrent infiltration by $CD8^+$ T cells and $CD4^+$ T cells is a favourable prognostic factor in non-small-cell lung carcinoma. *Br J Cancer* 2006;**94**:275–80.
174. Finak G, Bertos N, Pepin F, Sadekova S, Souleimanova M, Zhao H, et al. Stromal gene expression predicts clinical outcome in breast cancer. *Nat Med* 2008;**14**:518–27.
175. Fontaine P, Roy-Proulx G, Knafo L, Baron C, Roy DC, Perreault C. Adoptive transfer of T lymphocytes targeted to a single immunodominant minor histocompatibility antigen eradicates leukemia cells without causing graft-versus-host disease. *Nat Med* 2001;**7**:789–94.
176. Finn OJ. Cancer vaccines: between the idea and the reality. *Nat Rev Immunol* 2003;**3**:630–41.
177. Bleakley M, Riddell SR. Molecules and mechanisms of the graft-versus-leukaemia effect. *Nat Rev Cancer* 2004;**4**:371–80.
178. Meunier MC, Delisle JS, Bergeron J, Rineau V, Baron C, Perreault C. T cells targeted against a single minor histocompatibility antigen can cure solid tumors. *Nat Med* 2005;**11**:1222–9.
179. Gray A, Raff AB, Chiriva-Internati M, Chen SY, Kast WM. A paradigm shift in therapeutic vaccination of cancer patients: the need to apply therapeutic vaccination strategies in the preventive setting. *Immunol Rev* 2008;**222**:316–27.
180. Hunder NN, Wallen H, Cao J, Hendricks DW, Reilly JZ, Rodmyre R, et al. Treatment of metastatic melanoma with autologous $CD4^+$ T cells against NY-ESO-1. *N Engl J Med* 2008;**358**:2698–703.
181. Kenter GG, Welters MJ, Valentijn AR, Lowik MJ, Berends-van der Meer DM, Vloon AP, et al. Vaccination against HPV-16 oncoproteins for vulvar intraepithelial neoplasia. *N Engl J Med* 2009;**361**:1838–47.
182. Vella LA, Yu M, Fuhrmann SR, El-Amine M, Epperson DE, Finn OJ. Healthy individuals have T-cell and antibody responses to the tumor antigen cyclin B1 that when elicited in mice protect from cancer. *Proc Natl Acad Sci USA* 2009;**106**:14010–5.
183. Rosenberg SA, Restifo NP, Yang JC, Morgan RA, Dudley ME. Adoptive cell transfer: a clinical path to effective cancer immunotherapy. *Nat Rev Cancer* 2008;**8**:299–308.
184. Rolff J, Siva-Jothy MT. Invertebrate ecological immunology. *Science* 2003;**301**:472–5.
185. Pancer Z, Cooper MD. The evolution of adaptive immunity. *Annu Rev Immunol* 2006;**24**:497–518.

Questions Arising from "The Origin and Role of MHC Class I-Associated Self-Peptides"

Masanori Kasahara

Department of Pathology, Hokkaido University Graduate School of Medicine, Sapporo, Japan

1. Generation of some CD8 T cell epitopes is dependent on IFN-γ-inducible proteasome activators known as PA28.[1,2] Could the author comment on the contribution of PA28 to the formation of SMII?
2. The focus of this chapter is on the composition of SMII displayed on cells in peripheral tissue. Could the author comment on the importance of SMII displayed on the surface of thymic epithelial cells in CD8 T cell development?[3-5]

References

1. Murata S, Udono H, Tanahashi N, Hamada N, Watanabe K, Adachi K, et al. Immunoproteasome assembly and antigen presentation in mice lacking both PA28α and PA28β. *EMBO J* 2001;**20**:5898–907.
2. Sun Y, Sijts AJ, Song M, Janek K, Nussbaum AK, Kral S, et al. Expression of the proteasome activator PA28 rescues the presentation of a cytotoxic T lymphocyte epitope on melanoma cells. *Cancer Res* 2002;**62**:2875–82.
3. Nil A, Firat E, Sobek V, Eichmann K, Niedermann G. Expression of housekeeping and immunoproteasome subunit genes is differentially regulated in positively and negatively selecting thymic stroma subsets. *Eur J Immunol* 2004;**34**:2681–9.
4. Osterloh P, Linkemann K, Tenzer S, Rammensee HG, Radsak MP, Busch DH, et al. Proteasomes shape the repertoire of T cells participating in antigen-specific immune responses. *Proc Natl Acad Sci USA* 2006;**103**:5042–7.
5. Murata S, Sasaki K, Kishimoto T, Niwa S, Hayashi H, Takahama Y, et al. Regulation of $CD8^+$ T cell development by thymus-specific proteasomes. *Science* 2007;**316**:1349–53.

Progress in Molecular Biology
and Translational Science, Vol. 92
DOI: 10.1016/S1877-1173(10)92014-0

Response to Questions

Claude Perreault

Institute for Research in Immunology and Cancer, Université de Montréal, Montréal, Quebec, Canada

I. Response to Question #1

I am not aware of recent studies on the impact of PA28 on the SMII. However, using a high-throughput mass spectrometry based method,[1] we recently analyzed the transcriptome and the SMII of dendritic cells (DCs) expressing or not immunoproteasome (IP) subunits MECL1 and LMP7.[2] Of the 417 peptides eluted from wild-type (WT) DCs, 212 were expressed at similar levels in IP-deficient DCs. However, 199 peptides were overexpressed in WT relative to IP-deficient DCs. Among those 199 peptides, 60 were detected exclusively in WT DCs. Only six peptides were slightly overexpressed (three- to fivefold) in IP-deficient relative to WT DCs and none were unique to dKO DCs. We therefore concluded that IPs increase the abundance and diversity of MIPs at the surface of DCs. In addition, we found that IPs have a nonredundant impact on expression of a selected set of immune genes clustered primarily on chromosomes 4 and 8. We concluded that IPs have more than one nonredundant role. They have a dramatic impact on the MIP repertoire and a heretofore unrecognized impact on expression of immune-related genes. Both of these effects may be of great importance in adaptive immune responses.

II. Response to Question #2

Although commonly taken for granted, the conservation of the thymus as the primary T lymphoid organ in all animals with an adaptative immune system is remarkable. In contrast, about 10 different organs have served as primary sites of hematopoiesis (including "B cell poiesis") in jawed vertebrates.[3] At face value, conservation of the thymus as the primary T lymphoid organ suggests

DOI: 10.1016/S1877-1173(10)92015-2

1877-1173/10 $35.00

that intrathymic development endows T cells with elusive evolutionarily selected functional attributes.[4] Could T cells produced extrathymically substitute for classic T cells? We addressed this question in Oncostatin M (OM)-transgenic mice. Transgenic expression (or injection) of OM induces total thymic atrophy that is compensated by a massive production of T cells in lymph nodes (LN) that ectopically recapitulates all the development stages normally found in the thymus.[5,6] OM amplifies a cryptic T cell development pathway that is operative in LNs of nontransgenic mice.[7] Because OM-transgenic mice have no quantitative CD4 or CD8 T cell deficit, they provide a unique model to evaluate the function of T cells generated extrathymically. We found that extrathymic T cells cannot substitute for conventional thymus-derived T cells.[8,9] Extrathymic CD4 T cells contained an exceedingly high proportion of Treg cells and were unable to help B and CD8 T cells, whereas extrathymic CD8 T cells behave like innate T cells. A common feature of extrathymic CD8 T cells and other types of innate T cells is that they are positively selected by hematopoietic cells as opposed to thymic epithelial cells (TEC).[4,10] Why do TECs have the unique ability to induce generation of conventional T cells? At least in part because TECs express a unique form of proteasome (the thymoproteasome) that likely generates low-affinity MHC class I ligands compared with other proteasomes.[11] Thymoproteasome-dependent self-peptide production is required for the development of an immunocompetent repertoire of CD8 T cells.[12] Biochemical characterization of the SMII presented by TECs represents an interesting but challenging endeavor.

References

1. Fortier MH, Caron E, Hardy MP, Voisin G, Lemieux S, Perreault C, et al. The MHC class I peptide repertoire is molded by the transcriptome. *J Exp Med* 2008;**205**:595–610.
2. de Verteuil D, Muratore-Schroeder TL, Granados DP, Fortier MH, Hardy MP, Bramoullé A, et al. Deletion of immunoproteasome subunits imprints on the transcriptome and has a broad impact on peptides presented by major histocompatibility complex I molecules. *Mol. Cell Proteomics*, Epub May 19, 2010, PMID: 20484733.
3. Zapata A, Amemiya CT. Phylogeny of lower vertebrates and their immunological structures. *Curr Top Microbiol Immunol* 2000;**248**:67–107.
4. Heinonen KM, Perreault C. Development and functional properties of thymic and extrathymic T lymphocytes. *Crit Rev Immunol* 2008;**28**:441–66.
5. Clegg CH, Rulffes JT, Wallace PM, Haugen HS. Regulation of an extrathymic T-cell development pathway by oncostatin M. *Nature* 1996;**384**:261–3.
6. Boileau C, Houde M, Dulude G, Clegg CH, Perreault C. Regulation of extrathymic T cell development and turnover by Oncostatin M. *J Immunol* 2000;**164**:5713–20.
7. Guy-Grand D, Azogui O, Celli S, Darche S, Nussenzweig MC, Kourilsky P, et al. Extrathymic T cell lymphopoiesis: ontogeny and contribution to gut intraepithelial lymphocytes in athymic and euthymic mice. *J Exp Med* 2003;**197**:333–41.

8. Blais ME, Gérard G, Martinic MM, Roy-Proulx G, Zinkernagel RM, Perreault C. Do thymically and strictly extrathymically developing T cells generate similar immune responses? *Blood* 2004;**103**:3102–10.
9. Blais ME, Brochu S, Giroux M, Bélanger MP, Dulude G, Sékaly RP, et al. Why T cells of thymic versus extrathymic origin are functionally different. *J Immunol* 2008;**180**:2299–312.
10. Berg LJ. Signalling through TEC kinases regulates conventional versus innate $CD8^+$ T-cell development. *Nat Rev Immunol* 2007;**7**:479–85.
11. Murata S, Sasaki K, Kishimoto T, Niwa SI, Hayashi H, Takahama Y, et al. Regulation of $CD8^+$ T cell development by thymus-specific proteasomes. *Science* 2007;**316**:1349–53.
12. Nitta T, Murata S, Sasaki K, Fujii H, Ripen AM, Ishimaru N, et al. Thymoproteasome shapes immunocompetent repertoire of $CD8^+$ T cells. *Immunity* 2010;**32**:29–40.

Functional Development of the T Cell Receptor for Antigen

Peter J.R. Ebert,* Qi-Jing Li,† Johannes B. Huppa,‡ and Mark M. Davis*,‡

*The Department of Microbiology and Immunology, Stanford University School of Medicine, Stanford, California, USA
†The Department of Immunology, Duke University Medical Center, Durham, North Carolina, USA
‡Howard Hughes Medical Institute, Stanford University School of Medicine, Stanford, California, USA

For over three decades now, the T cell receptor (TCR) for antigen has not ceased to challenge the imaginations of cellular and molecular immunologists alike. T cell antigen recognition transcends every aspect of adaptive immunity: it shapes the T cell repertoire in the thymus and directs T cell-mediated effector functions in the periphery, where it is also central to the induction of peripheral tolerance. Yet, despite its central position, there remain many questions unresolved: how can one TCR be specific for one particular peptide-major histocompatibility complex (pMHC) ligand while also binding other pMHC ligands with an immunologically relevant affinity? And how can a T cell's extreme specificity (alterations of single methyl groups in their ligand can abrogate a response) and sensitivity (single agonist ligands on a cell surface are sufficient to trigger a measurable response) emerge from TCR–ligand interactions that are so low in affinity? Solving these questions is intimately tied to a fundamental understanding of molecular recognition dynamics within the many different contexts of various T cell–antigen presenting cell (APC)

Progress in Molecular Biology
and Translational Science, Vol. 92
DOI: 10.1016/S1877-1173(10)92004-8

1877-1173/10 $35.00

contacts: from the thymic APCs that shape the TCR repertoire and guide functional differentiation of developing T cells to the peripheral APCs that support homeostasis and provoke antigen responses in naïve, effector, memory, and regulatory T cells. Here, we discuss our recent findings relating to T cell antigen recognition and how this leads to the thymic development of foreign-antigen-responsive αβT cells.

I. Introduction

With the ongoing identification and characterization of new T cell lineages with distinctive immunological functions, we increasingly appreciate how diverse and dynamic the T cell compartment really is. Such attributes are not surprising in view of the daunting task T cells are entrusted with: to distinguish "self" from "foreign" based on protein composition and then neutralize "foreign" with little time left before pathogens breach the first lines of innate defense. To manage this job, education, memory, networking, and fine-tuning capabilities are of the essence. As we now know, most of these predicates of adaptive immunity depend *per se* on T cell antigen recognition. Yet, though considerable strides have been made, many of the parameters by which antigen recognition operates are not fully understood.

Here, we principally discuss the mechanisms by which the peripheral T cell receptor (TCR) repertoire arises, so we must first set the scene. The T cell's schoolhouse is the thymus, an organ whose bulk is comprised of the thymocytes that it educates, but whose structure is imparted by a spider's web of stromal cells that includes the epithelial cells that act as teachers. The thymus is separated anatomically into an outer cortex and an inner medulla, both of which are composed of distinct epithelial cells[1] and further populated with bone-marrow-derived macrophages and dendritic cells (DCs; although these are relatively more abundant in the medulla).[2] T cell precursors are delivered to the junction between these layers and then migrate outward into the cortex, eventually reaching the outermost subcapsular zone (SCZ).[3] These are double-negative (DN) thymocytes, so called because they are devoid of the coreceptors CD4 and CD8, and at this stage, they also lack a TCR.

The molecular genetic construction of the TCR is intimately and causally linked with T cell development. DNs in the outer cortex[3] initiate recombination of the TCRβ locus, and are allowed to develop further only upon successful generation of TCRβ protein,[4] as read out by TCRβ's capacity to generate TCR signals in tandem with an invariant pre-TCRα partner.[5] Once past this checkpoint, immature T cells form a TCRα gene and upregulate CD4 and CD8 to become CD4+CD8+ double-positive (DP) thymocytes. This represents the first T cell population that can receive signals from pMHC,[6] and these cells spend the next several days navigating the web of cortical and then medullary antigen presenting cells (APCs), testing their newly minted TCRs against their host's self-determinants.

II. TCR Structure and Diversity

TCRs are expressed on the surfaces of thymic and peripheral T cells as a disulfide-linked heterodimer composed of two glycosylated type I membrane proteins: TCRα (40–45 kDa) and TCRβ (40–50 kDa). Quite similar in structure to a Fab antibody fragment, the extracellular portion contains two variable (Vα, Vβ) immunoglobulin (Ig)-like domains and then two constant Ig domains (Cα, Cβ), which are followed by a transmembrane region and a short cytoplasmic tail. Relatively random VJ (TCRα) and VDJ (TCRβ) gene rearrangement at the DN stage of thymic development allows for a potential diversity on the order of $\sim 10^4$ TCR combinations,[7] while N-nucleotide and P-nucleotide additions within the junctional CDR3 regions of TCRα and TCRβ increase the potential diversity enormously. In theory, roughly 10^{15} different αβTCR heterodimers could be produced,[8] exceeding the number of T cells in a mouse ($\sim 2 \times 10^8$) or human ($\sim 2 \times 10^{11}$).

The concentration of diversity in the CDR3s and the relative lack in the germline V-region-encoded CDR1s and CDR2s seems to be tailor-made for recognizing diverse peptides in comparatively monomorphic MHC molecules.[8] Studies with mutated peptides[9,10] and experiments with transplanted CDR3s[11] showed a close relationship between CDR3s and peptide specificity. Structural studies of TCR–pMHC complexes have also generally borne this out by showing that the CDR3 loops form a large share of contacts with the peptide, while CDRs1 and 2 more often contact predominantly the helices that comprise the MHC's peptide binding cleft[12,13] (though exceptions do exist to this generality).[14] This is not to say that the mode of binding for a particular pair of V regions to a particular MHC is the same across all TCR–pMHC pairs. Rather, the CDRs1 and 2 appear to adopt one of a handful of discrete docking angles over the MHC.[15] As another line of evidence, a study of the binding between the 2B4 TCR and its pMHC ligand showed that mutations in the contact area between TCR and MHC led primarily to defects in the initial association of the two, while mutations in the contact area between TCR and peptide led primarily to defects in the stability of the bound complex. We suggested that this might be by design: that the CDRs1 and 2 might allow the 2B4 TCR to associate readily with pMHCs bearing many different peptides, but that the more stable association that leads to TCR triggering would be dictated more by the peptide

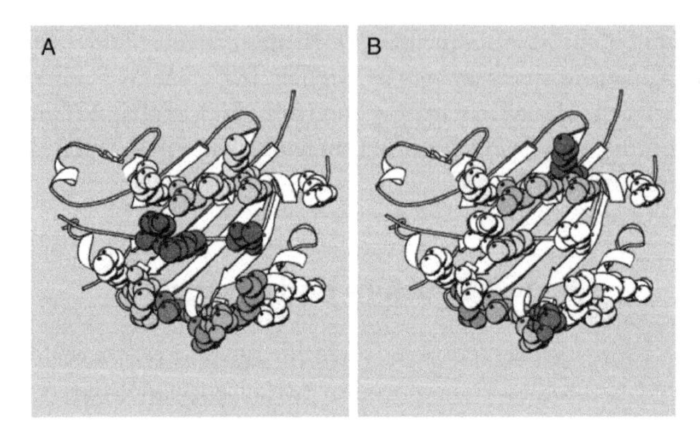

Fig. 1. ϕ analysis of the contribution of MCC–I-E^k surface residues to 2B4 TCR–pMHC association versus stabilization. The peptide-binding pocket of I-E^k and the MCC peptide are shown as white ribbons. Residues studied are shown as space-filling representations and are color-coded with respect to their effect on stability and association. (A) Contributions of residues to complex stability ($\Delta\Delta G$). Residues are shaded white (no effect, $\Delta\Delta G = 0$ kcal mol^{-1}) to gray (maximum effect in this study, $\Delta\Delta G > 3.14$ kcal mol^{-1}). (B) Contributions of residues to complex association (ϕ). Residues are shaded white (no interaction, $\phi = 0$) to gray (maximum interaction in this study, $\phi = 0.73$). Figure originally from Ref. 16.

and CDR3s[16] (Fig. 1). As no preselection TCR has been examined in this way, it is impossible to say whether this property is a result of evolutionary pressure on the germline V regions or is the result of thymic selection.[17]

In reality, many fewer than 10^{15} distinct functional specificities are apparent in the peripheral T cell repertoire. Studies in mice have shown that the frequency of naïve CD8+ or CD8+ T cells that are specific for a particular foreign pMHC tetramer lies in the range of dozens to a few hundred per 10^8.[18,19] Our own studies of naïve human CD8+ T cells yield a very similar answer (Yu and Davis, unpublished data), giving an estimate of functional ligand diversity that can be distinguished by naïve αβT cells of 10^6–10^7. The relationship of TCRs with ligands is not remotely one-to-one, as TCRs can exhibit substantial polyspecificity for pMHCs,[20] and distinct TCRs can recognize the same pMHC.[21] Rather, this relationship appears to be many-to-many[22] and, as there has not been any estimation of the clonality of these tetramer-positive populations, it is difficult to estimate the repertoire's actual diversity in terms of TCR protein sequences based on this method. Earlier estimates of the diversity based on TCR gene sequencing arrived at an answer of 10^7–10^8 distinct TCRs among human peripheral blood T cells.[23] Taken together, this might lead us to guess, very roughly, that each T cell in mice has on average ~10 identical twins. This would accord with still earlier results that found that it was common to observe identical TCRs arising in clones raised against a particular antigen from immunization of different mice.[24] That is, there is typically a significant TCR overlap between individual mice, so each antigen reactivity is likely represented by a handful of distinct TCRs in mice, and each of these TCRs is likely to be represented by multiple naïve T cells.

Even though these estimates are imprecise, they leave a substantial gap between the number of theoretically distinct TCRs and the number that occupy the peripheral lymphatics of mice and humans. While some of this loss in diversity can be explained by the mechanics of TCR recombination[4] or the inability of particular TCRαs to pair with particular TCRβs, more than 90% of successfully rearranged TCRs are lost in the thymus along with the T cells that bore them.[25,26] In the following sections we discuss the knowledge we have gained in recent years regarding which self-pMHCs guide the selection of these TCRs, how immature T cells access this self-antigenic information, and how a microRNA helps guarantee proper self-tolerization. And finally, we speculate on which TCRs are selected and why they are chosen.

III. Thymic Self-Peptides Involved in T Cell Selection

Cognate antigen, that is, the pMHC that acts as an agonist for a particular TCR and activates mature T cells bearing that TCR, also generally induces apoptosis in immature DP thymocytes *in vitro* and *in vivo*.[27] These agonist interactions often occur with affinities in the range of 1–40 μM in terms of their dissociation constants (K_D's).[28,29] For mature T cells, whether the dissociation constant better predicts the biological potency of the pMHC, as opposed to the half-life of the TCR–pMHC interaction or a combination of thermodynamic parameters, is a topic of contemporary debate and is described elsewhere.[29] However, for immature T cells, across three distinct class I-restricted TCR systems, the threshold separating positively from negatively selecting pMHC was found to be consistent and very narrow in terms of the K_D of TCR–pMHC interaction (T1, S14, and OT-I TCRs).[30] This analysis was performed using soluble photo-crosslinkable pMHCs on the surfaces of T cells and so describes the threshold in terms of apparent K_D (which includes the small contribution of CD8 to binding),[31] but the stronger pMHC ligands examined bound tightly enough to be measureable in terms of monomeric affinity and could be used for comparison. The positive/negative selection threshold thus appears to lie very close to the range of affinities found for agonists. Furthermore, earlier studies had measured the monomeric affinities of two positively selecting altered peptide ligands, E1 and R4, in the OT-I system (i.e., variants of the agonist Ova peptide). At 20 μM[28] and 40–57 μM[28,32] at 25 °C, respectively, these measurements also support the idea of a narrow affinity gap between positive and negative selection.

There is a snare, however. When E1 was coexpressed alongside the OT-I TCR transgene, in otherwise peptide-deficient mice, OT-I T cells were positively selected and exported to the periphery. But these cells were functionally compromised and hyporesponsive,[33] similar to the way in which the P14 T cells[34] selected by very low doses of agonist peptide in fetal thymic organ cultures (FTOCs) had earlier been found to be compromised.[35] Whether E1-selected cells bearing the OT-I TCR would contribute to the peripheral repertoire in normal mice, then, is in question.

Two naturally occurring H-2K^b-binding peptides that promote positive selection of OT-I T cells have also been known for some time.[36,37] Both these peptides bore structural similarities to the nominal Ova antigen, prompting the authors to suggest that structural similarity might underlie the potential of self-peptides to promote selection. Until recently, however, we did not know the affinity of the OT-I receptor for these peptide-H-2K^b. We have now determined their affinities and, somewhat surprisingly, they are much weaker than the aforemeasured E1 and R4.[38] Indeed, they bound so poorly at 25 °C that they could be measured only at 10 °C, where their dissociation was sufficiently slow to calculate an accurate K_D. In order to more accurately compare them, we therefore also measured the K_D of Ova-H-2K^b at this temperature. At a K_D of 136 mM for Catnb-H-2K^b (15.6-fold lower than Ova-H-2K^b at 10 °C) and 211 mM for Cappa1-H-2K^b (24.3-fold lower than Ova-H-2K^b at 10 °C), these endogenous selecting ligands bind OT-I two- to sixfold more weakly than R4 (8.8-fold lower than Ova-H-2K^b at 25 °C) and E1 (4.4-fold lower than Ova-H-2K^b at 25 °C). These results significantly broaden the spectrum of self-peptides, in terms of affinity, that are relevant for thymic selection. Also, while the threshold between positive and negative selection appears quite sharp *in vitro*,[30] for the admittedly few naturally occurring selecting peptides, it appears much broader. This in turn might also help us understand the TCR specificity fostered by the thymus: it seems quite a task to accurately distinguish pMHCs that differ by only twofold in affinity, but if the thymus effectively removes this "middle ground" by negative selection or attenuation,[33] or by some other mechanism, then the task becomes much easier.

With respect to the identification of selecting peptides, the study of class II-restricted systems has lagged behind, as no selecting peptides have been described until recently. But we have now begun to close that gap. Recently, our group[39] as well as that of Paul Allen and colleagues[40] reported the identification of endogenous I-E^k-bound peptides that can drive the *in vitro* selection of functionally mature 5C.C7 or AND TCR transgenic T cells on the B10.BR background, respectively (Fig. 2). Among the 95 tested, 6 different peptides were found with this property for 5C.C7, the most potent among them being an endogenous retroviral Gag-Pol-polyprotein-derived peptide (which we refer to as Gag-Pol or GP), whereas for AND, 1 peptide, a gp250-derived peptide, was

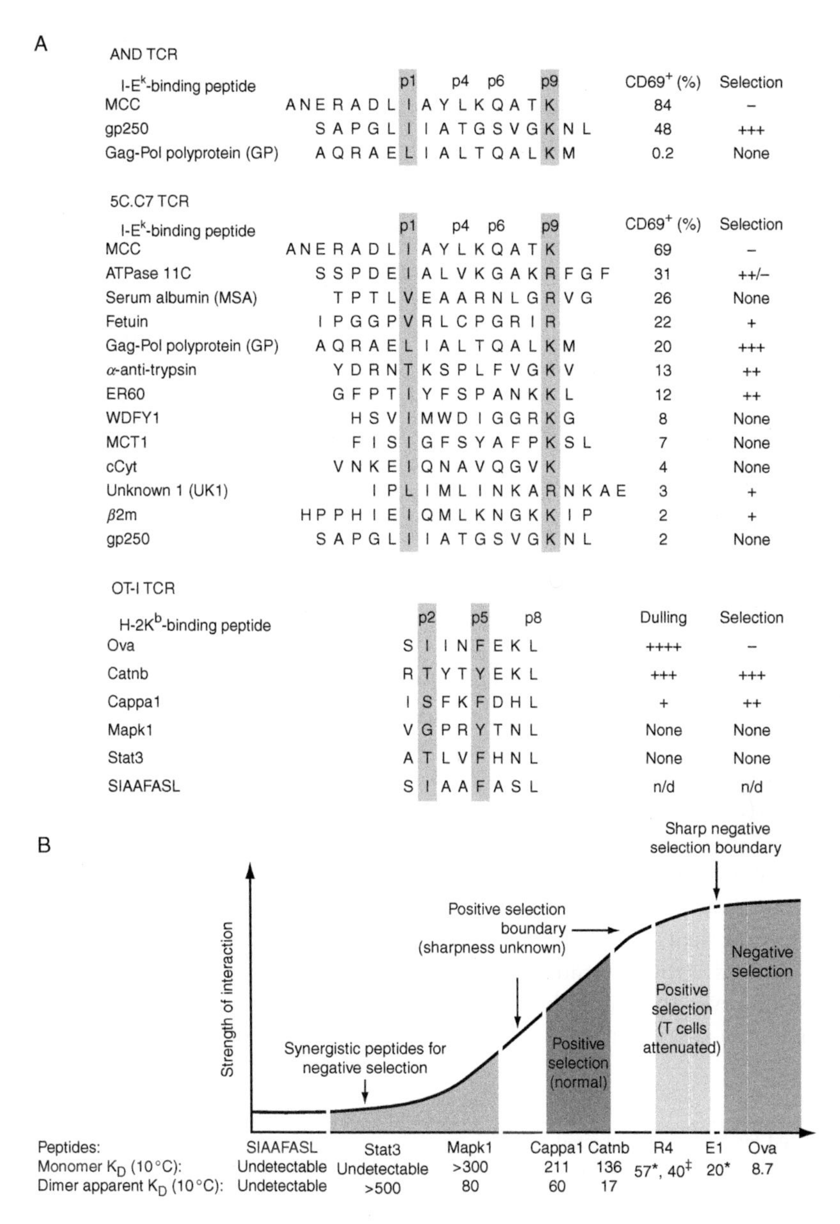

A

AND TCR

I-E^k-binding peptide	Sequence (p1, p4, p6, p9)	CD69$^+$ (%)	Selection
MCC	ANERADLIAYLKQATK	84	–
gp250	SAPGLIIATGSVGKNL	48	+++
Gag-Pol polyprotein (GP)	AQRAELIALTQALKM	0.2	None

5C.C7 TCR

I-E^k-binding peptide	Sequence (p1, p4, p6, p9)	CD69$^+$ (%)	Selection
MCC	ANERADLIAYLKQATK	69	–
ATPase 11C	SSPDEIALVKGAKRFGF	31	++/–
Serum albumin (MSA)	TPTLVEAARNLGRVG	26	None
Fetuin	IPGGPVRLCPGRIR	22	+
Gag-Pol polyprotein (GP)	AQRAELIALTQALKM	20	+++
α-anti-trypsin	YDRNTKSPLFVGKV	13	++
ER60	GFPTIYFSPANKKL	12	++
WDFY1	HSVIMWDIGGRKG	8	None
MCT1	FISIGFSYAFPKSL	7	None
cCyt	VNKEIQNAVQGVK	4	None
Unknown 1 (UK1)	IPLIMLINKARNKAE	3	+
β2m	HPPHIEIQMLKNGKKIP	2	+
gp250	SAPGLIIATGSVGKNL	2	None

OT-I TCR

H-2K^b-binding peptide	Sequence (p2, p5, p8)	Dulling	Selection
Ova	SIINFEKL	++++	–
Catnb	RTYTYEKL	+++	+++
Cappa1	ISFKFDHL	+	++
Mapk1	VGPRYTNL	None	None
Stat3	ATLVFHNL	None	None
SIAAFASL	SIAAFASL	n/d	n/d

FIG. 2. Summary of sequence and thermodynamic data for the self-peptides with reactivity toward the AND, 5C.C7, and OT-I TCRs. (A) Amino acid sequences of a subset of the known I-E^k-bound self-peptides[41] and H-2K^b-bound self-peptides (Ref. 37; the Cappa1 peptide was originally described in Ref. 36). I-E^k-binding peptides are shown in their predicted I-E^k-binding register; MHC pocket residues are highlighted.[42,43] The peptides' reactivity toward the indicated TCR was measured by CD69 upregulation of DP thymocytes bearing that TCR toward the indicated peptide

found among the same pool. In its putative I-E^k-binding register, GP harbors four residues that are identical to those in the agonist Moth Cytochrome C (MCC) and a fifth residue that is a chemically conservative substitution (the P1 leucine, which is an isoleucine in MCC). These identical and similar residues occupy the I-E^k-contacting positions,[42] while the TCR-contacting side chains are very different. This strengthens the argument that positively selecting ligands for a given TCR are structurally similar to that of TCR's nominal antigen[37,45] and/or that the selecting ligand might help to enforce an MHC conformation that is similar to that of the nominal antigen-MHC.[46] The ability of antibodies raised against the complex of nominal antigen-MHC to block positive selection in two different TCR systems further supported this case.[47,48] Again, there is a snare, though: the third data point does not follow this rule. The gp250 peptide that positively selects AND T cells is structurally very distinct from the agonist MCC: only the P1 and P9 anchor residues are conserved, while the P4 and P6 anchors are chemically quite distinct (L → T and Q → S, respectively, from MCC to gp250), and the TCR contacts are likewise quite distinct. Nor does this represent degenerate recognition of gp250 by AND because single mutations of the gp250 peptide destroy its ability to select AND T cells. Even a substitution at P2 that restores the "MCC-like" alanine residue abrogates rather than strengthens AND reactivity. To levy a parting defense of the role for structural similarity, however, we found that an ATPase 11c-derived peptide could weakly promote positive selection of 5C.C7 thymocytes and that this peptide could also negatively select 5C.C7 thymocytes at high doses. The ATPase 11c peptide is highly similar to the agonist MCC and, unlike GP, it is chemically identical to MCC at known TCR-contacting residues (P2, P5, and P7).[42]

Speaking more generally in terms of the thymic peptide repertoire, prior work has shown that the bulk of the quantity of self-peptides presented in the thymus represents a limited number of distinct species and these are roughly the same species as those found on splenic APCs in the periphery.[49] It must also be noted, however, that multiple genetic programs that are unique to thymic APCs and result in alteration of the peptide repertoire presented by those APCs have

---

presented by APCs or platebound I-E^k.[39,40] For OT-I, reactivity was measured in terms of CD4 CD8 coreceptor downregulation, or dulling.[37] The ability of these peptides to positively select T cells bearing the indicated TCR was assessed: for AND, in reaggregate thymus cultures with the ANV41.2 cTEC cell line as APCs; for 5C.C7, in 5C.C7 Ii$^{-/-}$ fetal thymus organ cultures (FTOCs); and for OT-I, in OT-I TAP$^{-/-}$ FTOCs. (B) A summary of affinities of the OT-I TCR for different peptide–H-2K^b and a schematic of the relation between affinity and selection outcome. Positive selection data for E1 (EIINFEKL) and R4 (SIIRFEKL) are from Ref. 44, positive selection data for the self-peptides are from Ref. 37. Affinity measurements are from Ref. 38 except for * (Ref. 28), at 25 °C, and ‡ (Ref. 32), also at 25 °C.

been identified, presumably to better select maturing T cells. For example, the autoimmune regulator, or AIRE, directs the expression of tissue-restricted proteins (e.g., pancreatic-, salivary gland-, ovary-specific proteins) in a set of medullary epithelial cells,[50] which are then presented directly[51] and/or cross-presented by nearby phagocytic APCs[52] to achieve tolerance against those tissues. Lymphotoxin-β receptor regulates a similar expression of tissue-restricted antigens (TRAs) in a distinct subset of medullary thymic epithelial cells (mTECs).[53] While many thymic DCs, both cortical and medullary, are autochthonous, DCs also migrate into the thymus from the skin and blood, and potentially from other sites. These DCs reside primarily at the corticomedullary junction and their role in promoting central tolerance has also been confirmed.[54] With respect to positive selection, cortical thymic epithelial cells (cTECs) are unique among epithelial cells in that they express class II, which is obviously an advantage for educating CD4+ T cells. They also have adaptations in their MHC loading pathways that are thought to facilitate selection. For example, cTECs express cathepsin L, which diversifies and/or alters the repertoire of self-peptides presented by class II molecules.[55] In the class I loading pathway, cTECs constitutively express a thymus-restricted β5t proteasome subunit along with a β5i subunit that is IFNγ-inducible in somatic cells, which similarly results in a more diverse array of class I-presented peptides.[56] Both these adaptations are necessary for a fully diverse T cell repertoire, although whether their necessity stems simply from the increased peptide diversity (as suggested by the classic experiments of Hogquist)[57] or from the presentation of uniquely cTEC-restricted peptides remains to be seen.[58] In either case, these mechanisms all illustrate the pains taken by the thymus to broaden the scope of the immunological "self" presented to developing T cells.

IV. Synaptic and Nonsynaptic Interactions in the Thymus

As discussed above, the thymus contains a wealth of information regarding the host's antigenic self. To optimize its education, a thymocyte would be best served by referencing all of it against its TCRs. But how to accomplish this? The thymus is vast, on a cellular scale, and the thymocyte's task is great. Since the APCs cannot go to the thymocytes, the thymocytes must go to the APCs. And this is what they do, in a manner well suited to their purpose.

The first clues pointing to a potential mechanism came from the earliest studies of TCR transgenic mice. Here, it was found that a relatively linear increase in the total number of selected T cells resulted from an increase in the surface density of MHC on the selecting epithelium[59](but no increase was observed with increased TCR density on the developing T cells).[60] This contrasted with negative selection, which could be induced by minute amounts of agonist ligand and which titrated precipitously at these very low doses.[35] And, although this positive selection "niche" can be saturated in mice expressing TCR transgenes, who therefore have abnormally large numbers of "selectable" immature T cells, it does not appear to be saturated in normal mice.[59,61] The idea of a niche for positive selection was expanded by Merkenschlager and colleagues,[62] who generated reaggregate thymus cultures in which only some cTECs expressed the selecting ligand. In this case, a dose-dependent and fairly linear increase in T cell selection could be observed as the proportion of selecting to nonselecting cTECs increased. Again, negative selection differed, in that only very few negatively selecting APCs were required to obliterate nearly all autoresponsive DP thymocytes. A physical niche characterized by a static interaction of thymocytes with positively selecting cTECs was suggested as the explanation for these data.[63]

With the benefit of new studies exploring the migration of maturing T cells within the thymus, we can reexamine this interpretation. While DN thymocytes arrived at the SCZ by virtue of directed migration in response to CXCR4 ligands,[64] DPs do not appear initially to follow a directed pattern of migration on their journey back through the cortex. Rather, they adopt a pattern of motility best described as a random walk,[65] where nonresponsiveness to cues for directed migration is actively maintained by a repressor of chemokine receptor signaling, the G protein-coupled receptor kinase-interactor 2 (GIT2).[66] Upon encountering a positively selecting cTEC, signaling ensues and the DP arrests its motility.[67,68] However, this arrest is short-lived, on the order of 5–10 min, and the DP does not form a stable contact or "synapse" with the stimulating cTEC,[69] as a mature T cell would in response to a stimulatory APC in most[70,71] but not all cases.[72] After this brief interlude, motility resumes. But this transient contact alone is insufficient for commitment to positive selection: the DP must find new selecting cTECs, or return to an old one, in order to maintain active Nuclear Factor of Activated T cells (NFAT) and other chromosome remodeling machinery long enough[73,74] to continue its maturation.[69] Thus, it seems more likely that the selecting niche is actually scattered along the DP's path and is measured not in terms of cellular space available to swaddle DPs but in terms of the probability of finding sufficiently many distinct sources of selecting pMHCs as the DP meanders its way through the cortex. This hypothesis, which we have described as a "gauntlet" model, can explain how selection could be linear with both the number of selecting APCs and with the total density of selecting MHC, as both affect the probability of encountering a selecting cTEC. In further support of this, positive selection is attenuated in the absence of GIT2 and the random walk behavior that it promotes.[66] The majority of thymocytes likely do not have sufficiently self-peptide-self-MHC-compatible TCRs, and these fail to survive for want of promaturation TCR signals.

The cortex holds additional perils for developing thymocytes as well.[75] We know that negative selection is brutally efficient both in terms of the number of APCs required to achieve maximal clonal deletion in thymus cultures (Ref. 62 and our unpublished data) and the number of agonist pMHC ligands required to achieve deletion *in vitro*.[69,76,77] Cortical DCs are potent negative selectors[2] and, although cTECs generally lack costimulatory molecules that are required for direct induction of negative selection,[78–80] at least a subset of cTECs also can contribute to deletional tolerance.[81] It seems likely that, for a thymocyte burdened with an autoreactive TCR, these APCs will be very hard to avoid. After all, the thymocyte is already compelled to bounce around the cortex to seek out new brief cTEC contacts from which to glean positively selecting signals. Upon encounter with a thymic DC or TEC that presents negatively selecting ligands in reaggregate thymus cultures, rather than arrest and resume its motility, the thymocyte stops and forms a stable contact, typified by TCR accumulation at the interface with the APC as in a mature T cell synapse.[82,83] In thymic slices, negatively selecting thymocyte–DC interactions appear less stable, with the thymocyte restricted in its movements to ~30 μM, but whether synapses form in this situation could not be addressed.[84] *In vitro*, this negatively selecting synapse can be formed in response to as few as two agonist pMHCs, and apoptosis ensues within 90 minutes.[69] In the case of cTEC-mediated negative selection, simultaneous or subsequent interaction with distinct APCs may also be important,[85] but this has not yet been addressed by the microscopy studies. Finally, we can show that this synapse formation is necessary for efficient negative selection *in vitro* and in thymus cultures, as shown by antibody blockade of LFA-1 and the ensuing inability of thymocytes to form stable contacts with agonist-presenting APCs (Ref. 69 and our unpublished data). This is similar to mature T cells, which require the prolonged contact with APCs that is afforded by synapse formation to commit to expansion and effector function.[86]

We have likened this process to a gauntlet because to have any chance of survival and to be initiated into the peripheral repertoire, the developing thymocyte must "run the gauntlet" of the thymic cortex (Fig. 3). It must continually find suitable low-affinity self-pMHCs to maintain active NFAT in the nucleus, yet at any moment it may encounter an APC bearing a number of negatively selecting peptides that is sufficient to induce synapse formation and apoptosis.

A thymocyte that escapes deletion and receives sufficiently many positively selecting signals continues its maturation program. This includes upregulation of the chemokine receptor CCR7, which prompts migration toward the medulla[89,91] in a directed fashion.[65] Entry into the medulla is then further gated by a pertussis-toxin-sensitive mechanism.[92] Once migrated there, the thymocyte is exposed to new APCs with new self-antigens that can delete autoreactive thymocytes whose cognate self-antigens are not present in the cortex[88,93]: AIRE-expressing medullary TECs, for example,[50,94] and

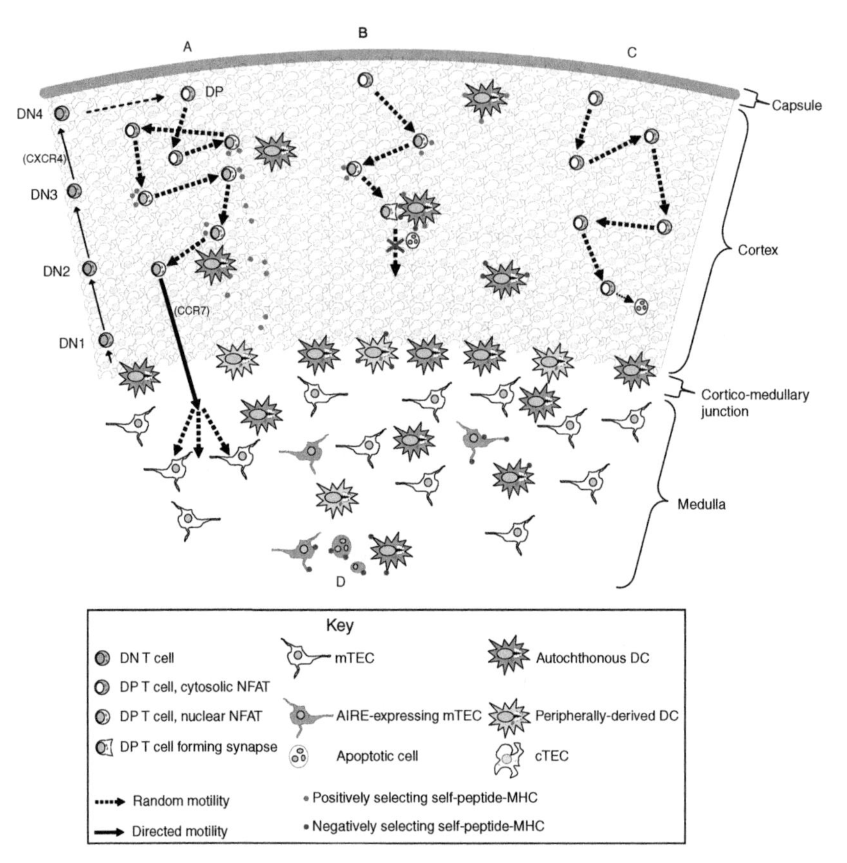

FIG. 3. The gauntlet model of thymic T cell maturation. A Schematic of thymocyte migration in response to developmental and TCR-triggered cues. (A) A thymocyte whose eventual fate is positive selection and maturation. Common lymphoid progenitors (CLPs) initially seed the thymus[87] and migrate outward in a directed fashion in response to chemokine signaling,[3] while discrete anatomic locations within the cortex support TCR recombination and maturation to the DP stage. Once these DN cells become DPs at the SCZ, they instead adopt a random-walk pattern of migration. When a DP encounters positively selecting pMHCs, it pauses briefly and initiates signaling, which includes NFAT nuclear translocation and activation, but does not include synapse formation with the stimulating cTEC. Individually, these contacts are too brief to support the sustained signaling that is required for maturation to proceed,[73,74] and so reencounter with additional positively selecting cTECs presumably is required. When sufficient signaling has occurred, the DP transitions to the late DP/semi-mature SP stage,[88] which includes upregulation of CCR7[89] and downregulation of GIT2,[66] allowing directed migration toward the medulla. Additional signals within the medulla support further maturation,[90] while encounters with medullary APCs (which can present antigens not found in the cortex) run the risk of negative selection.[84] (B) A thymocyte whose eventual fate is negative selection in response to a cortically presented antigen. DPs' random walk motility increases the probability that they will encounter fairly sparse cortical DCs that can present negatively selecting self-antigens. Once encountered, the DP forms an immunological synapse that prevents it from leaving the APC,[82] providing sufficient time to

peripherally derived DCs.[54] The necessity of this step is evidenced by the failure of central tolerance in CCR7-deficient mice.[95,96] Alternatively, if the thymocyte's TCR passes this final test, it is fostered and eventually exported to the periphery as a naïve T cell.[90]

V. miR-181a, A MicroRNA That Enhances T Cell Sensitivity and Promotes Tolerance

Another feature of developing thymocytes is their heightened sensitivity toward pMHC ligands, over and above the already remarkable sensitivity of mature T cells, both in terms of the quality[97] and number[76,77] of ligands they can recognize. For example, while mature CD4+[98] and CD8+[99] T cells require contact with ~3–10 agonist pMHCs to mount a full effector response, DP thymocytes require only 2 to trigger apoptosis.[69] Alternative sialylation of cell surface molecules could partially explain this heightened sensitivity,[100] as could lower CD5 expression in an independent study[101] and more recently high CD45 expression,[102] but other factors clearly are at play.

One such factor is miR-181a, a microRNA previously identified as playing a role in influencing the B/T lineage decision.[103] microRNAs are ~20–22 nucleotide RNAs in their mature form and exert their influence in combination with proteins of the RNA-induced silencing complex (RISC) by targeting mRNAs for degradation or translational repression.[104] We previously showed[105] that miR-181a suppresses multiple phosphatases that negatively regulate the TCR signaling cascade: Ptpn22, which targets the activating phosphorylation sites of lck and ZAP70[106,107]; Shp2, an effector for inhibitory transmembrane adapter proteins (TRAPs) such as SIT and TRIM, which have also been shown to influence thymocyte sensitivity and tolerance[108]; DUSP5, which dephosphorylates activated Erk in the nucleus; and DUSP6, which targets cytosolic Erk.[109] While each of these four targets was repressed only moderately, from two- to fivefold, overexpression of miR-181a in mature T cells resulted in a dramatic shift in their responsiveness, in the form of both higher absolute calcium signals, an ability to respond to fewer agonist pMHCs than are normally required, and the ability to respond to pMHC ligands that normally are too weak to stimulate a response.[105] This resulted from increases in both steady-

---

complete the apoptotic response.[69] (C) A thymocyte whose eventual fate is neglect. (D) Distinct medullary APCs present antigens derived from disparate sources. DCs can migrate from the blood and skin, or develop intrathymically. Subsets of mTECs possess genetic programs that lead to expression of tissue-restricted antigens. These TRAs might be presented directly by mTECs,[51] or cross-presented by other medullary APCs such as DCs.[52]

state and TCR-signaling-induced levels of activated lck and Erk. This mirrors the similar and elegantly delineated mechanism by which the balance of the lck-activatory phosphatase CD45 and the lck-inhibitory kinase Csk modulates TCR sensitivity in developing T cells.[102]

With regard to thymic selection, we also found that miR-181a was expressed at about four- to eightfold higher levels in DP thymocytes than in mature T cells. When miR-181a was inhibited in 5C.C7 transgenic thymus cultures by use of an anti-miR-181a antagomir,[110] a reduction in the number of positively selected T cells could be observed. A similar diminution of the TCR response could be seen in negatively selecting thymus cultures, where a relatively high number of 5C.C7 thymocytes survived in the miR-181a-inhibited thymi. These results led us to suggest that miR-181a was at least in part responsible for thymocyte hyperreactivity toward pMHC and that the developmentally regulated expression of miR-181a at the DP stage might represent a mechanism favoring central tolerance.

This presumably would operate by raising the sensitivity of DP cells and allowing the deletion of thymocytes bearing TCRs with moderate affinity for self-pMHC. As the remaining T cells mature, they would reduce their miR-181a expression and lose the ability to respond to these moderate-affinity pMHCs. This might create a buffer zone in terms of affinity for self, presumably representing K_D's in the range of $\sim$50–100 μM (Fig. 1). In order to test whether such a buffer exists and whether it is important for self-tolerance, we subsequently made use of the various I-E^k-bound self-peptides that we had characterized in the 5C.C7 system.[39] First, to test whether miR-181a is important for efficient induction of central tolerance, we allowed wild type T cells to mature in miR-181a-inhibited thymus cultures and then challenged these T cells with syngeneic APCs. Compared to the T cells from unmanipulated thymi, which remained quiescent, miR-181a inhibition resulted in a small but appreciable population of autoreactive T cells as measured by CD69 upregulation and the production of inflammatory cytokines. To test whether these autoresponsive T cells were responding toward particular self-peptides of especially high affinity, we made use of the two self-peptides that we had previously found to have the highest capacity to stimulate the 5C.C7 TCR, GP, and ATPase 11c (Fig. 4). By fixing the 5C.C7β chain but allowing the TCRα chain to vary as normal, we hoped to generate a preselection repertoire that would be biased toward recognizing these two peptides. We then allowed the 5C.C7β-transgenic thymocytes to mature in thymi in which miR-181a was inhibited. Using GP-I-E^k and ATPase-I-E^k tetramers, we could then see that the miR-181a-inhibited thymi had allowed the maturation of many more GP- and ATPase-reactive T cells, as compared to otherwise unmanipulated 5C.

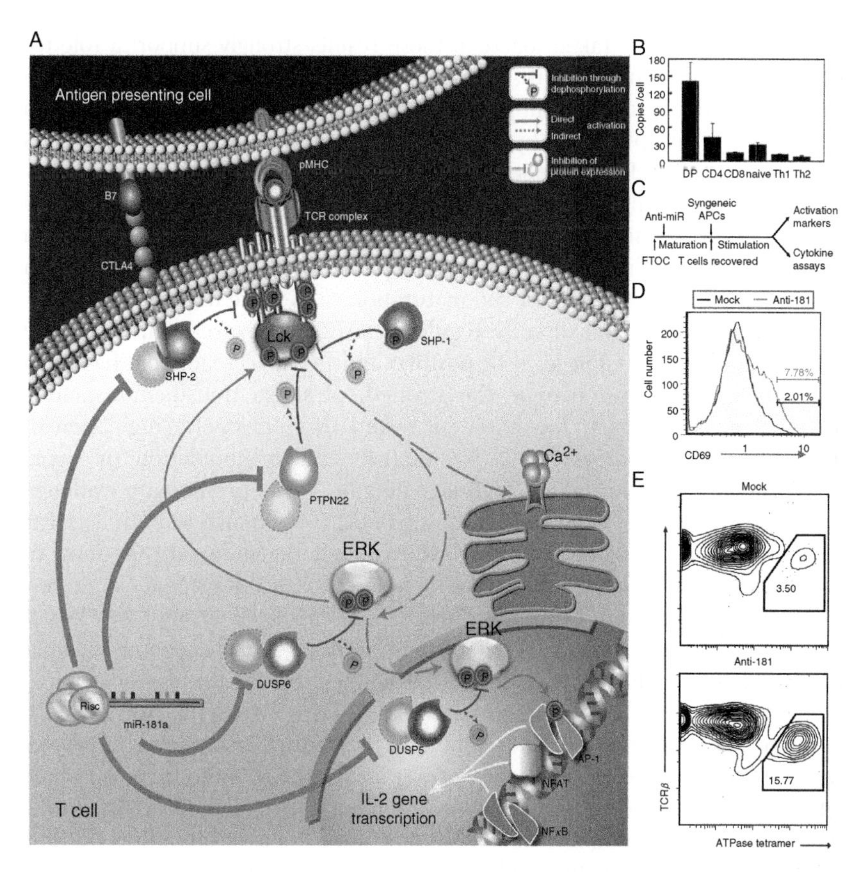

FIG. 4. miR-181a-mediated hypersensitivity in thymic selection. (A) A schematic of miR-181a's targets and their positions in a mature T cell's TCR signaling cascade. (B) Expression level of miR-181a in developing T cell populations, as measured by quantitative real-time PCR. (C) Schematic of the protocol used to test the effect of miR-181a inhibition on T cell development. E16 FTOC were established and treated with antagomir specific for miR-181a, or with seed-region-scrambled antagomir which does not reduce miR-181a levels. After 5–6 days, mature T cells were harvested from these FTOCs, washed to remove any residual antagomirs, and then cocultured with syngeneic (B10.BR) splenic APCs. (D) B10.BR FTOC (in which T cells possess a normally diverse TCR repertoire) were treated in this way, and the resulting mature T cell populations were assayed for expression of the activation marker CD69 by FACS. Gated CD4+ T cells are shown: miR-181a inhibition in the FTOC resulted in a small population of autoreactive T cells. (E) 5C.C7β-transgenic B10.BR FTOC were treated as in part (C), and the resulting T cells were assayed using an ATPase 11c-I-E^k tetramer by FACS. Gated CD4+ T cells are shown: miR-181a inhibition in the FTOC resulted in an increased proportion of T cells that recognize the ATPase 11c self-peptide. Parts (A, B) are originally from Ref. 105 parts (C–E) are originally from Ref. 39.

C7β-transgenic thymi. Taken together, these results strongly support a role for developmentally regulated miR-181a-dependent hypersensitivity in enforcing central tolerance. This in turn supports the notion that, in addition to the spectrum of self-determinants presented by disparate APCs and the affinity of a thymocyte's TCR for those self-pMHCs, the sensitivity of a thymocyte's TCR signaling network is also a critical parameter determining the outcome of selection.

How, then, is miR-181a regulated during T cell maturation? miR-181a downregulation might simply be part and parcel of one of the many known genetic programs that support T cell maturation.[111] A more intriguing possibility is that TCR signaling directly regulates miR-181a expression. Stochastic variation in the expression level of positive and negative regulators has been shown to lead to differences in the strength of signal transduction among individual cells,[112] and it has been suggested that epigenetic regulation of other negative regulators might accomplish similar modulation of signal strength in response to thymic TCR signaling.[113,114] As preliminary evidence for this possibility, we found that TCR signaling in response to both agonists and positively selecting ligands could dramatically downregulate the level of mature miR-181a within 1 hour.[39] We do not yet know the mechanism underlying this unusually rapid miRNA downregulation, however, and we do not know whether it is this direct mechanism or another entirely separate mechanism that results in the low level of miR-181a in mature T cells. It is tempting to speculate, however, that this might also explain why T cells selected by high-affinity positively selecting ligands appear to be attenuated in their responsiveness in the periphery.[33] If especially strong TCR signals in the thymus induce higher than normal miR-181a downregulation and if this low miR-181a expression is maintained during the remainder of the T cell's maturation, then this low miR-181a might render the resulting T cell population hyporesponsive. Also, interestingly, miR-181a is even more highly expressed in DN T cells than in DPs, and miR-181a enhances progression of T cells from the DN to the DP stage in OP9-DL1 cocultures.[105,115] This even higher level of miR-181a expression might in turn cause an even higher degree of TCR hypersensitivity, which might allow DNs to receive maturation cues from the transient dimerization of pre-TCR.[5,116] Whether the ensuing miR-181a downregulation then results directly from these pre-TCR signals, or as the indirect result of the DN-to-DP developmental program, is also unknown.

VI. The Selecting Ligand and the Purpose of Selection

The critical role for self-MHC in the positive selection of developing T cells has been well accepted,[117–119] and when mice bearing monoclonal T cell populations became available, this requirement was also found in most cases to be specific to the particular MHC allele to which that TCR was restricted, in terms of its ability to respond to foreign-peptide-self-MHC.[21,120] This dovetailed beautifully with the established phenomenon of self-restriction of polyclonal T cell populations to thymic MHC, and self-MHC has been credited with self-restriction ever since.[121] More recently, evidence for the importance of self-MHC in selection has also come from the TCR side of the equation. In the context of a polyclonal repertoire, mutations of residues within CDR2β that contact invariant portions of I-A^b reduced the number of selected T cells in I-A^b+ mice.[122] While we have been forced to admit that there is also a requirement for specific self-peptides,[37,48,57] this requirement has been generally characterized as permissive, in that the peptide might allow the invariant portions of the MHC to adopt a particular conformation that is reminiscent of the agonist-MHC[46] or might itself act as a structural mimic of the agonist.[37]

Self-peptides have taken a back seat to self-MHC in terms of their influence on the peripheral TCR repertoire, and it is not immediately clear why this should be. A strong case can be made for the role of self-peptides, based on evidence that is very parallel to the case for self-MHC. Absence of class I or class II molecules results in the near absence of CD8+ or CD4+ T cells, respectively. Mutations in the peptide-binding cleft of I-E^k were found to have a similar effect in mice transgenic for an I-E^k-restricted TCR.[123] Reducing the repertoire of self-peptides from its usual diversity to a handful of distinct species also results in a $>95\%$ deficit in T cell numbers in a polyclonal setting: whether by β2m or TAP mutation in class I systems[33,57] or combined invariant chain/HLA-DM mutation in class II systems.[124,125] In a beautiful demonstration of the peptide's importance, the size of the selected CD8+ T cell repertoire in b2m-deficient thymi could be reconstituted in a dose-dependent fashion by the addition of increasingly complex mixtures of exogenous peptide.[57] Upon later examination of thymi with monoclonal expression of the OT-I TCR, only 2 out of 27 tested endogenously derived H-$2K^b$-binding peptides were capable of supporting positive selection.[37] And, in fact, this fraction overestimates the actual proportion of selecting peptides among the self-peptide repertoire by roughly twofold, as the peptides were split into two fractions prior to mass spectrometric identification and only the active half was assayed further. More recently, we[39] and Lo *et al.*[40] have mirrored this result using the class II-restricted 5C.C7 and AND TCRs, respectively: 6 of 95 naturally occurring I-E^k-bound peptides could reconstitute positive selection of 5C.C7 (5 of them only very weakly) and 1 of 95 peptides could drive selection of AND T cells. As an additional demonstration of the critical role for specific peptides in selection, it should be noted that, although 5C.C7 and AND recognize the same nominal antigen (a pigeon or moth cytochrome *c*-derived peptide presented by I-E^k), the self-peptides that select these two TCRs are distinct. So, while self-MHC is necessary for positive selection, it is also clearly not sufficient. These data, both old and new, are substantial hurdles for the idea that positive selection confers self-MHC-restriction.

It is certainly a compelling idea that T cells should be self-referential on the level of an individual animal. But how necessary is it that TCRs reference the invariant portions of self-MHC in each successive generation? Recently, the possibility that MHC restriction is encoded at least partially in the germline elements of the TCR's V regions has received new attention. If TCRs are predisposed to bind MHC by virtue of their natural history, then the critical question for self-restriction is whether thymic selection can improve the likelihood that a selected T cell will be able to recognize a foreign-peptide-self-MHC. Some 90% of thymocytes are thought to be discarded because they fail positive selection.[25,26] The present model of self-restriction posits that they fail because they could not bind to foreign-peptide-self-MHC or at least that they are less able to bind foreign-peptide-self-MHC complexes than those thymocytes that survive. This has been difficult to address directly, as it requires an extensive evaluation of the preselection TCR repertoire not only in terms of TCR sequence but also in terms of pMHC binding. But the indirect evidence that is available points to a different answer. Earlier work had shown that T cells selected by a TCR-nonspecific stimulus (either anti-TCR antibody[126] or Ly6A.2[127]) were as reactive toward a third-party MHC as T cells from a normal mouse. But as this third-party MHC was not "self" (as indeed these T cells had no "self" to speak of), it remained possible that positive selection would in fact focus the repertoire on the particular MHC allele(s) present in the thymus. This possibility became less likely, however, in light of more recent results showing that T cells that arise in negative-selection-deficient mice are cross-reactive toward multiple MHC alleles.[128] While negative selection presumably reigns in this cross-reactivity in a normal animal,[20] positive selection appears to generate a repertoire that is not particularly focused on the selecting MHC allele. Also compelling in terms of the TCR side of the relationship was the finding that this cross-reactivity depended on conserved residues within the CDRs1 and 2.[129] As another test of the germline compatibility of TCRs with MHC, the class II homolog HLA-DM was expressed on the cell surface in MHC-deficient mice, and this could not restore T cell selection.[130] The question remains open, however, whether this failure resulted from the lack of positive evolutionary pressure on the invariant MHC-like features of HLA-DM or the inability of HLA-DM to present a diverse peptide repertoire (just as the monolithic class II pMHC of Ii/H2M-deficient mice failed to select T cells).[125]

To build a parallel TCR-centric argument, one might consider the impact on positive selection of fixing CDR1 and 2 and allowing only CDR3 to vary. If T cells are selected for their compatibility with the invariant portions of MHC, then CDR1 and 2 should be sufficient to drive selection irrespective of the CDR3 sequence. If, on the other hand, they are selected for their ability to bind specific pMHC combinations, then the CDR3 should prove critical. We can find an answer in the classic study of Sant'Angelo.[131] Here, the authors utilized a transgenic mouse bearing the beta chain of a class II-restricted TCR and examined the sequences of TCRα chains from T cells at various stages of development. By further restricting their analysis to TCRαs using Vα2.3, they effectively considered a set of T cells where five of six CDRs were fixed and only CDR3α could be diverse. In this case, only 6% of DP cells possessed TCRα chains that could be found among peripheral CD4+ T cells. So fixing favorable CDRs1 and 2 certainly did not give these TCRs a free pass in terms of selection; in fact, they conferred a selection efficiency that was barely distinguishable from the selection efficiency we estimate from the proportion of DPs to CD4+ SPs in a normal mouse. By generating mice that are limited to Vα2.3/Vβ8.2 but whose CDR3s are subject to normal diversification, we find a similar inefficiency of positive selection on an H-2^b background, even though both Vα2.3 and Vβ8.2 are utilized by large numbers of T cells in b-haplotype mice (and also accounting for the more limited opportunity for receptor editing due to the Vα transgene; Campbell and Davis, unpublished data). Similarly, several different transgenic mice bearing TCRs with identical Vα11/Vβ3 usage displayed markedly different efficiencies of positive selection,[61] as did TCRs distinguished by a single mutation in the V–J junction.[132] But, remarkably, the Sant'Angelo study also showed that the TCRs that they found among CD4+ SP T cells still were sufficient to very accurately confer appropriate CD4+ lineage choice in the thymus. This further underscores the point that, rather than being driven by degenerate binding of CDRs1 and 2 with MHC, positive selection relies upon specific interactions of CDR3s with particular self-pMHC(s).

Rather than foretelling the reactivity with an as-yet-unknown peptide, previous work has told us that the same self-pMHC that selected a T cell is operative in the periphery, in that it signals the TCR-dependent homeostasis of naïve CD4+[133] and CD8+ T cells.[134,135] A later investigation also demonstrated a role for the selecting self-pMHC in potentiating T cell triggering, in this case by maintaining naïve T cells' CD3 chains in a state of semi-phosphorylation and thus suspending them in a state of near-activation.[136] Lately, we have presented data that could provide a more causal link between positive selection and peripheral activation. It has been known that soluble monomeric agonist pMHC ligands are generally incapable of initiating signaling events in T cells and that dimeric agonist pMHC was sufficient.[137–140] In the case of activated 5C.C7 CD4+ T cells, we could show that heterodimeric engagement of TCRs with a strong agonist pMHC paired with particular self-pMHC (in the form of a covalently linked heterodimer of pMHCs) could trigger calcium and cytokine signaling.[141] We further hypothesized that those particular self-pMHCs that could partner with antigenic peptides to support mature 5C.C7 T cell activation would be operative in positive selection.[142,143]

There are now two studies that support this hypothesis, one from our own work with the 5C.C7 TCR[39] and another from the work of Lo *et al.*[40] with the AND TCR. In the case of 5C.C7, an endogenous retrovirus-derived GP polyprotein peptide was the strongest of 6 positively selecting peptides among the 95 self-peptides tested, and this GP peptide could enhance naïve 5C.C7 T cell activation in response to minute amounts of exogenously added agonist MCC peptide while other self-peptides could not. In the case of AND, one positively selecting peptide was found among the same test set of 95 self-peptides, and this peptide could enhance naïve AND T cell activation in the context of platebound agonist MCC–self-peptide heterodimers, while again other self-peptides could not. Additional studies will certainly be needed to properly evaluate the model of selection-for-coagonist. Nonetheless, we feel it better explains the results of recent years. As one particularly striking example, Chu *et al.*[144] examined the populations of naïve T cells that bound to a particular antigen-I-A^b tetramer, in b-haplotype and in d-haplotype mice. Contrary to what one would expect based on the selection-for-self-restriction model, the two mice harbored very similar numbers of these cells. However, the tetramer-binding cells from the d-haplotype mouse had a fivefold reduced capacity for activation. Thus positive selection had seemingly improved the responsiveness of those T cells without improving their ability to bind agonist peptide-self-MHC. We would suggest that these cells have instead been educated with respect to which self-peptide-I-A^b could support TCR triggering when paired with the agonist-I-A^b. In this case, the critical feature imparted by thymic selection would not be the TCRs' affinity for the agonist pMHC but their fine specificity for much lower affinity self-peptide-MHCs.

There are also hurdles to overcome. One wrinkle in this story is that one of the 5C.C7 "coactivating" self-peptides found in the initial heterodimer study, ER60, was not found to coactivate naïve 5C.C7 T cells when stimulation was done using APCs. This could possibly be due to the different activation states of the T cells used, as the prior study had utilized preactivated T cells which have been shown to be more sensitive to restimulation.[145] Alternatively, this might have been due to a difference in the agonist peptides used, as the 2005 study used a stronger K5 variant of the MCC agonist peptide, which might relax the affinity requirement for the partnering self-peptide. This is plausible, as the later study showed ER60 to be among the five weakly positively selecting peptides for 5C.C7, and so its affinity for the 5C.C7 TCR is likely only slightly weaker than GP's. Finally, this discrepancy might have resulted from a difference between the covalently linked heterodimeric pMHCs and cell-surface-bound pMHCs, in terms of how they trigger TCRs. This possibility will be discussed further in the following section.

VII. On the Role of Self-Peptides in the Activation of CD4+ Versus CD8+ T Cells

Another wrinkle is the disparate role of self-pMHC so far observed in the activation of CD4+ and CD8+ T cells. One report suggested that self-pMHC did not influence OT-I T cell activation, based on the comparison of the ability of RMA versus RMA-S cells to act as APCs.[146] As RMA-S cells lack TAP, they lack the capacity to process their own self-peptides for presentation by class I molecules. However, they do retain a small but significant amount of surface class I molecules[147] and so this data could be interpreted as suggesting only that self-pMHC are not required in large numbers to observe an enhancement of agonist signaling. A second series of reports[148,149] also used RMA-S cells, but in those studies added the various self-peptides known to bind H-2K^b exogenously[37] in order to stabilize and increase H-2K^b surface expression. In this case, an enhancing effect on agonist signaling was observed when H-2K^b expression was rescued, but the identity of the self-peptide made no difference in the magnitude of signaling, even though a subset of the self-peptides tested are known to positively select the OT-I TCR.[37] This strongly suggested that the invariant portions of the self-H-2K^b were responsible for the enhancement of TCR signaling. This could occur through simultaneous binding of TCR to the agonist pMHC and local enrichment of CD8 by binding to the self-pMHC. And, because we know that CD8 contributes to the stabilization of TCR–pMHC[31,150] while CD4 does not,[151] this result seemed to represent a very plausible explanation for the differential role of specific self-peptides in activating the two types of T cell. However, it represents a serious blow to the idea that positive selection selects T cells for their ability to utilize specific self-peptides in the periphery. This is because regardless of their eventual lineage decision, positive selection initially operates on the same population of DP T cells, is triggered by the same TCR signals,[152] and depends on very particular and comparably scarce self-pMHCs in the case of both class I[37] and class II.[40,153] How, then, could their mechanisms of signaling be so different? And if the identity of the selecting peptide is so important for both types of T cell, why do CD4+ T cells remember those peptides in the periphery, while CD8+ T cells forsake them?

To answer these questions, we examined signaling by the OT-I TCR using the soluble heterodimer technique that had initially illuminated the role of self-peptides in triggering CD4+ T cells.[154] Our results using Ova-K^b–self-peptide-K^b heterodimers essentially recapitulated the results of Yachi *et al.*[149]: the agonist Ova-K^b could partner with any of the self-peptide-K^b tested to provoke a calcium response in OT-I DP thymocytes or negative selection in OT-I H-$2K^{-/-}$ FTOCs. This was true irrespective of whether the self-peptide-K^b was a positively selecting peptide or a nonselecting peptide such as Mapk1 or Stat3. However, a different picture began to emerge in our examination of positive selection. Here, a calcium response or a positive selection response was observed only when the positively selecting Catnb-K^b was partnered with another Catnb-K^b or with the other positively selecting ligand, Cappa1-K^b. Catnb-K^b could not support signaling when partnered with the nonselecting Mapk1-K^b or Stat3-K^b. This suggested that both partners in the pMHC heterodimer needed to bind TCR in order to cause a signal. And, in fact, we could show that Mapk1-K^b and Stat3-K^b do bind to the OT-I TCR, albeit very weakly, by measuring the interaction of dimeric Mapk1-K^b or Stat3-K^b with OT-I. Finally, we could abrogate this OT-I binding by generating a TCR-contact-mutated version of the Ova antigen, SIAAFASL, or by sterically blocking Ova's central TCR-binding residue with a biotin. Heterodimers of these peptide-K^bs paired with Ova-K^b were unable to stimulate calcium flux or selection of OT-I thymocytes even when used at a concentration 100-fold higher than was sufficient to observe signaling with the Ova-K^b–Mapk1-K^b or Ova-K^b–Stat3-K^b. Thus we could show that TCR engagement by both partners of the heterodimeric ligand was necessary for thymocyte signaling in this class I-restricted system, as had previously been observed for class II-restricted T cells.

These data also allow us to comment on how faithfully soluble heterodimers recapitulate the natural situation of membrane-bound pMHCs. From our studies of the self-peptide–H-$2K^b$s, it is clear that the covalently linked heterodimers do increase the operational avidity, or apparent affinity, of the ligands involved. This might explain why we can observe an enhancement of 5C.C7 T cell activation by ER60 in the context of the heterodimers,[154] but not in the context of APC-presented ligands.[39] But, critically, the dimers do preserve the rank order of affinities of the individual component pMHCs[38] (Fig. 5). And, further, the increased avidity is not sufficient to explain why the heterodimeric ligand signals while the monomeric ligand does not. This is because we now have a direct measurement of the dimer's apparent K_D, and so we can directly compare the dimers to monomers in terms of how many receptors they are binding. For example, we have measured the calcium response of OT-I DP T cells when their TCRs are engaged by monomeric Ova/H-$2K^b$ at a concentration above the K_D of this interaction (at concentrations of 17 μM; $K_D = 6–10$ μM).[28,38] Remarkably, when more than half of the T cell's TCRs are bound by the monomeric agonist pMHC, no signaling whatsoever results. We can then compare this to the result

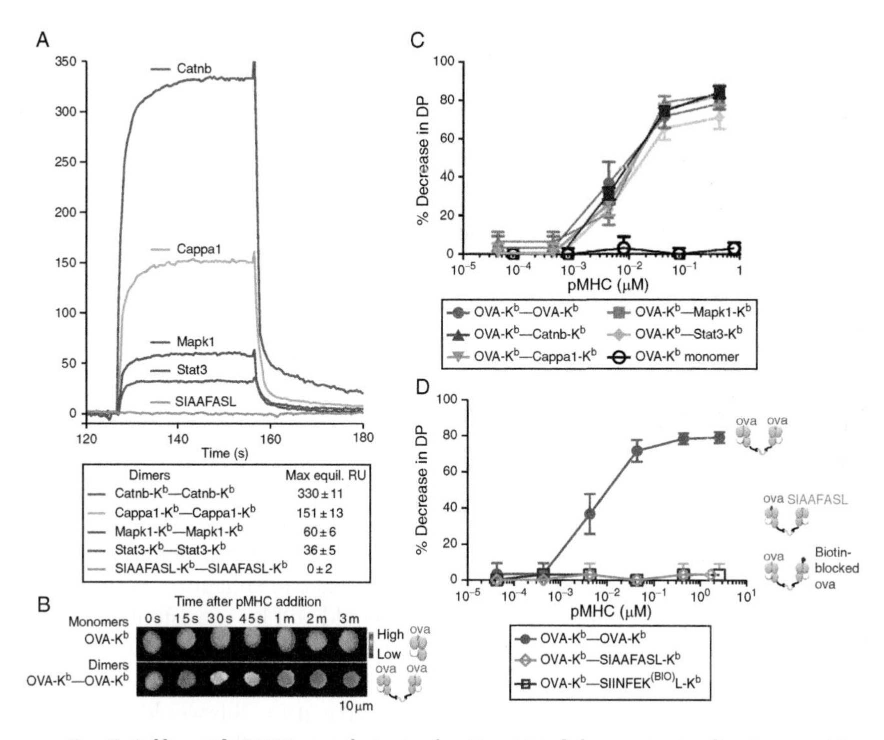

FIG. 5. Self-peptide-MHC contributes to class I-restricted thymocyte signaling in a peptide-specific manner. (A) Surface Plasmon Resonance (SPR) measurement of binding of soluble dimeric complexes of the indicated self-peptide-K^bs to immobilized OT-I TCR. The Mapk1 and Stat3 self-peptide-K^bs bound OT-I with a measurable affinity, while the engineered "null" SIAAFASL-K^b did not. (B) Dimeric but not monomeric Ova-K^b drives calcium signaling by OT-I DP thymocytes. Thymocytes were taken from OT-I H-$2K^{-/-}$ H-$2D^{-/-}$ mice, loaded with the ratiometric calcium dye Fura2am, and exposed to high concentrations of soluble monomeric or dimeric Ova-K^b. Fura2 ratios (340/380 nm) are shown in a false-color scale. (C) Homodimeric or heterodimeric Ova-K^b, but not monomeric Ova-K^b, drives negative selection in whole thymi. E16 FTOC were established from OT-I H-$2K^{-/-}$ H-$2D^{-/-}$ mice, and the indicated soluble monomeric or dimeric H-$2K^b$s were added at varying concentrations. After 3–4 days, FTOC were disrupted, and the resulting thymocyte populations were assayed by FACS: the percentage of surviving DP thymocytes, relative to untreated control FTOCs, are shown. Ova-K^b monomers had no effect on thymocyte selection, even when added at concentrations above the K_D of the Ova-K^b/OT-I interaction, where more than 50% of the thymocytes' TCRs are expected to be bound by the Ova-K^b. Ova-K^b homodimers, or Ova-K^b–self-peptide-K^b heterodimers, could drive negative selection of the OT-I thymocytes to a similar extent, and at an equally low dose. (D) The ability of Ova-K^b–self-peptide-K^b to negatively select depends on the self-peptide-K^b's affinity for the OT-I TCR. FTOC were established and assayed as in part (C). Negative selection did not result from even very high doses of the Ova-K^b–SIAAFASL-K^b (in which the peptide's OT-I-contacting residues are mutated), or the Ova-K^b–SIINFEK$^{(BIO)}$L-K^b (in which a primary predicted OT-I contacting residue in the Ova peptide, the p7 K, is modified with a biotin at the lysine's free amine group). Data are originally from Ref. 38.

of engaging TCRs with a soluble homodimeric Cappa1-H-2K^b–Cappa1-H-2K^b. OT-I's apparent affinity for this dimeric ligand is ~ 60 μM, yet at concentrations well below this (e.g., 8.7 μM), the homodimeric Cappa1-H-2K^b induces robust calcium responses in OT-I DP T cells. The same picture emerges when the OT-I DP cells are exposed to the monomeric or dimeric pMHCs in the context of the whole thymus: agonist monomers are completely nonstimulatory even at very high concentrations (in terms of positive selection, negative selection, or upregulation of activation markers such as CD69), while the Cappa1-H-2K^b homodimers and Catnb-H-2K^b homodimers which have lower apparent affinities, can mediate positive selection at lower concentrations.

Finally, although we believe that we have provided evidence for a more unified mechanism of self-peptide-supported TCR triggering in CD4+ and CD8+ T cells, it is harder to make the connection between positively selecting ligands and peripheral activation for the CD8+ T cells. Whereas signaling by the mature 5C.C7 and AND CD4+ T cells could only be enhanced by the selecting peptide, we now have multiple independent sets of data that suggest that CD8+ T cells overshoot this mark. That is, in the OT-I system, positive selection is driven only by a high-affinity subset of the self-peptides that can enhance mature T cell responses. On the one hand, this observation could still be consistent with our hypothesis that positive selection produces a repertoire of T cells that can respond to agonist pMHC paired with self-pMHC. By forcing the OT-I TCR to recognize a ligand with a K_D of 100–200 μM affinity in order to mature, this would guarantee that the OT-I TCR has coactivating peptides available to it in the periphery, even if weaker pMHCs (~ 300 μM) are also sufficient. In this case, it seems that thymic selection would waste potentially useful TCRs. And, as every additional naïve CD8+ T cell appears to be useful in combating infection,[19] this would seem to be a less than optimal mechanism. On the other hand, we know that naïve CD8+ T cells require self-pMHC for their homeostatic survival, and that this peripheral maintenance is dependent on the same self-pMHCs that selected those T cells in the thymus.[135] In this case, perhaps the weaker ~ 300 μM self-pMHCs are insufficient for OT-I's peripheral maintenance. None of these speculations is particularly satisfying, and it is clear that further investigation of other class I-restricted TCRs will be needed to resolve this.

VIII. Conclusions

The T cell repertoire is molded by a wealth of information about the self. Here we have briefly summarized recent findings relating to how this information is packaged in the thymus for developing T cells' consideration, how thymocytes go about accessing it, and the nature of the information that gives a developing T cell its imprimatur to join the peripheral repertoire.

On the first point, we have accumulated an excellent description of the movement of thymocytes within and between the different anatomic compartments of the thymus,[65,155] and we have begun to understand the molecular queues that underlie this movement, both developmentally regulated[66,92,156] and TCR-triggered.[68,60,82] We also have excellent evidence suggesting that the APCs in these distinct physical locations present nonoverlapping sets of self-peptides, in addition to being phenotypically different in terms of their ability to promote positive versus negative selection.[58] This evidence is mostly indirect, coming from studies of mutations in the MHC loading pathway and their influence on the efficiency of T cell selection.[56,124,157,158] For example, we know that positive selection depends on thymocytes' ability to "scan" the cortex in a wide-ranging and randomly directed fashion,[66,69] and we know that positive selection depends on self-peptides that are quite rare.[37,39,40] While the indirect evidence is compelling, we can only infer the TCR–pMHC interaction that connects these. We also know that negative selection depends on APC-specific mechanisms of TRA presentation,[50] and that limiting thymocytes' access to the medulla results in autoimmunity.[95] But did this occur specifically because thymocytes were denied access to self-peptides, as opposed to another tolerizing property of the medulla?

Clearly these questions would benefit from a more complete identification of the peptides that actually occupy the MHCs on the surfaces of thymic APCs.[58] But making the connection between these peptides and selection events will be a daunting task. Positive selection relies on the diversity of the self-peptide repertoire, such that a very small but especially diverse subset of the total mass of self-pMHC can support positive selection of a large diversity of T cells.[125] We do not know how many peptides this subset might represent: antibodies that block positive selection bound ~3–5% of self-pMHCs,[47,48] and this can serve as a reasonable upper limit, but we have no good sense of the lower limit. In the extreme case, it is quite possible that two phenotypically indistinguishable APCs might present nonoverlapping sets of peptides due only to stochastic variation, combined perhaps with epigenetic variegation.[159] This might offer an explanation for why thymocytes' cortical "scanning" motility is beneficial for their selection: perhaps multiple cTECs must be examined in order to find a selecting interaction.[66,69] But it would also make it incredibly difficult to isolate the particular peptides that are relevant for any given TCR. Negatively selecting self-peptides present a similar conundrum. Thymocytes are capable of responding toward two high-affinity pMHCs,[69] and can seek out such peptides presented by one in 20 APCs.[62] If we conservatively estimate that the APCs involved express ~25,000–50,000 MHC molecules each, then we are talking about a detection limit of between two and four parts per million. Only radical improvements in mass spectrometry, or entirely new technology, will allow us to directly see the peptides that T cells can see.

T cells require only a week to develop a profound understanding of the self, and so we can continue to learn about the nature of self from the T cells themselves. A renewed interest and ability to detect very rare pMHC tetramer-binding T cell populations has yielded new information regarding the diversity of the naïve TCR repertoire and by extension has placed constraints on the number of distinct TCRs that emerge from the thymus.[17,160] Comparing these results with a similar survey of the preselection repertoire would be very valuable in assessing how positive and negative selection shape the mature T cell pool in a normal individual. There are obstacles to overcome here as well, as thymocytes' low surface expression of TCR makes tetramer analysis difficult,[161] and multiple genetic programs conspire to make TCR ligation a lethal proposition.[111,162,163] The tetramer approach also a priori forces us to select peptide targets prior to data collection, and so such an analysis could not be unbiased without a very large sample size. This is a critical difficulty, as the affinity model of selection predicts that tetramers made from positively selecting pMHCs (e.g., $K_D > 50$ μM) should not detect the T cells they select,[38] whereas the altered self model predicts the opposite.[58] Thus a more global analysis may be useful. Advances in high-throughput DNA sequencing technology could be leveraged here, as they have been in the study of antibody diversity.[164] On a sufficient scale, such studies have the potential not only to precisely gauge the diversity of pre- and postselection TCR repertoires but also on a population level to assess the degree of coevolution of TCR and MHC genes.[165] But, as valuable as they are, such techniques will have difficulty connecting what they discover about TCRs to the self-peptides that are relevant for those TCRs' selection and survival.[135]

If we are to understand the mechanics of positive selection without a complete understanding of the self-peptides involved, then we must answer this critical question: what is the selecting parameter, and what is the selected property? Does a TCR's weak affinity for self-peptide-self-MHC predict high affinity for an as-yet-unknown peptide-self-MHC? We do not favor this model, because in this case the selecting parameter must be the affinity afforded by the TCR's contact with genetically encoded MHC residues. Several lines of evidence indicate that this property is satisfied prior to selection,[20,126] and that peptide-specific contacts are crucial for selection.[57,131] Perhaps the positive selection parameter is the breadth of the TCR's reactivity toward self-pMHC, and the selected property is a high degree of polyspecificity.[20,22] This might in turn confer the potential for each selected T cell to recognize multiple distinct antigens and thereby improve the chance that any one T cell will be useful in combating infection. We know, however, that positive selection can proceed when a particular TCR is afforded an abundance of a single pMHC species.[37,39,40] Still, we can only remark this property in the case of three distinct

TCRs among many thousands or millions, and so this may represent the exception rather than the rule. If these do represent an exception, then we might also discard the coinciding observation that selection can be driven by low-affinity peptides presented ubiquitously throughout the thymus. This would allow us to consider the possibility that thymocytes are selected by high-affinity interactions with a unique set of "altered self" peptides on cTECs, and that autoimmunity can be averted because these peptides will never be seen again by mature T cells.[58,166] The selecting parameter might then be a "training set" of agonist cortical peptides, with a completely distinct "test set" due to peptide-selective differences in the MHC loading pathway among peripheral and somatic cells. In this case, the selected property is agonist reactivity with ostensibly "nonself"-peptide-self-MHC and therefore could perfectly predict self-restriction of the peripheral repertoire. But, as noted earlier, the uniqueness of cTECs' peptide repertoire is inferred from mutations of the MHC loading pathway and the ensuing impact on T cell selection. The extent to which peripheral APCs can successfully hide this very dangerous "training set" of agonist self-peptides is far from clear. The fidelity required for this separation of self to be a successful strategy is enormous, because we know that allowing T cells even the tiniest sniff of those cortical agonist peptides could spell disaster.[98,142]

Finally, we will discuss the possibility that we currently favor. Perhaps the TCR's weak affinity for self-peptide-self-MHC predicts improved responsiveness toward as-yet-unknown peptide-self-MHCs specifically because the self-peptide-self-MHCs participate in the latter process. We have suggested a mechanism for TCR triggering based on heterodimeric engagement of TCR with agonist pMHC together with self-pMHC, which could explain the basis for the selected parameter in this case.[143] More recently, we have also shown that this dimer-based mechanism of TCR triggering can be extended to explain how self-peptide-MHCs trigger selection in the thymus.[38] The triggering mechanism can be conserved, despite the weaker ligands involved, at least in part because developmental regulation of miR-181a lowers the signaling threshold in DP thymocytes.[105] Lastly, there are now two peptides that can demonstrate that the property of self-pMHC contributing to peripheral responses is a correlate of the selection process.[39,40] Further studies in additional TCR systems will certainly be needed to support the universality of this observation, however, and additional evidence is still required to settle the precise molecular mechanism governing TCR signaling in the context of cell–cell interactions.[167] More advanced molecular imaging techniques will be invaluable in addressing this last point,[151] while cell and tissue imaging will continue to give us clues as to which APCs are relevant to developing T cells and how the twain meet.[66] And, while significant advances in the detection of macromolecules will be required to directly assess the self-peptide repertoire, we are hopeful that high-throughput and systems-level analyses will allow us to test the predictions of the above models against the actual polyclonal TCR repertoire in healthy individuals.

References

1. Rodewald HR. Thymus organogenesis. *Annu Rev Immunol* 2008;**26**:355–88.
2. McCaughtry TM, Baldwin TA, Wilken MS, Hogquist KA. Clonal deletion of thymocytes can occur in the cortex with no involvement of the medulla. *J Exp Med* 2008;**205**:2575–84.
3. Petrie HT. Cell migration and the control of post-natal T-cell lymphopoiesis in the thymus. *Nat Rev Immunol* 2003;**3**:859–66.
4. Nemazee D. Receptor editing in lymphocyte development and central tolerance. *Nat Rev Immunol* 2006;**6**:728–40.
5. Yamasaki S, Saito T. Molecular basis for pre-TCR-mediated autonomous signaling. *Trends Immunol* 2007;**28**:39–43.
6. Singer A, Bosselut R. CD4/CD8 coreceptors in thymocyte development, selection, and lineage commitment: analysis of the CD4/CD8 lineage decision. *Adv Immunol* 2004;**83**:91–131.
7. Gascoigne NR, Chien Y, Becker DM, Kavaler J, Davis MM. Genomic organization and sequence of T-cell receptor beta-chain constant- and joining-region genes. *Nature* 1984;**310**:387–91.
8. Davis MM, Bjorkman PJ. T-cell antigen receptor genes and T-cell recognition. *Nature* 1988;**334**:395–402.
9. Sorger SB, Paterson Y, Fink PJ, Hedrick SM. T cell receptor junctional regions and the MHC molecule affect the recognition of antigenic peptides by T cell clones. *J Immunol* 1990;**144**:1127–35.
10. Ehrich EW, Devaux B, Rock EP, Jorgensen JL, Davis MM, Chien YH. T cell receptor interaction with peptide/major histocompatibility complex (MHC) and superantigen/MHC ligands is dominated by antigen. *J Exp Med* 1993;**178**:713–22.
11. Katayama CD, Eidelman FJ, Duncan A, Hooshmand F, Hedrick SM. Predicted complementarity determining regions of the T cell antigen receptor determine antigen specificity. *EMBO J* 1995;**14**:927–38.
12. Garcia KC, Degano M, Stanfield RL, Brunmark A, Jackson MR, Peterson PA, et al. An alphabeta T cell receptor structure at 2.5 A and its orientation in the TCR–MHC complex. *Science* 1996;**274**:209–19.
13. Garboczi DN, Ghosh P, Utz U, Fan QR, Biddison WE, Wiley DC. Structure of the complex between human T-cell receptor, viral peptide and HLA-A2. *Nature* 1996;**384**:134–41.
14. Rudolph MG, Stanfield RL, Wilson IA. How TCRs bind MHCs, peptides, and coreceptors. *Annu Rev Immunol* 2006;**24**:419–66.
15. Garcia KC, Adams JJ, Feng D, Ely LK. The molecular basis of TCR germline bias for MHC is surprisingly simple. *Nat Immunol* 2009;**10**:143–7.
16. Wu LC, Tuot DS, Lyons DS, Garcia KC, Davis MM. Two-step binding mechanism for T-cell receptor recognition of peptide MHC. *Nature* 2002;**418**:552–6.
17. Jenkins MK, Chu HH, McLachlan JB, Moon JJ. On the composition of the preimmune repertoire of T cells specific for peptide-major histocompatibility complex ligands. *Annu Rev Immunol* 2010;**28**:275–94.

18. Moon JJ, Chu HH, Pepper M, McSorley SJ, Jameson SC, Kedl RM, et al. Naive CD4(+) T cell frequency varies for different epitopes and predicts repertoire diversity and response magnitude. *Immunity* 2007;**27**:203–13.
19. Obar JJ, Khanna KM, Lefrancois L. Endogenous naive CD8+ T cell precursor frequency regulates primary and memory responses to infection. *Immunity* 2008;**28**:859–69.
20. Huseby ES, White J, Crawford F, Vass T, Becker D, Pinilla C, et al. How the T cell repertoire becomes peptide and MHC specific. *Cell* 2005;**122**:247–60.
21. Berg LJ, Pullen AM, de St Fazeka, Groth B, Mathis D, Benoist C, et al. Antigen/MHC-specific T cells are preferentially exported from the thymus in the presence of their MHC ligand. *Cell* 1989;**58**:1035–46.
22. Wucherpfennig KW, Allen PM, Celada F, Cohen IR, De Boer R, Garcia KC, et al. Polyspecificity of T cell and B cell receptor recognition. *Semin Immunol* 2007;**19**:216–24.
23. Arstila TP, Casrouge A, Baron V, Even J, Kanellopoulos J, Kourilsky P. A direct estimate of the human alphabeta T cell receptor diversity. *Science* 1999;**286**:958–61.
24. Sant'Angelo DB, Waterbury PG, Cohen BE, Martin WD, Van Kaer L, Hayday AC, et al. The imprint of intrathymic self-peptides on the mature T cell receptor repertoire. *Immunity* 1997;**7**:517–24.
25. Scollay RG, Butcher EC, Weissman IL. Thymus cell migration. Quantitative aspects of cellular traffic from the thymus to the periphery in mice. *Eur J Immunol* 1980;**10**:210–8.
26. Surh CD, Sprent J. T-cell apoptosis detected in situ during positive and negative selection in the thymus. *Nature* 1994;**372**:100–3.
27. Kisielow P, Bluthmann H, Staerz UD, Steinmetz M, von Boehmer H. Tolerance in T-cell-receptor transgenic mice involves deletion of nonmature CD4+8+ thymocytes. *Nature* 1988;**333**:742–6.
28. Alam SM, Travers PJ, Wung JL, Nasholds W, Redpath S, Jameson SC, et al. T-cell-receptor affinity and thymocyte positive selection. *Nature* 1996;**381**:616–20.
29. Armstrong KM, Piepenbrink KH, Baker BM. Conformational changes and flexibility in T-cell receptor recognition of peptide-MHC complexes. *Biochem J* 2008;**415**:183–96.
30. Naeher D, Daniels MA, Hausmann B, Guillaume P, Luescher I, Palmer E. A constant affinity threshold for T cell tolerance. *J Exp Med* 2007;**204**:2553–9.
31. Garcia KC, Scott CA, Brunmark A, Carbone FR, Peterson PA, Wilson IA, et al. CD8 enhances formation of stable T-cell receptor/MHC class I molecule complexes. *Nature* 1996;**384**:577–81.
32. Alam SM, Davies GM, Lin CM, Zal T, Nasholds W, Jameson SC, et al. Qualitative and quantitative differences in T cell receptor binding of agonist and antagonist ligands. *Immunity* 1999;**10**:227–37.
33. Stefanski HE, Mayerova D, Jameson SC, Hogquist KA. A low affinity TCR ligand restores positive selection of CD8+ T cells in vivo. *J Immunol* 2001;**166**:6602–7.
34. Ashton-Rickardt PG, Van Kaer L, Schumacher TN, Ploegh HL, Tonegawa S. Peptide contributes to the specificity of positive selection of CD8+ T cells in the thymus. *Cell* 1993;**73**:1041–9.
35. Hogquist KA, Jameson SC, Bevan MJ. Strong agonist ligands for the T cell receptor do not mediate positive selection of functional CD8+ T cells. *Immunity* 1995;**3**:79–86.
36. Hogquist KA, Tomlinson AJ, Kieper WC, McGargill MA, Hart MC, Naylor S, et al. Identification of a naturally occurring ligand for thymic positive selection. *Immunity* 1997;**6**:389–99.
37. Santori FR, Kieper WC, Brown SM, Lu Y, Neubert TA, Johnson KL, et al. Rare, structurally homologous self-peptides promote thymocyte positive selection. *Immunity* 2002;**17**:131–42.
38. Juang J, Ebert PJ, Feng D, Garcia KC, Krogsgaard M, Davis MM. Peptide-MHC heterodimers show that thymic positive selection requires a more restricted set of self-peptides than negative selection. *J Exp Med* 2010;**207**:1223–34.

39. Ebert PJ, Jiang S, Xie J, Li QJ, Davis MM. An endogenous positively selecting peptide enhances mature T cell responses and becomes an autoantigen in the absence of microRNA miR-181a. *Nat Immunol* 2009;**10**:1162–9.
40. Lo W-L, Felix NJ, Walters JJ, Rohrs H, Gross ML, Allen PM. An endogenous peptide-MHC ligand positively selects and augments peripheral CD4+ T cell activation and survival. *Nat Immunol* 2009;**10**:1155–61.
41. Felix NJ, Donermeyer DL, Horvath S, Walters JJ, Gross ML, Suri A, et al. Alloreactive T cells respond specifically to multiple distinct peptide-MHC complexes. *Nat Immunol* 2007;**8**:388–97.
42. Reay PA, Kantor RM, Davis MM. Use of global amino acid replacements to define the requirements for MHC binding and T cell recognition of moth cytochrome *c* (93–103). *J Immunol* 1994;**152**:3946–57.
43. Fremont DH, Stura EA, Matsumura M, Peterson PA, Wilson IA. Crystal structure of an H-2Kb-ovalbumin peptide complex reveals the interplay of primary and secondary anchor positions in the major histocompatibility complex binding groove. *Proc Natl Acad Sci USA* 1995;**92**:2479–83.
44. Hogquist KA, Jameson SC, Heath WR, Howard JL, Bevan MJ, Carbone FR. T cell receptor antagonist peptides induce positive selection. *Cell* 1994;**76**:17–27.
45. Starr TK, Jameson SC, Hogquist KA. Positive and negative selection of T cells. *Annu Rev Immunol* 2003;**21**:139–76.
46. Stefanski HE, Jameson SC, Hogquist KA. Positive selection is limited by available peptide-dependent MHC conformations. *J Immunol* 2000;**164**:3519–26.
47. Baldwin KK, Reay PA, Wu L, Farr A, Davis MM. A T cell receptor-specific blockade of positive selection. *J Exp Med* 1999;**189**:13–24.
48. Viret C, He X, Janeway Jr. CA. On the self-referential nature of naive MHC class II-restricted T cells. *J Immunol* 2000;**165**:6183–92.
49. Marrack P, Ignatowicz L, Kappler JW, Boymel J, Freed JH. Comparison of peptides bound to spleen and thymus class II. *J Exp Med* 1993;**178**:2173–83.
50. Anderson MS, Venanzi ES, Klein L, Chen Z, Berzins SP, Turley SJ, et al. Projection of an immunological self shadow within the thymus by the aire protein. *Science* 2002;**298**:1395–401.
51. Aschenbrenner K, D'Cruz LM, Vollmann EH, Hinterberger M, Emmerich J, Swee LK, et al. Selection of Foxp3+ regulatory T cells specific for self antigen expressed and presented by Aire+ medullary thymic epithelial cells. *Nat Immunol* 2007;**8**:351–8.
52. Gray D, Abramson J, Benoist C, Mathis D. Proliferative arrest and rapid turnover of thymic epithelial cells expressing Aire. *J Exp Med* 2007;**204**:2521–8.
53. Seach N, Ueno T, Fletcher AL, Lowen T, Mattesich M, Engwerda CR, et al. The lymphotoxin pathway regulates Aire-independent expression of ectopic genes and chemokines in thymic stromal cells. *J Immunol* 2008;**180**:5384–92.
54. Bonasio R, Scimone ML, Schaerli P, Grabie N, Lichtman AH, von Andrian UH. Clonal deletion of thymocytes by circulating dendritic cells homing to the thymus. *Nat Immunol* 2006;**7**:1092–100.
55. Honey K, Nakagawa T, Peters C, Rudensky A. Cathepsin L regulates CD4+ T cell selection independently of its effect on invariant chain: a role in the generation of positively selecting peptide ligands. *J Exp Med* 2002;**195**:1349–58.
56. Nitta T, Murata S, Sasaki K, Fujii H, Ripen AM, Ishimaru N, et al. Thymoproteasome shapes immunocompetent repertoire of CD8+ T cells. *Immunity* 2010;**32**:29–40.
57. Hogquist KA, Gavin MA, Bevan MJ. Positive selection of CD8+ T cells induced by major histocompatibility complex binding peptides in fetal thymic organ culture. *J Exp Med* 1993;**177**:1469–73.

58. Klein L, Hinterberger M, Wirnsberger G, Kyewski B. Antigen presentation in the thymus for positive selection and central tolerance induction. *Nat Rev Immunol* 2009;**9**:833–44.
59. Berg LJ, Frank GD, Davis MM. The effects of MHC gene dosage and allelic variation on T cell receptor selection. *Cell* 1990;**60**:1043–53.
60. Labrecque N, Whitfield LS, Obst R, Waltzinger C, Benoist C, Mathis D. How much TCR does a T cell need? *Immunity* 2001;**15**:71–82.
61. Yelon D, Berg LJ. Structurally similar TCRs differ in their efficiency of positive selection. *J Immunol* 1997;**158**:5219–28.
62. Merkenschlager M, Benoist C, Mathis D. Evidence for a single-niche model of positive selection. *Proc Natl Acad Sci USA* 1994;**91**:11694–8.
63. Merkenschlager M. Tracing interactions of thymocytes with individual stromal cell partners. *Eur J Immunol* 1996;**26**:892–6.
64. Plotkin J, Prockop SE, Lepique A, Petrie HT. Critical role for CXCR4 signaling in progenitor localization and T cell differentiation in the postnatal thymus. *J Immunol* 2003;**171**:4521–7.
65. Witt CM, Raychaudhuri S, Schaefer B, Chakraborty AK, Robey EA. Directed migration of positively selected thymocytes visualized in real time. *PLoS Biol* 2005;**3**:e160.
66. Phee H, Dzhagalov I, Mollenauer M, Wang Y, Irvine DJ, Robey E, et al. Regulation of thymocyte positive selection and motility by GIT2. *Nat Immunol* 2010;**11**:503–11.
67. Bousso P, Bhakta NR, Lewis RS, Robey E. Dynamics of thymocyte–stromal cell interactions visualized by two-photon microscopy. *Science* 2002;**296**:1876–80.
68. Bhakta NR, Oh DY, Lewis RS. Calcium oscillations regulate thymocyte motility during positive selection in the three-dimensional thymic environment. *Nat Immunol* 2005;**6**:143–51.
69. Ebert PJ, Ehrlich LI, Davis MM. Low ligand requirement for deletion and lack of synapses in positive selection enforce the gauntlet of thymic T cell maturation. *Immunity* 2008;**29**:734–45.
70. Monks CR, Freiberg BA, Kupfer H, Sciaky N, Kupfer A. Three-dimensional segregation of supramolecular activation clusters in T cells. *Nature* 1998;**395**:82–6.
71. Grakoui A, Bromley SK, Sumen C, Davis MM, Shaw AS, Allen PM, et al. The immunological synapse: a molecular machine controlling T cell activation. *Science* 1999;**285**:221–7.
72. Gunzer M, Schafer A, Borgmann S, Grabbe S, Zanker KS, Brocker EB, et al. Antigen presentation in extracellular matrix: interactions of T cells with dendritic cells are dynamic, short lived, and sequential. *Immunity* 2000;**13**:323–32.
73. Kisielow P, Miazek A. Positive selection of T cells: rescue from programmed cell death and differentiation require continual engagement of the T cell receptor. *J Exp Med* 1995;**181**:1975–84.
74. Liu X, Bosselut R. Duration of TCR signaling controls CD4-CD8 lineage differentiation in vivo. *Nat Immunol* 2004;**5**:280–8.
75. Viret C, Sant'Angelo DB, He X, Ramaswamy H, Janeway Jr. CA. A role for accessibility to self-peptide-self-MHC complexes in intrathymic negative selection. *J Immunol* 2001;**166**:4429–37.
76. Davey GM, Schober SL, Endrizzi BT, Dutcher AK, Jameson SC, Hogquist KA. Preselection thymocytes are more sensitive to T cell receptor stimulation than mature T cells. *J Exp Med* 1998;**188**:1867–74.
77. Peterson DA, DiPaolo RJ, Kanagawa O, Unanue ER. Cutting edge: negative selection of immature thymocytes by a few peptide-MHC complexes: differential sensitivity of immature and mature T cells. *J Immunol* 1999;**162**:3117–20.
78. Punt JA, Osborne BA, Takahama Y, Sharrow SO, Singer A. Negative selection of CD4+CD8+ thymocytes by T cell receptor-induced apoptosis requires a costimulatory signal that can be provided by CD28. *J Exp Med* 1994;**179**:709–13.

79. Laufer TM, DeKoning J, Markowitz JS, Lo D, Glimcher LH. Unopposed positive selection and autoreactivity in mice expressing class II MHC only on thymic cortex. *Nature* 1996;**383**:81–5.
80. Laufer TM, Fan L, Glimcher LH. Self-reactive T cells selected on thymic cortical epithelium are polyclonal and are pathogenic in vivo. *J Immunol* 1999;**162**:5078–84.
81. Ahn S, Lee G, Yang SJ, Lee D, Lee S, Shin HS, et al. TSCOT+ thymic epithelial cell-mediated sensitive CD4 tolerance by direct presentation. *PLoS Biol* 2008;**6**:e191.
82. Richie LI, Ebert PJ, Wu LC, Krummel MF, Owen JJ, Davis MM. Imaging synapse formation during thymocyte selection: inability of CD3zeta to form a stable central accumulation during negative selection. *Immunity* 2002;**16**:595–606.
83. Hailman E, Burack WR, Shaw AS, Dustin ML, Allen PM. Immature CD4(+)CD8(+) thymocytes form a multifocal immunological synapse with sustained tyrosine phosphorylation. *Immunity* 2002;**16**:839–48.
84. Le Borgne M, Ladi E, Dzhagalov I, Herzmark P, Liao YF, Chakraborty AK, et al. The impact of negative selection on thymocyte migration in the medulla. *Nat Immunol* 2009;**10**:823–30.
85. Punt JA, Havran W, Abe R, Sarin A, Singer A. T cell receptor (TCR)-induced death of immature CD4+CD8+ thymocytes by two distinct mechanisms differing in their requirement for CD28 costimulation: implications for negative selection in the thymus. *J Exp Med* 1997;**186**:1911–22.
86. Huppa JB, Gleimer M, Sumen C, Davis MM. Continuous T cell receptor signaling required for synapse maintenance and full effector potential. *Nat Immunol* 2003;**4**:749–55.
87. Serwold T, Ehrlich LI, Weissman IL. Reductive isolation from bone marrow and blood implicates common lymphoid progenitors as the major source of thymopoiesis. *Blood* 2008;**113**:807–15.
88. Kishimoto H, Sprent J. Negative selection in the thymus includes semimature T cells. *J Exp Med* 1997;**185**:263–71.
89. Ueno T, Saito F, Gray DH, Kuse S, Hieshima K, Nakano H, et al. CCR7 signals are essential for cortex-medulla migration of developing thymocytes. *J Exp Med* 2004;**200**:493–505.
90. Park JH, Adoro S, Guinter T, Erman B, Alag AS, Catalfamo M, et al. Signaling by intrathymic cytokines, not T cell antigen receptors, specifies CD8 lineage choice and promotes the differentiation of cytotoxic-lineage T cells. *Nat Immunol* 2010;**11**:257–64.
91. Misslitz A, Pabst O, Hintzen G, Ohl L, Kremmer E, Petrie HT, et al. Thymic T cell development and progenitor localization depend on CCR7. *J Exp Med* 2004;**200**:481–91.
92. Ehrlich LI, Oh DY, Weissman IL, Lewis RS. Differential contribution of chemotaxis and substrate restriction to segregation of immature and mature thymocytes. *Immunity* 2009;**31**:986–98.
93. Cho HJ, Edmondson SG, Miller AD, Sellars M, Alexander ST, Somersan S, et al. Cutting edge: identification of the targets of clonal deletion in an unmanipulated thymus. *J Immunol* 2003;**170**:10–3.
94. Liston A, Lesage S, Wilson J, Peltonen L, Goodnow CC. Aire regulates negative selection of organ-specific T cells. *Nat Immunol* 2003;**4**:350–4.
95. Kurobe H, Liu C, Ueno T, Saito F, Ohigashi I, Seach N, et al. CCR7-dependent cortex-to-medulla migration of positively selected thymocytes is essential for establishing central tolerance. *Immunity* 2006;**24**:165–77.
96. Nitta T, Nitta S, Lei Y, Lipp M, Takahama Y. CCR7-mediated migration of developing thymocytes to the medulla is essential for negative selection to tissue-restricted antigens. *Proc Natl Acad Sci USA* 2009;**106**:17129–33.
97. Pircher H, Rohrer UH, Moskophidis D, Zinkernagel RM, Hengartner H. Lower receptor avidity required for thymic clonal deletion than for effector T-cell function. *Nature* 1991;**351**:482–5.

98. Irvine DJ, Purbhoo MA, Krogsgaard M, Davis MM. Direct observation of ligand recognition by T cells. *Nature* 2002;**419**:845–9.
99. Purbhoo MA, Irvine DJ, Huppa JB, Davis MM. T cell killing does not require the formation of a stable mature immunological synapse. *Nat Immunol* 2004;**5**:524–30.
100. Starr TK, Daniels MA, Lucido MM, Jameson SC, Hogquist KA. Thymocyte sensitivity and supramolecular activation cluster formation are developmentally regulated: a partial role for sialylation. *J Immunol* 2003;**171**:4512–20.
101. Tarakhovsky A, Kanner SB, Hombach J, Ledbetter JA, Muller W, Killeen N, et al. A role for CD5 in TCR-mediated signal transduction and thymocyte selection. *Science* 1995;**269**:535–7.
102. Zikherman J, Jenne C, Watson S, Doan K, Raschke W, Goodnow CC, et al. CD45-Csk phosphatase-kinase titration uncouples basal and inducible T cell receptor signaling during thymic development. *Immunity* 2010;**32**:342–54.
103. Chen CZ, Li L, Lodish HF, Bartel DP. MicroRNAs modulate hematopoietic lineage differentiation. *Science* 2004;**303**:83–6.
104. Bartel DP, Chen CZ. Micromanagers of gene expression: the potentially widespread influence of metazoan microRNAs. *Nat Rev* 2004;**5**:396–400.
105. Li QJ, Chau J, Ebert PJ, Sylvester G, Min H, Liu G, et al. miR-181a is an intrinsic modulator of T cell sensitivity and selection. *Cell* 2007;**129**:147–61.
106. Wu J, Katrekar A, Honigberg LA, Smith AM, Conn MT, Tang J, et al. Identification of substrates of human protein-tyrosine phosphatase PTPN22. *J Biol Chem* 2006;**281**:11002–10.
107. Zikherman J, Hermiston M, Steiner D, Hasegawa K, Chan A, Weiss A. PTPN22 deficiency cooperates with the CD45 E613R allele to break tolerance on a non-autoimmune background. *J Immunol* 2009;**182**:4093–106.
108. Koelsch U, Schraven B, Simeoni L. SIT and TRIM determine T cell fate in the thymus. *J Immunol* 2008;**181**:5930–9.
109. Theodosiou A, Ashworth A. MAP kinase phosphatases. *Genome Biol* 2002;**3**: (REVIEWS3009).
110. Soutschek J, Akinc A, Bramlage B, Charisse K, Constien R, Donoghue M, et al. Therapeutic silencing of an endogenous gene by systemic administration of modified siRNAs. *Nature* 2004;**432**:173–8.
111. Jones ME, Zhuang Y. Acquisition of a functional T cell receptor during T lymphocyte development is enforced by HEB and E2A transcription factors. *Immunity* 2007;**27**:860–70.
112. Feinerman O, Veiga J, Dorfman JR, Germain RN, Altan-Bonnet G. Variability and robustness in T cell activation from regulated heterogeneity in protein levels. *Science* 2008;**321**:1081–4.
113. Schmedt C, Saijo K, Niidome T, Kuhn R, Aizawa S, Tarakhovsky A. Csk controls antigen receptor-mediated development and selection of T-lineage cells. *Nature* 1998;**394**:901–4.
114. Wong P, Barton GM, Forbush KA, Rudensky AY. Dynamic tuning of T cell reactivity by self-peptide-major histocompatibility complex ligands. *J Exp Med* 2001;**193**:1179–87.
115. Mao TK, Chen CZ. Dissecting microRNA-mediated gene regulation and function in T-cell development. *Methods Enzymol* 2007;**427**:171–89.
116. Yamasaki S, Ishikawa E, Sakuma M, Ogata K, Sakata-Sogawa K, Hiroshima M, et al. Mechanistic basis of pre-T cell receptor-mediated autonomous signaling critical for thymocyte development. *Nat Immunol* 2006;**7**:67–75.
117. Bevan MJ, Fink PJ. The influence of thymus H-2 antigens on the specificity of maturing killer and helper cells. *Immunol Rev* 1978;**42**:3–19.
118. Zinkernagel RM, Callahan GN, Althage A, Cooper S, Klein PA, Klein J. On the thymus in the differentiation of "H-2 self-recognition" by T cells: evidence for dual recognition? *J Exp Med* 1978;**147**:882–96.
119. Zuniga-Pflucker JC, Longo DL, Kruisbeek AM. Positive selection of CD4-CD8+ T cells in the thymus of normal mice. *Nature* 1989;**338**:76–8.

120. Kisielow P, Teh HS, Bluthmann H, von Boehmer H. Positive selection of antigen-specific T cells in thymus by restricting MHC molecules. *Nature* 1988;**335**:730–3.
121. Hogquist KA, Jameson SC, Bevan MJ. The ligand for positive selection of T lymphocytes in the thymus. *Curr Opin Immunol* 1994;**6**:273–8.
122. Scott-Browne JP, White J, Kappler JW, Gapin L, Marrack P. Germline-encoded amino acids in the alphabeta T-cell receptor control thymic selection. *Nature* 2009;**458**:1043–6.
123. Bhayani HR, Hedrick SM. The role of polymorphic amino acids of the MHC molecule in the selection of the T cell repertoire. *J Immunol* 1991;**146**:1093–8.
124. Barton GM, Rudensky AY. Evaluating peptide repertoires within the context of thymocyte development. *Semin Immunol* 1999;**11**:417–22.
125. Barton GM, Rudensky AY. Requirement for diverse, low-abundance peptides in positive selection of T cells. *Science* 1999;**283**:67–70.
126. Zerrahn J, Held W, Raulet DH. The MHC reactivity of the T cell repertoire prior to positive and negative selection. *Cell* 1997;**88**:627–36.
127. Henderson SC, Berezovskaya A, English A, Palliser D, Rock KL, Bamezai A. CD4+ T cells mature in the absence of MHC class I and class II expression in Ly-6A.2 transgenic mice. *J Immunol* 1998;**161**:175–82.
128. Huseby ES, Crawford F, White J, Kappler J, Marrack P. Negative selection imparts peptide specificity to the mature T cell repertoire. *Proc Natl Acad Sci USA* 2003;**100**:11565–70.
129. Rubtsova K, Scott-Browne JP, Crawford F, Dai S, Marrack P, Kappler JW. Many different Vbeta CDR3s can reveal the inherent MHC reactivity of germline-encoded TCR V regions. *Proc Natl Acad Sci USA* 2009;**106**:7951–6.
130. Kim HJ, Guo D, Sant'Angelo DB. Coevolution of TCR–MHC interactions: conserved MHC tertiary structure is not sufficient for interactions with the TCR. *Proc Natl Acad Sci USA* 2005;**102**:7263–7.
131. Sant'Angelo DB, Lucas B, Waterbury PG, Cohen B, Brabb T, Goverman J, et al. A molecular map of T cell development. *Immunity* 1998;**9**:179–86.
132. Kaye J, Vasquez NJ, Hedrick SM. Involvement of the same region of the T cell antigen receptor in thymic selection and foreign peptide recognition. *J Immunol* 1992;**148**:3342–53.
133. Ernst B, Lee DS, Chang JM, Sprent J, Surh CD. The peptide ligands mediating positive selection in the thymus control T cell survival and homeostatic proliferation in the periphery. *Immunity* 1999;**11**:173–81.
134. Goldrath AW, Bevan MJ. Low-affinity ligands for the TCR drive proliferation of mature CD8 + T cells in lymphopenic hosts. *Immunity* 1999;**11**:183–90.
135. Goldrath AW, Bevan MJ. Selecting and maintaining a diverse T-cell repertoire. *Nature* 1999;**402**:255–62.
136. Stefanova I, Dorfman JR, Germain RN. Self-recognition promotes the foreign antigen sensitivity of naive T lymphocytes. *Nature* 2002;**420**:429–34.
137. Boniface JJ, Rabinowitz JD, Wulfing C, Hampl J, Reich Z, Altman JD, et al. Initiation of signal transduction through the T cell receptor requires the multivalent engagement of peptide/MHC ligands [corrected]. *Immunity* 1998;**9**:459–66.
138. Stone JD, Stern LJ. CD8 T cells, like CD4 T cells, are triggered by multivalent engagement of TCRs by MHC-peptide ligands but not by monovalent engagement. *J Immunol* 2006;**176**:1498–505.
139. Cochran JR, Cameron TO, Stern LJ. The relationship of MHC-peptide binding and T cell activation probed using chemically defined MHC class II oligomers. *Immunity* 2000;**12**:241–50.
140. Cochran JR, Cameron TO, Stone JD, Lubetsky JB, Stern LJ. Receptor proximity, not intermolecular orientation, is critical for triggering T-cell activation. *J Biol Chem* 2001;**276**:28068–74.

141. Krogsgaard M, Li QJ, Sumen C, Huppa JB, Huse M, Davis MM. Agonist/endogenous peptide-MHC heterodimers drive T cell activation and sensitivity. *Nature* 2005;**434**:238–43.
142. Davis MM, Krogsgaard M, Huse M, Huppa JB, Lillemeier BF, Li QJ. T cells as a self-referential, sensory organ. *Annu Rev Immunol* 2007;**25**:681–95.
143. Krogsgaard M, Juang J, Davis MM. A role for "self" in T-cell activation. *Semin Immunol* 2007;**19**:236–44.
144. Chu HH, Moon JJ, Takada K, Pepper M, Molitor JA, Schacker TW, et al. Positive selection optimizes the number and function of MHCII-restricted CD4+ T cell clones in the naive polyclonal repertoire. *Proc Natl Acad Sci USA* 2009;**106**:11241–5.
145. Stefanova I, Dorfman JR, Tsukamoto M, Germain RN. On the role of self-recognition in T cell responses to foreign antigen. *Immunol Rev* 2003;**191**:97–106.
146. Sporri R, Reis e Sousa C. Self peptide/MHC class I complexes have a negligible effect on the response of some CD8+ T cells to foreign antigen. *Eur J Immunol* 2002;**32**:3161–317.
147. Ljunggren HG, Stam NJ, Ohlen C, Neefjes JJ, Hoglund P, Heemels MT, et al. Empty MHC class I molecules come out in the cold. *Nature* 1990;**346**:476–80.
148. Yachi PP, Ampudia J, Gascoigne NR, Zal T. Nonstimulatory peptides contribute to antigen-induced CD8-T cell receptor interaction at the immunological synapse. *Nat Immunol* 2005;**6**:785–92.
149. Yachi PP, Lotz C, Ampudia J, Gascoigne NR. T cell activation enhancement by endogenous pMHC acts for both weak and strong agonists but varies with differentiation state. *J Exp Med* 2007;**204**:2747–57.
150. Wooldridge L, van den Berg HA, Glick M, Gostick E, Laugel B, Hutchinson SL, et al. Interaction between the CD8 coreceptor and major histocompatibility complex class I stabilizes T cell receptor-antigen complexes at the cell surface. *J Biol Chem* 2005;**280**:27491–501.
151. Huppa JB, Axmann M, Mortelmaier MA, Lillemeier BF, Newell EW, Brameshuber M, et al. TCR–peptide-MHC interactions in situ show accelerated kinetics and increased affinity. *Nature* 2010;**463**:963–7.
152. Bosselut R, Feigenbaum L, Sharrow SO, Singer A. Strength of signaling by CD4 and CD8 coreceptor tails determines the number but not the lineage direction of positively selected thymocytes. *Immunity* 2001;**14**:483–94.
153. Ebert PJ, Baker JF, Punt JA. Immature CD4+CD8+ thymocytes do not polarize lipid rafts in response to TCR-mediated signals. *J Immunol* 2000;**165**:5435–42.
154. Krogsgaard M, Davis MM. How T cells 'see' antigen. *Nat Immunol* 2005;**6**:239–45.
155. Robey EA, Bousso P. Visualizing thymocyte motility using 2-photon microscopy. *Immunol Rev* 2003;**195**:51–7.
156. Griffith AV, Fallahi M, Nakase H, Gosink M, Young B, Petrie HT. Spatial mapping of thymic stromal microenvironments reveals unique features influencing T lymphoid differentiation. *Immunity* 2009;**31**:999–1009.
157. Tourne S, Nakano N, Viville S, Benoist C, Mathis D. The influence of invariant chain on the positive selection of single T cell receptor specificities. *Eur J Immunol* 1995;**25**:1851–6.
158. Ignatowicz L, Kappler J, Marrack P. The repertoire of T cells shaped by a single MHC/peptide ligand. *Cell* 1996;**84**:521–9.
159. Venanzi ES, Melamed R, Mathis D, Benoist C. The variable immunological self: genetic variation and nongenetic noise in Aire-regulated transcription. *Proc Natl Acad Sci USA* 2008;**105**:15860–5.
160. Day CL, Seth NP, Lucas M, Appel H, Gauthier L, Lauer GM, et al. Ex vivo analysis of human memory CD4 T cells specific for hepatitis C virus using MHC class II tetramers. *J Clin Invest* 2003;**112**:831–42.
161. Savage PA, Davis MM. A kinetic window constricts the T cell receptor repertoire in the thymus. *Immunity* 2001;**14**:243–52.

162. Bouillet P, Purton JF, Godfrey DI, Zhang LC, Coultas L, Puthalakath H, et al. BH3-only Bcl-2 family member Bim is required for apoptosis of autoreactive thymocytes. *Nature* 2002;**415**:922–6.
163. Kovalovsky D, Pezzano M, Ortiz BD, Sant'Angelo DB. A novel TCR transgenic model reveals that negative selection involves an immediate, Bim-dependent pathway and a delayed, Bim-independent pathway. *PLoS ONE* 2010;**5**:e8675.
164. Weinstein JA, Jiang N, White 3rd RA, Fisher DS, Quake SR. High-throughput sequencing of the zebrafish antibody repertoire. *Science* 2009;**324**:807–10.
165. Jiang X, Fares MA. Identifying coevolutionary patterns in human leukocyte antigen (Hla) molecules. *Evolution; Int J Org Evol* 2009;**64**:1429–45.
166. Marrack P, Kappler J. The T cell receptor. *Science* 1987;**238**:1073–9.
167. Ma Z, Sharp KA, Janmey PA, Finkel TH. Surface-anchored monomeric agonist pMHCs alone trigger TCR with high sensitivity. *PLoS Biol* 2008;**6**:e43.

Section II

Thymic Requirements for T Cell Immunity

Transcriptional Regulation of Thymus Organogenesis and Thymic Epithelial Cell Differentiation

NANCY R. MANLEY AND
BRIAN G. CONDIE

Department of Genetics, University of Georgia, Athens, Georgia, USA

Transcriptional regulatory networks are the central regulatory mechanisms that control organ identity, patterning, and differentiation. In the case of the thymus, several key transcription factors have been identified that are critical for various aspects of thymus organogenesis and thymic epithelial cell (TEC) differentiation. The thymus forms from the third pharyngeal pouch endoderm during embryogenesis. Organ development progresses from initial thymus cell fate specification, through multiple stages of TEC differentiation and cortical (cTEC) and medullary (mTEC) formation. Transcription factors have been identified for each of these stages: a Hoxa3-dependent cascade at initial fate specification, Foxn1 for early (and later) TEC differentiation, and NF-κB for mTEC differentiation. As important as these factors are, their interrelationships are not understood, and many more transcription factors are likely required for complete thymus organogenesis to occur. In this chapter, we review the literature on these known genes, as well as identify gaps in our knowledge for future studies.

Progress in Molecular Biology
and Translational Science, Vol. 92
DOI: 10.1016/S1877-1173(10)92005-X

I. Introduction

The vertebrate thymus is the primary lymphoid organ responsible for the production of self-restricted, self-tolerant T cells. In the mature thymus, T cell precursors (thymocytes) undergo a series of developmental stages via interactions with thymic stromal cells, resulting in the generation of mature T lymphocytes that are exported into the periphery. The thymus is composed of resident stromal cells that compose the microenvironments that promote the various steps of thymocyte differentiation. Thymic epithelial cells (TECs) are a major component of the thymic stroma, and are required both for thymus organogenesis and for the promotion of most if not all stages of thymocyte maturation.[1–9]

Essentially all aspects of postnatal thymocyte proliferation, survival, migration, and differentiation require interactions with stromal microenvironments. It is also well established that thymocytes require signals from and interactions with TECs for all currently defined stages of thymocyte differentiation, even if we do not know the molecular identity of all of the signals involved.[10,11] While recent efforts to generate "T cells in a dish" have made some progress toward cell culture-based generation of T cells,[12–14] none of these methods as yet approach the efficacy of the thymus in making T cells. Therefore, understanding the formation, composition, and structure of the thymic stroma is essential to understanding how the thymus promotes T cell development *in vivo*, and may provide essential information for efforts to generate T cells *in vitro*. Errors in early patterning and organogenesis can also lead to a variety of organ defects, from complete failure of organ formation to ectopic, hypo- or hyperplastic, or subfunctional organs.

This review will focus on the current understanding of the transcriptional networks that drive thymus organogenesis and fetal TEC differentiation (Table I). Organogenesis is a complex process, requiring interactions between multiple cell types, and often directed by transcription factor hierarchies that direct and coordinate organ formation and cellular differentiation. In the case of the thymus, there is still surprisingly little that is known about the identity and function of the transcriptional pathways required for correct organ formation.

A. Coordinated Early Organogenesis of the Thymus and Parathyroids in Mice

The thymus and parathyroid glands are bilateral organs that develop from the third pharyngeal pouch endoderm and surrounding neural crest cells (NCCs).[2] The third pouch forms during embryonic days (E) 9–10.5 in mice. The pouch endoderm is patterned into separate parathyroid and thymus-fated domains during early formation, based on two criteria. First, the parathyroid-fated domain is marked by the expression of the tissue-specific transcription factor Gcm2 as early as E9.5.[15,16] Furthermore, transplanting the E9 third

TABLE I
Transcription Factors Important for Thymus Organogenesis and TEC Differentiation

Name	Relevant expression pattern[a]	Mutant phenotype[a]
Hoxa3	E10.5 third pharyngeal pouch endoderm and surrounding mesenchyme; later in thymic stroma	Athymia; failure of pouch patterning into organ domains, and of initial organ formation.
Pax1	E10.5 third pharyngeal pouch endoderm	Normal initial organogenesis, then mild thymic hypoplasia; may have redundancy with Pax9.
Pax9	E10.5 third pharyngeal pouch endoderm	Failure to separate from pharynx, extreme hypoplasia.
Eya1	E10.5 third pharyngeal pouch endoderm and surrounding mesenchyme; later in thymic stroma	Athymia; similar to Hoxa3, except Pax1 expression normal (see text).
Six1	E10.5 third pharyngeal pouch endoderm and surrounding mesenchyme; later in thymic stroma	Thymus organogenesis initiates, Foxn1 is expressed, then undergoes apoptosis by E12.
Six4	Coexpressed with Six1 in pharyngeal pouch endoderm	Single mutants normal; Six1; Six4 double mutants similar to Six1 mutant, but with smaller initial primordium and no Foxn1 expression.
Tbx1	E9.5 pouch endoderm; after E10.5 excluded from thymus domain	Athymia secondary to failure of pouch formation
Foxn1	In all TECs from E11 to E11.5, then expression is modulated in different TEC subsets	Null mutant arrests after initial primordium formation at TEC stem/progenitor stage; hypomorphic alleles have milder phenotypes. Required in both fetal and postnatal stages.
NF-κB2	Postnatal mTECs; fetal stages not well defined	Reduced UEA-1+ and Aire+ mTECs; failure to support NKT cell differentiation.
RelB	In UEA-1+ mTECs from E14	Failure of mTEC differentiation and medullary formation.

[a]Expression patterns and mutant phenotypes listed are only those relevant to the thymus; all of these transcription factors have roles in other locations as well.

pouch endoderm to an ectopic location results in ectopic thymus formation in both chick[17] and mouse,[18] a classic test of cell fate specification. This patterning appears to be endoderm-intrinsic, both from the transplant experiments, and because neural crest-deficient Splotch mutants have normal initial pouch patterning.[19]

After initial organ domains are patterned, subsequent epithelial-mesenchymal interactions are required both to establish the location of the final boundary between the thymus and parathyroid-fated domains (and thus influencing organ size), and to promote endoderm proliferation and formation of the bilateral primordia. Each primordium contains the epithelial precursors to one thymus lobe and one parathyroid gland[20] and is surrounded by condensing neural crest mesenchyme, particularly around the thymus domain. The two organ-specific domains are not morphologically distinguishable until they begin to physically separate, at about E12. Subsequently, the primordia undergo a series of morphological changes resulting in the formation of separate thymus and parathyroid organs in distinct locations. The thymic rudiments migrate ventrally and medially to their final position above the heart. The molecular and cellular mechanisms underlying this migration are largely unknown, although current evidence does suggest that it is primarily driven by neural crest-intrinsic mechanisms.

B. Organization and Structure of Thymic Compartments and Diversity of TEC Subsets

Once organogenesis is initiated, a complex set of cellular interactions between TECs, surrounding mesenchyme, and immigrating lymphoid and endothelial progenitor cells results in the development of a complex cellular environment. The postnatal thymus thus consists of the developing T cells, or thymocytes, and the nonlymphoid thymic stromal elements that in aggregate compose the microenvironments that promote the different stages of thymocyte differentiation: TECs, mesenchyme, endothelium, and nonlymphoid hematopoietic-derived cells (dendritic cells, macrophages). Among these stromal elements, TECs are the primary functional cell type, and are required for promoting essentially all stages of thymocyte differentiation. After birth, the thymus continues to develop and organize its compartmental structure, and is periodically seeded with hematopoietic progenitors,[21] expanding in size and increasing thymic output of naïve T cells to fill the "empty" peripheral environment.[22–24]

The postnatal thymus is organized into regions or compartments that contain different populations of TECs and developing thymocytes. The outer and inner compartments are termed the cortex and medulla respectively, and the zone where they meet is the corticomedullary junction (CMJ). The stereotypical migration of thymocytes through the complex and highly organized postnatal thymus, combined with the functional interdependence of thymocyte and TEC differentiation provide strong evidence that TEC compartments and the TEC subtypes within them are functionally diverse. The thymic stroma, especially TECs, provide discrete environments that promote specific stages of thymocyte development, and provide both the migration signals and the structural pathways for their migration through the thymus.[10,25,26] The correct structure of the thymus is likely required for optimal thymocyte production, demonstrating the importance of understanding the cellular and molecular interactions required for both TEC differentiation and compartment organization. The classically defined TEC subsets are, however, still poorly defined by molecular and functional criteria; one could make the argument that we do not really know how many functionally distinct subsets of TECs exist, much less what the molecular mechanisms are that control their development and function.

II. Transcription Factors Controlling Organ Patterning and Initial Organogenesis

A. Hoxa3—The Earliest Player

The data suggest that the earliest regulator of organ patterning and development after pouch formation is the transcription factor Hoxa3 (Table I). *Hoxa3* mutants have numerous defects in the pharyngeal region, including abnormal or absent development of all pharyngeal-derived organs.[27–36] Hoxa3 is not required for pouch formation, but in its absence initial thymus and parathyroid fate are not specified, and the shared organ primordia do not form. Hoxa3 is expressed in both the third pouch endoderm and surrounding NCCs, but our data suggests that it is *Hoxa3* expression in pouch endoderm that is required for normal third pouch patterning and initial organ development. In mice, *Hoxa3* is the only Hox3 paralog expressed in the pharyngeal endoderm, and the only one required for thymus and parathyroid initial organogenesis, as mice with only a single wild-type *Hoxa3* allele ($Hoxa3^{+/-}Hoxb3^{-/-}Hoxd3^{-/-}$) still have a thymus and parathyroids, albeit ectopically located.[30] Furthermore, deletion of Hoxa3 specifically from NCCs results in failure of organ separation and migration,[37] but normal initial stages of patterning and organ formation (Masuda *et al.*, unpublished data). This conclusion is consistent with functional data discussed above showing that establishment of thymus fate is endoderm-intrinsic. Furthermore, although Hoxa3 is expressed in TECs during later organogenesis,[32] deletion of Hoxa3 in TECs after initial organ formation does not result in any obvious defects in TEC differentiation (Masuda *et al.*, unpublished data), so any function at later stages remains unclear.

Taken together, the current data show that thymus organogenesis in the *Hoxa3* null mutant fails after third pharyngeal pouch formation, but prior to either organ specification or initial organogenesis. Based on these data, we propose that Hoxa3 acts first to specify the identity of the third pouch, laying the groundwork for induction of organ-specific identities. However, it is not clear whether Hoxa3 directly regulates thymus and parathyroid organ fate. Thus, the precise relationship between Hoxa3 and initial establishment of thymus fate remains to be determined.

B. The Pax–Eya–Six Gene Network may Act Downstream of Hoxa3

While specific downstream targets for Hoxa3 have not been verified, the effects of Hoxa3 in the endoderm are likely mediated at least in part via a Pax1/9-Eya1–Six1/4 network acting downstream of Hoxa3 (Table I). Recent data in mouse have suggested that the Pax-Eya–Six–Dach regulatory network, originally identified in *Drosophila*, has been evolutionarily conserved in vertebrates, and co-opted for roles in development of several organs and tissues, including the thymus and parathyroids.[38–43] All of these genes are at least initially expressed throughout the pouch and null mutants for each of these genes have mild to severe defects in initial thymus (and parathyroid) organogenesis,[6,32,43–47] although none precisely phenocopies the *Hoxa3* mutant phenotype.

Pax1 and *Pax9* are related by duplication, and constitute the Group I Pax transcription factors.[48,49] *Pax1* and *Pax9* are both broadly expressed in the pharyngeal pouches, with *Pax1* being upregulated in the endoderm of the third pharyngeal pouch prior to organ formation,[29] and both continue to be expressed in TECs. While these genes' expression in the pouch is initiated independently of Hoxa3, the continued expression of at least *Pax1* depends on Hoxa3.[32] *Pax1* mutants have a mild thymus phenotype primarily characterized by low thymocyte cellularity.[50] *Pax9* mutants have a more severe phenotype, with severe thymic hypoplasia and failure to separate from the pharyngeal endoderm.[47] Given their high degree of similarity and overlapping expression, it is likely that these genes provide redundant function for one another; however, while this has been shown in skeletal development, analysis of pharyngeal organogenesis has not been reported in *Pax1;Pax9* double mutants.[51,52]

The Manley lab has also shown that Hoxa3 and Pax1 function in the same genetic pathway controlling both thymus and parathyroid organogenesis by analyzing *Hoxa3*$^{+/-}$*Pax1*$^{-/-}$ compound mutants.[6,32] These mutants have morphogenesis defects including delayed separation from the pharynx, and thymic hypoplasia associated with increased TEC apoptosis and decreased proliferation, and reduced MHC Class II expression. Thymocyte differentiation was only mildly affected, but total numbers were reduced. These genetic interactions combined with the gene expression results indicate that Pax1 (and also probably Pax9) act downstream of Hoxa3 during fetal thymus organogenesis.

Eya1, Six1, and Six4 have all been shown to be expressed in the third pouch and surrounding mesenchyme, and to have effects on early thymus organogenesis when mutated. *Eya1* null mutants have a phenotype most similar to *Hoxa3* null mutants, with an apparent failure to initiate thymus and parathyroid organogenesis.[39] Differences include some hypoplasia of the pharyngeal arches, which could lead to a smaller than normal pouch, and normal *Pax1* gene expression. Six1 is expressed in both the third pouch endoderm and surrounding mesenchyme, and in scattered cells in the fetal thymus at the newborn stage, presumably TECs.[53] *Six1* single and *Six1;Six4* double mutants are also athymic and aparathyroid, although in both cases organogenesis initiates at E11, but the primordium undergoes apoptosis by E12.[43] Double mutant and gene expression analyses led to the conclusion that the order of gene function is Hoxa3–Pax1/9–Eya1–Six1/4, with each gene being required for the expression of the next in a gene expression regulatory cascade.[43] However, this cascade still affects both thymus and parathyroid organogenesis equally, and so does not yet account for the establishment of organ-specific domains and thymus fate.

C. Tbx1—Conflicting Theories

Another transcription factor that is expressed in the pouch endoderm and has been associated with thymus organogenesis is Tbx1 (Table I). The *Tbx1* gene is located in the DiGeorge critical region, and is expressed throughout the third pouch during initial pouch formation.[54] Tbx1 deletion in mice results in phenotypes consistent with DiGeorge Syndrome in humans, including athymia and aparathyroidism,[55] and temporal-specific deletion between E9.5 and E11.5 results in either athymia or hypoplastic thymus.[54] As a result, Tbx1 is often suggested to play an important and direct role in thymus organogenesis.[54,56] However, we have shown that Tbx1 is specifically expressed in the parathyroid domain at E10.5 and may act upstream of Gcm2 to specify parathyroid fate, but is excluded from the thymus domain and the thymic primordia as early as E10.5.[57,58] These seemingly contradictory results can be explained by an early role for Tbx1 in pouch formation, with only an indirect role in thymus organogenesis. Deleting Tbx1 before E9.5 (the null mutant) causes failure of pouch formation, and thus indirectly causing athymia; deleting Tbx1 during pouch formation results in a smaller pouch, indirectly reducing the size of the thymus-fated domain resulting in thymic hypoplasia. Thus, the most likely model is that Tbx1 is required for pouch formation, and subsequently plays role in parathyroid, but not thymus, organogenesis; indeed, it is even possible that Tbx1 down regulation in the distal pouch is required to establish thymus fate.

III. Patterning Events Lead to Induction of the Thymus-Specific Transcription Factor Foxn1

A. Genetic Analysis of Foxn1 Function

Although both the exact timing of thymic specification and the gene(s) that specify thymus organ identity within the third pouch are still uncertain, by E11.5 it appears that all cells within the thymus-specific domain express the forkhead transcription factor Foxn1 (Table I). Perhaps the best-known mouse mutants affecting TEC carry mutations at the nude locus, which encodes Foxn1.[59–61] A recent study in which Foxn1 was activated stochastically in individual TEC progenitors within the nude thymic primordium showed that Foxn1 expression induced cTEC and mTEC differentiation and compartment formation.[62] Thus, *Foxn1* expression is both necessary and sufficient to induce TEC differentiation in presumptive TEC progenitors; however, given its role in skin, it is unlikely to be sufficient to direct TEC differentiation in cells not already fated to become thymus.

Analysis of multiple mutant alleles of Foxn1 has shown that it is required for multiple stages of TEC differentiation, but does not specify thymus organ identity.[61,63,64] Although commonly considered "athymic," the thymic rudiment forms in nude embryos but is not populated by LPCs,[65,66] and TEC differentiation is blocked at an early stage. Consistent with this phenotype, expression of the chemokine CCL25 is largely dependent on Foxn1.[67] This timing is also consistent with our data showing that *Foxn1* is not highly expressed in the wild type thymic rudiment until E11.25, after initial primordium formation but prior to significant LPC infiltration.[15] Presumptive TECs in the nude thymus do not express differentiation markers such as MHC Class I or II.[68] Instead, they express markers typical of normal TECs from the earliest thymic rudiment, which have been shown to characterize a multipotent TEC progenitor.[69,70] These results suggest that the presumptive TECs in nude mice are arrested at a very early progenitor stage.

Although some TEC differentiation occurs in the absence of thymocytes, complete TEC differentiation and regional patterning within the thymus require stage-dependent interactions with differentiating thymocytes.[71–73] These interactions begin primarily between E15 and birth, and result in the organization of cortical and medullary regions within the thymic rudiment and the elaboration of TEC subsets. Our lab has shown that mice homozygous for a hypomorphic allele of *Foxn1*, *Foxn1*$^{\Delta}$, have relatively normal initial organogenesis, but are blocked in TEC differentiation at about E13.5,[64] with a phenotype consistent with a failure to respond to TEC-thymocyte interaction signals.[73] The phenotype is also tissue-specific, severely affecting the thymus but with only mild, low penetrance effects on hair and skin. These data show that Foxn1 is critical for multiple aspects of fetal TEC differentiation, and that different tissues may have different sensitivities to Foxn1 dosage. Foxn1 has also been implicated in promoting proliferation during initial thymus organogenesis,[65] raising the possibility that Foxn1 functions to balance or coordinate differentiation and proliferation, as has been proposed in keratinocytes.[74,75]

B. Foxn1 in the Postnatal Thymus

Because of its central role in multiple aspects of fetal TEC differentiation, Foxn1 is also an obvious candidate for a key regulator of TEC maintenance and function in the postnatal thymus. *Foxn1* is widely expressed in both fetal and postnatal TECs. The early studies identifying the *Foxn1* gene showed that it is extensively expressed in both cTECs and mTECs in the postnatal thymus.[60,61] However, a recent study using a Foxn1 antibody concluded that at later stages Foxn1 is not maintained at high levels in all TECs,[76] although analysis of Foxn1 transcription suggests a broader expression pattern in postnatal TECs.[37] Analysis of Foxn1 mRNA levels in sorted TEC subpopulations also showed that Foxn1 transcription levels vary by at least 10-fold in different TEC subsets.[37] This range of Foxn1 expression suggests that dynamic regulation of Foxn1 levels may be functionally important in TEC differentiation.

Foxn1 has also been implicated in maintaining the postnatal thymus, an in aging-associated thymic involution. The first evidence in favor of this possibility was the identification of *Foxn1* down regulation as an early event in aging-associated thymic involution.[77] We recently generated an allele of Foxn1 that has normal fetal expression, but is downregulated to about 30% of normal levels at 1 week after birth.[37] This down regulation results in a thymic degeneration phenotype resembling premature aging-associated involution, with decreased proliferation in and loss of specific TEC subsets. Furthermore, by combining this allele with the *nude* allele, we showed a clear quantitative response to different Foxn1 expression levels.[37] A subsequent study using conditional deletion of Foxn1 confirmed a requirement for Foxn1 in postnatal TECs.[78] These recent studies, combined with the critical role Foxn1 plays in both cross talk dependent and independent TEC differentiation at fetal stages, suggest that differential regulation of Foxn1 in TEC subsets plays critical roles throughout the lifespan of the organ.

C. Structural and Functional Analysis of the Foxn1 Gene and Protein

The *Foxn1* gene is encoded by a total of 10 exons spanning ~30 kb of genomic sequence.[79] The first two exons, 1a and 1b, are short, noncoding exons that are differentially transcribed by two predicted promoters.[79] Transcripts initiating at exon 1b are found only in skin, while those initiating at exon 1a are found in both skin and thymus. Consistent with this data, a YAC transgene containing genomic sequences from exons 1b-9 rescued only the skin and hair defects in nude mice,[80] while a significantly larger 110-kb transgene rescued both phenotypes.[81] These results indicate that a thymus-specific promoter for Foxn1 may exist, although its exact location has not yet been identified. Transcripts from both promoters are predicted to encode the same protein.

Studies by two labs have mapped functional domains within the Foxn1 protein using genetic analysis, transfection studies and DNA-binding assays. These studies confirmed the locations of the forkhead class DNA binding domain (DBD) and a C-terminal acidic activation domain (CTD), both of which were predicted by amino acid (aa) sequence homology.[75,82] Transfection studies also mapped a nuclear localization signal (NLS) to within 120 aa immediately N-terminal to the DBD. However, these studies were unable to identify any other function for the N-terminal 285 aa of the protein.[75,82] Transfection studies performed in both primary keratinocytes[75] and BHK cells[82] demonstrated that this N-terminal domain was dispensable for Foxn1 to activate known keratinocyte targets. However, similar studies were not done with TEC lines,[79] leaving open the possibility that this domain could have a TEC-specific activity.

Our work with the *Foxn1*$^{\Delta}$ allele also identified a transcript that splices from exon 2 directly to exon 4, encoding for a protein that deletes the majority of the N-terminal domain but retains the NLS, DBD, and CTD.[64] This transcript is present in both skin and thymus, as early as E12.5 in the wild-type thymus primordium (our unpublished data). Although the normal function of the Δ2–4 transcript is still unclear, the Foxn1$^{\Delta/\Delta}$ mutant that expresses only this transcript has a strong thymus-specific hypomorphic phenotype, suggesting that the NTD may have a thymus-specific function.[64] This data is particularly intriguing in light of studies from the Boehm lab comparing the sequence and function of Foxn1-related genes from mouse, zebrafish, and amphioxus.[83] This analysis showed that the DNA binding and C-terminal domains of these three proteins have similar function in an assay activating hair keratin genes in HeLa cells, even though zebrafish and amphioxus do not have hair. The functional similarity of the zebrafish gene is particularly striking since the level of sequence conservation outside the DBD is quite low; also, this analysis was performed with a gene later determined not to be the true Foxn1 ortholog, but a closely related Foxn gene.[84] However, the NTD of amphioxus specifically blocked this function. This chapter concluded that the NTD "contains a functionally distinct domain(s) that has undergone functionally relevant changes early in vertebrate evolution."

IV. NFκb Pathway and Medullary Formation

Efficient generation of functional T cells requires an elaborated TEC meshwork and organized thymic compartmental structure,[10,11] which in turn requires thymocyte-derived signals for its development (reviewed in Refs. 85,86). This well-established mutually inductive process is termed "cross talk,"[4,87,88] and contributes to the regulation of thymus organogenesis, homeostasis, and involution. Investigation of the molecular basis for these critical cross-inductive interactions is ongoing. Several key inductive interactions have been identified, including a requirement for RANKL and CD40L from thymocytes for the differentiation of specific mTEC subtypes, and a requirement for the NF-κB pathway for compartment formation and expansion (reviewed in Ref. 89). To date, the NF-κB family constitute the only transcription factors specifically required for mTEC differentiation (Table I).

The NF-κB family of transcription factors consists of five members, Nfkb1 (NF-κB1, p150/p50), Nfkb2 (NF-κB2, p100/p52), Rel (c-rel), Rela (p65), and Relb. These factors are expressed in a wide range of fetal and adult tissues and play important roles in the development and function of multiple cell types including lymphocytes.[90] *In situ* hybridization analysis of adult mouse thymus detected the expression of all five NF-κB factors.[91] Intriguingly, different family members were localized to different regions within the thymus with Nfkb1 and Rela detected in cortical areas while Nfkb2, Rel, and Relb were expressed in medullary regions.[91] In contrast, studies of NF-κB factor expression or function during the fetal stages of thymus development and differentiation have been very limited even though generic NF-κB mediated transcriptional activity has been detected in the mouse thymus at E13.5.[92]

The expression of RelB and Rel has been defined in fetal thymus. *In situ* hybridization analysis has shown that RelB is expressed in E15 thymus and in the thymic medulla of E17 day old mouse embryos.[93] RelB appears to be coexpressed with the medullary TEC marker UEA-1 in E14 thymus and RelB expression at this stage is dependent on Traf6.[94] No UEA-1^{+} cells were detected at in Traf6 mutants at E16 and keratin 5^{+} cells were dispersed, suggesting an early defect in mTEC differentiation.[94] In addition, Rel mRNA was detected in thymus at E13.5 and was predominantly expressed in medulla by E17.5.[95] However, there is no published data regarding the embryonic thymus phenotype of RelB or Rel mutants.

Overall, there is ample evidence from studies of postnatal thymus to hint at a role for several NF-κB factors in the normal development and differentiation of TECs. For example, it has been shown that Nfkb2 is required in thymic stroma for the normal postnatal differentiation of NKT cells,[96] presumably secondary to changes in TEC differentiation and/or function. For example, the numbers of both UEA-1$^+$ and Aire expressing mTECs are markedly reduced in Nfkb2 mutants.[97,98] In addition, it is well documented that Relb function is required for the formation of well-defined and functional medullary regions in the postnatal thymus.[90,99] In this context, it is surprising that fetal thymus differentiation has not been extensively examined in these mutants.

V. Future Directions

Perhaps the most striking characteristic of our current knowledge about the transcription factors required for thymus organogenesis and TEC differentiation is how little we know. As is clear from this review, very few of them have been identified. Of those that have, only Foxn1 and the Nfkb family have been clearly shown to play a role in TEC differentiation, and even in those cases little or nothing is known about their downstream targets. However, the information we do have points to key roles for transcriptional regulators at multiple stages of thymus organogenesis. Recent data further shows that at least some of these factors are likely to play essential roles for organ homeostasis in the postnatal thymus as well. Thus, it is clear that much further work remains to be done, and that many key elements in the transcriptional control of thymus remain to be identified.

Given this paucity of information, there are many questions that remain unanswered; here, we highlight a few key areas that we believe to be of critical importance in understanding the transcriptional regulation of thymus organogenesis and TEC differentiation:

1. *Identifying the transcription factor(s) that specify thymus identity*: As in other organs, transcription factors undoubtedly play a critical role in specifying thymus identity in the third pharyngeal pouch. The current data suggest that thymus fate may be specified by one or more transcription factors acting downstream of Hoxa3 and upstream of Foxn1, and that its expression is induced by endoderm-intrinsic mechanisms, perhaps as early as E9.5. The identity of the transcription factor(s) that specify thymus fate and the molecular mechanisms that control their expression is a major unanswered question, and is critical information both for understanding thymus organogenesis, and for efforts to induce thymus differentiation in cultured ES or iPS cells for therapeutic purposes.
2. *Identifying transcription factor(s) that regulate cTEC differentiation*: Although current data indicates that Foxn1 is required broadly for both mTEC and cTEC differentiation, and the NFκB factors are specifically required for mTEC differentiation, no cTEC-specific

transcription factors have yet been identified. While it is possible that Foxn1 itself directs cTEC differentiation, this remains to be determined, and other transcription factors are almost certainly required either downstream of or in parallel to Foxn1.

3. *Delineation of the lineage relationships between different TEC subsets*: While not a "missing transcription factor" *per se*, a major impediment to understanding the regulation of TEC differentiation is the lack of a well-supported, commonly accepted hierarchy of TEC differentiation. While markers of different TEC subsets are accumulating, and some lineage relationships have been proposed, there are clearly many more unknowns than verified facts. Are cTECs and mTECs independent differentiation lineages derived from a common bipotent progenitor cell? Are different mTEC subsets lineally related, or independently specified? How many different stem/progenitor cells are there, and do they differ in the fetal and postnatal thymus? Are TECs in the postnatal thymus generated by reiteration of the same differentiation pathways used in initial organogenesis and fetal TEC differentiation? Identifying new transcription factors that play specific roles in TEC differentiation could help to dissect these lineage relationships, and provide key information regarding both the molecular and cellular mechanisms controlling these processes.

References

1. Markowitz JS, Auchincloss Jr. H, Grusby MJ, Glimcher LH. Class II-positive hematopoietic cells cannot mediate positive selection of CD4+ T lymphocytes in class II-deficient mice. *Proc Natl Acad Sci USA* 1993;**90**:2779–83.
2. Manley NR. Thymus organogenesis and molecular mechanisms of thymic epithelial cell differentiation. *Semin Immunol* 2000;**12**:421–8.
3. Le Douarin NM, Dieterlen-Lièvre F, Oliver PD. Ontogeny of primary lymphoid organs and lymphoid stem cells. *Am J Anat* 1984;**170**:261–99.
4. Ritter MA, Boyd RL. Development in the thymus: it takes two to tango. *Immunol Today* 1993;**14**:462–9.
5. Boyd RL, Tucek CL, Godfrey DI, Izon DJ, Wilson TJ, Davidson NJ, et al. The thymic microenvironment. *Immunol Today* 1993;**14**:445–59.
6. Su DM, Manley NR. Hoxa3 and pax1 transcription factors regulate the ability of fetal thymic epithelial cells to promote thymocyte development. *J Immunol* 2000;**164**:5753–60.
7. Anderson G, Jenkinson EJ, Moore NC, Owen JJ. MHC class II-positive epithelium and mesenchyme cells are both required for T-cell development in the thymus. *Nature* 1993;**362**:70–3.
8. Anderson G, Owen JJ, Moore NC, Jenkinson EJ. Thymic epithelial cells provide unique signals for positive selection of CD4+CD8+ thymocytes in vitro. *J Exp Med* 1994;**179**:2027–31.

9. Holländer GA, Wang B, Nichogiannopoulou A, Platenburg PP, van Ewijk W, Burakoff SJ, et al. Developmental control point in induction of thymic cortex regulated by a subpopulation of prothymocytes. *Nature* 1995;**373**:350–3.
10. Lind EF, Prockop SE, Porritt HE, Petrie HT. Mapping precursor movement through the postnatal thymus reveals specific microenvironments supporting defined stages of early lymphoid development. *J Exp Med* 2001;**194**:127–34.
11. Petrie HT, Tourigny M, Burtrum DB, Livak F. Precursor thymocyte proliferation and differentiation are controlled by signals unrelated to the pre-TCR. *J Immunol* 2000;**165**:3094–8.
12. De Smedt M, Hoebeke I, Plum J. Human bone marrow CD34+ progenitor cells mature to T cells on OP9-DL1 stromal cell line without thymus microenvironment. *Blood Cells Mol Dis* 2004;**33**:227–32.
13. Schmitt TM, de Pooter RF, Gronski MA, Cho SK, Ohashi PS, Zuniga-Pflucker JC. Induction of T cell development and establishment of T cell competence from embryonic stem cells differentiated in vitro. *Nat Immunol* 2004;**5**:410–7.
14. Schmitt TM, Zuniga-Pflucker JC. Induction of T cell development from hematopoietic progenitor cells by delta-like-1 in vitro. *Immunity* 2002;**17**:749–56.
15. Gordon J, Bennett AR, Blackburn CC, Manley NR. Gcm2 and Foxn1 mark early parathyroid- and thymus-specific domains in the developing third pharyngeal pouch. *Mech Dev* 2001;**103**:141–3.
16. Patel SR, Gordon J, Mahbub F, Blackburn CC, Manley NR. Bmp4 and Noggin expression during early thymus and parathyroid organogenesis. *Gene Expr Patterns* 2006;**6**:794–9.
17. LeDouarin N, Jotereau FV. Tracing of cells for the avian thymus through embryonic life in interspecific chimeras. *J Exp Med* 1975;**142**:17–40.
18. Gordon J, Wilson VA, Blair NF, Sheridan J, Farley A, Wilson L, et al. Functional evidence for a single endodermal origin for the thymic epithelium. *Nat Immunol* 2004;**5**:546–53.
19. Griffith AV, Cardenas K, Carter C, Gordon J, Iberg A, Engleka K, et al. Increased thymus- and decreased parathyroid-fated organ domains in Splotch mutant embryos. *Dev Biol* 2009;**327**:216–27.
20. Cordier AC, Haumont SM. Development of thymus, parathyroids, and ultimobranchial bodies in NMRI and Nude mice. *Am J Anat* 1980;**157**:227–63.
21. Goldschneider I. Cyclical mobilization and gated importation of thymocyte progenitors in the adult mouse: evidence for a thymus-bone marrow feedback loop. *Immunol Rev* 2006;**209**:58–75.
22. Adkins B. T-cell function in newborn mice and humans. *Immunol Today* 1999;**20**:330–5.
23. Adkins B. Peripheral CD4+ lymphocytes derived from fetal versus adult thymic precursors differ phenotypically and functionally. *J Immunol* 2003;**171**:5157–64.
24. Min B, McHugh R, Sempowski GD, Mackall C, Foucras G, Paul WE. Neonates support lymphopenia-induced proliferation. *Immunity* 2003;**18**:131–40.
25. Petrie HT. Role of thymic organ structure and stromal composition in steady-state postnatal T-cell production. *Immunol Rev* 2002;**189**:8–19.
26. Prockop SE, Palencia S, Ryan CM, Gordon K, Gray D, Petrie HT. Stromal cells provide the matrix for migration of early lymphoid progenitors through the thymic cortex. *J Immunol* 2002;**169**:4354–61.
27. Chisaka O, Capecchi MR. Regionally restricted developmental defects resulting from targeted disruption of the mouse homeobox gene *hox-1.5*. *Nature* 1991;**350**:473–9.
28. Kameda Y, Arai Y, Nishimaki T, Chisaka O. The role of Hoxa3 gene in parathyroid gland organogenesis of the mouse. *J Histochem Cytochem* 2004;**52**:641–51.
29. Manley NR, Capecchi MR. The role of Hoxa-3 in mouse thymus and thyroid development. *Development* 1995;**121**:1989–2003.

30. Manley NR, Capecchi MR. Hox group 3 paralogs regulate the development and migration of the thymus, thyroid, and parathyroid glands. *Dev Biol* 1998;**195**:1–15.
31. Manley NR, Capecchi MR. Hox group 3 paralogous genes act synergistically in the formation of somitic and neural crest-derived structures. *Dev Biol* 1997;**192**:274–88.
32. Su D, Ellis S, Napier A, Lee K, Manley NR. Hoxa3 and pax1 regulate epithelial cell death and proliferation during thymus and parathyroid organogenesis. *Dev Biol* 2001;**236**:316–29.
33. Gaufo GO, Thomas KR, Capecchi MR. Hox3 genes coordinate mechanisms of genetic suppression and activation in the generation of branchial and somatic motoneurons. *Development* 2003;**130**:5191–201.
34. Guidato S, Prin F, Guthrie S. Somatic motoneurone specification in the hindbrain: the influence of somite-derived signals, retinoic acid and Hoxa3. *Development* 2003;**130**:2981–96.
35. Kameda Y, Watari-Goshima N, Nishimaki T, Chisaka O. Disruption of the Hoxa3 homeobox gene results in anomalies of the carotid artery system and the arterial baroreceptors. *Cell Tissue Res* 2003;**311**:343–52.
36. Kameda Y, Nishimaki T, Takeichi M, Chisaka O. Homeobox gene hoxa3 is essential for the formation of the carotid body in the mouse embryos. *Dev Biol* 2002;**247**:197–209.
37. Chen L, Xiao S, Manley NR. Foxn1 is required to maintain the postnatal thymic microenvironment in a dosage-sensitive manner. *Blood* 2009;**113**:567–74.
38. Heanue TA, Reshef R, Davis RJ, Mardon G, Oliver G, Tomarev S, et al. Synergistic regulation of vertebrate muscle development by Dach2, Eya2, and Six1, homologs of genes required for *Drosophila* eye formation. *Genes Dev* 1999;**13**:3231–43.
39. Xu P-X, Zheng W, Laclef C, Maire P, Maas RL, Peters H, Xu X. Eya1 is required for the morphogenesis of mammalian thymus, parathyroid and thyroid. *Development* 2002;**129**:3033–44.
40. Manley NR, Capecchi MR. The role of Hoxa-3 in mouse thymus and thyroid development. *Development* 1995;**121**:1989–2003.
41. Su D-M, Manley NR. Hoxa3 and Pax1 transcription factors regulate the ability of fetal thymic epithelial cells to promote thymocyte development. *J Immunol* 2000;**164**:5753–60.
42. Su D-m, Ellis S, Napier A, Lee K, Manley NR. Hoxa3 and Pax1 regulate epithelial cell death and proliferation during thymus and parathyroid organogenesis. *Dev Biol* 2001;**236**:316–29.
43. Zou D, Silvius D, Davenport J, Grifone R, Maire P, Xu PX. Patterning of the third pharyngeal pouch into thymus/parathyroid by Six and Eya1. *Dev Biol* 2006;**293**:499–512.
44. Manley NR, Blackburn CC. A developmental look at thymus organogenesis: where do the non-hematopoietic cells in the thymus come from? *Curr Opin Immunol* 2003;**15**:225–32.
45. Xu PX, Zheng W, Laclef C, Maire P, Maas RL, Peters H, Xu X. Eya1 is required for the morphogenesis of mammalian thymus, parathyroid and thyroid. *Development* 2002;**129**:3033–44.
46. Peters H, Neubuser A, Kratochwil K, Balling R. Pax9-deficient mice lack pharyngeal pouch derivatives and teeth and exhibit craniofacial and limb abnormalities. *Genes Dev* 1998;**12**:2735–47.
47. Hetzer-Egger C, Schorpp M, Haas-Assenbaum A, Balling R, Peters H, Boehm T. Thymopoiesis requires Pax9 function in thymic epithelial cells. *Eur J Immunol* 2002;**32**:1175–81.
48. Dahl E, Koseki H, Balling R. Pax genes and organogenesis. *Bioessays* 1997;**19**:755–65.
49. Strachan T, Read AP. PAX genes. *Cur Opinion Gen Dev* 1994;**4**:427–38.
50. Deutsch U, Dressler GR, Gruss P. Pax 1, a member of a paired box homologous murine gene family, is expressed in segmented structures during development. *Cell* 1988;**53**:617–25.
51. Rodrigo I, Hill RE, Balling R, Munsterberg A, Imai K. Pax1 and Pax9 activate Bapx1 to induce chondrogenic differentiation in the sclerotome. *Development* 2003;**130**:473–82.
52. Peters H, Wilm B, Sakai N, Imai K, Maas R, Balling R. Pax1 and Pax9 synergistically regulate vertebral column development. *Development* 1999;**126**:5399–408.

53. Laclef C, Souil E, Demignon J, Maire P. Thymus, kidney and craniofacial abnormalities in Six1 deficient mice. *Mech Dev* 2003;**120**:669–79.
54. Xu H, Cerrato F, Baldini A. Timed mutation and cell-fate mapping reveal reiterated roles of Tbx1 during embryogenesis, and a crucial function during segmentation of the pharyngeal system via regulation of endoderm expansion. *Development* 2005;**132**:4387–95.
55. Jerome LA, Papaioannou VE. DiGeorge syndrome phenotype in mice mutant for the T-box gene, Tbx1. *Nat Genet* 2001;**27**:286–91.
56. Hollander G, Gill J, Zuklys S, Iwanami N, Liu C, Takahama Y. Cellular and molecular events during early thymus development. *Immunol Rev* 2006;**209**:28–46.
57. Manley NR, Selleri L, Brendolan A, Gordon J, Cleary ML. Abnormalities of caudal pharyngeal pouch development in Pbx1 knockout mice mimic loss of Hox3 paralogs. *Dev Biol* 2004;**276**:301–12.
58. Liu Z, Yu S, Manley NR. Gcm2 is required for the differentiation and survival of parathyroid precursor cells in the parathyroid/thymus primordia. *Dev Biol* 2007;**305**:333–46.
59. Kaestner KH, Knochel W, Martinez DE. Unified nomenclature for the winged helix/forkhead transcription factors. *Genes Dev* 2000;**14**:142–6.
60. Nehls M, Pfeifer D, Schorpp M, Hedrich H, Boehm T. New member of the winged-helix protein family disrupted in mouse and rat nude mutations. *Nature* 1994;**372**:103–7.
61. Nehls M, Kyewski B, Messerle M, Waldschütz R, Schüddekopf K, Smith AJ, Boehm T. Two genetically separable steps in the differentiation of thymic epithelium. *Science* 1996;**272**:886–9.
62. Bleul CC, Corbeaux T, Reuter A, Fisch P, Monting JS, Boehm T. Formation of a functional thymus initiated by a postnatal epithelial progenitor cell. *Nature* 2006;**441**:992–6.
63. Blackburn CC, Augustine CL, Li R, Harvey RP, Malin MA, Boyd RL, Miller JF, Morahan G. The nu gene acts cell-autonomously and is required for differentiation of thymic epithelial progenitors. *Proc Natl Acad Sci USA* 1996;**93**:5742–6.
64. Su DM, Navarre S, Oh WJ, Condie BG, Manley NR. A domain of Foxn1 required for crosstalk-dependent thymic epithelial cell differentiation. *Nat Immunol* 2003;**4**:1128–35.
65. Itoi M, Kawamoto H, Katsura Y, Amagai T. Two distinct steps of immigration of hematopoietic progenitors into the early thymus anlage. *Int Immunol* 2001;**13**:1203–11.
66. Bleul CC, Boehm T. Chemokines define distinct microenvironments in the developing thymus. *Eur J Immunol* 2000;**30**:3371–9.
67. Liu C, Saito F, Liu Z, Lei Y, Uehara S, Love P, et al. Coordination between CCR7- and CCR9-mediated chemokine signals in prevascular fetal thymus colonization. *Blood* 2006;**108**:2531–9.
68. Jenkinson EJ, Van Ewijk W, Owen JJ. Major histocompatibility complex antigen expression on the epithelium of the developing thymus in normal and nude mice. *J Exp Med* 1981;**153**:280–92.
69. Bennett AR, Farley A, Blair NF, Gordon J, Sharp L, Blackburn CC. Identification and characterization of thymic epithelial progenitor cells. *Immunity* 2002;**16**:803–14.
70. Gill J, Malin M, Hollander GA, Boyd R. Generation of a complete thymic microenvironment by MTS24(+) thymic epithelial cells. *Nat Immunol* 2002;**3**:635–42.
71. van Ewijk W, Kawamoto H, Germeraad WT, Katsura Y. Developing thymocytes organize thymic microenvironments. *Curr Top Microbiol Immunol* 2000;**251**:125–32.
72. van Ewijk W, Hollander G, Terhorst C, Wang B. Stepwise development of thymic microenvironments in vivo is regulated by thymocyte subsets. *Development* 2000;**127**:1583–91.
73. Klug DB, Carter C, Crouch E, Roop D, Conti CJ, Richie ER. Interdependence of cortical thymic epithelial cell differentiation and T-lineage commitment. *Proc Natl Acad Sci USA* 1998;**95**:11822–7.
74. Baxter RM, Brissette JL. Role of the nude gene in epithelial terminal differentiation. *J Invest Dermatol* 2002;**118**:303–9.

75. Brissette JL, Li J, Kamimura J, Lee D, Dotto GP. The product of the mouse nude locus, Whn, regulates the balance between epithelial cell growth and differentiation. *Genes Dev* 1996;**10**:2212–21.
76. Itoi M, Tsukamoto N, Amagai T. Expression of Dll4 and CCL25 in Foxn1-negative epithelial cells in the post-natal thymus. *Int Immunol* 2007;**19**:127–32.
77. Ortman CL, Dittmar KA, Witte PL, Le PT. Molecular characterization of the mouse involuted thymus: aberrations in expression of transcription regulators in thymocyte and epithelial compartments. *Int Immunol* 2002;**14**:813–22.
78. Cheng L, Guo J, Sun L, Fu J, Barnes PF, Metzger D, et al. Postnatal tissue-specific disruption of transcription factor FoxN1 triggers acute thymic atrophy. *J Biol Chem* 2010;**285**:5836–47.
79. Schorpp M, Hofmann M, Dear TN, Boehm T. Characterization of mouse and human nude genes. *Immunogenetics* 1997;**46**:509–15.
80. Kurooka H, Segre JA, Hirano Y, Nemhauser JL, Nishimura H, Yoneda K, et al. Rescue of the hairless phenotype in nude mice by transgenic insertion of the wild-type Hfh11 genomic locus. *Int Immunol* 1996;**8**:961–6.
81. Cunliffe VT, Furley AJ, Keenan D. Complete rescue of the nude mutant phenotype by a wild-type Foxn1 transgene. *Mamm Genome* 2002;**13**:245–52.
82. Schuddekopf K, Schorpp M, Boehm T. The whn transcription factor encoded by the nude locus contains an evolutionarily conserved and functionally indispensable activation domain. *Proc Natl Acad Sci USA* 1996;**93**:9661–4.
83. Schlake T, Schorpp M, Boehm T. Formation of regulator/target gene relationships during evolution. *Gene* 2000;**256**:29–34.
84. Schorpp M, Leicht M, Nold E, Hammerschmidt M, Haas-Assenbaum A, Wiest W, Boehm T. A zebrafish orthologue (whnb) of the mouse nude gene is expressed in the epithelial compartment of the embryonic thymic rudiment. *Mech Dev* 2002;**118**:179–85.
85. Anderson G, Jenkinson WE, Jones T, Parnell SM, Kinsella FA, White AJ, et al. Establishment and functioning of intrathymic microenvironments. *Immunol Rev* 2006;**209**:10–27.
86. Gill J, Malin M, Sutherland J, Gray D, Hollander G, Boyd R. Thymic generation and regeneration. *Immunol Rev* 2003;**195**:28–50.
87. van Ewijk W, Shores EW, Singer A. Crosstalk in the mouse thymus. *Immunol Today* 1994;**15**:214–7.
88. Anderson G, Jenkinson EJ. Lymphostromal interactions in thymic development and function. *Nat Rev Immunol* 2001;**1**:31–40.
89. Zhu M, Fu YX. Coordinating development of medullary thymic epithelial cells. *Immunity* 2008;**29**:386–8.
90. DeKoning J, DiMolfetto L, Reilly C, Wei Q, Harvan WL, Lo D. Thymic cortical epithelium is sufficient for the development of mature T cells in relB-deficient mice. *J Immunol* 1997;**158**:2558–66.
91. Kuo CT, Leiden JM. Transcriptional regulation of T lymphocyte development and function. *Annu Rev Immunol* 1999;**17**:149–87.
92. Schmidt-Ullrich R, Memet S, Lilienbaum A, Feuillard J, Raphael M, Israel A. NF-kappaB activity in transgenic mice: developmental regulation and tissue specificity. *Development* 1996;**122**:2117–28.
93. Carrasco D, Ryseck R-P, Bravo R. Expression of *relB* transcripts during lymphoid organ development: specific expression in dendritic antigen-presenting cells. *Development* 1993;**118**:1221–31.
94. Akiyama T, Maeda S, Yamane S, Ogino K, Kasai M, Kajiura F, et al. Dependence of self-tolerance on TRAF6-directed development of thymic stroma. *Science* 2005;**308**:248–51.
95. Carrasco D, Weih F, Bravo R. Developmental expression of the mouse c-rel proto-oncogene in hematopoietic organs. *Development* 1994;**120**:2991–3004.

96. Sivakumar V, Hammond KJ, Howells N, Pfeffer K, Weih F. Differential requirement for Rel/nuclear factor kappa B family members in natural killer T cell development. *J Exp Med* 2003;**197**:1613–21.
97. Zhu M, Chin RK, Christiansen PA, Lo JC, Liu X, Ware C, et al. NF-kappaB2 is required for the establishment of central tolerance through an Aire-dependent pathway. *J Clin Invest* 2006;**116**:2964–71.
98. Zhang B, Wang Z, Ding J, Peterson P, Gunning WT, Ding HF. NF-kappaB2 is required for the control of autoimmunity by regulating the development of medullary thymic epithelial cells. *J Biol Chem* 2006;**281**:38617–24.
99. Burkly L, Hession C, Ogata L, Reilly C, Marconi LA, Olson D, et al. Expression of *relB* is required for the development of thymic medulla and dendritic cells. *Nature* 1995;**373**:531–6.

Early T Cell Differentiation: Lessons from T-Cell Acute Lymphoblastic Leukemia

Cédric S. Tremblay,[*]
Thu Hoang,[†,✠] and
Trang Hoang[*,‡]

[*]*Institute of Research in Immunology and Cancer, University of Montreal, Montréal, Québec, Canada*
[†]*Mathématiques Appliquées Paris-5, Université René Descartes, Paris, France*
[‡]*Pharmacology, Biochemistry and Molecular Biology, Faculty of Medicine, University of Montréal, Québec, Canada*

T cells develop from bone marrow-derived self-renewing hematopoietic stem cells (HSC). Upon entering the thymus, these cells undergo progressive commitment and differentiation driven by the thymic stroma and the pre-T cell receptor (pre-TCR). These processes are disrupted in T-cell acute lymphoblastic leukemia (T-ALL). More than 70% of recurring chromosomal rearrangements in T-ALL activate the expression of oncogenic transcription factors,

✠Deceased

Progress in Molecular Biology
and Translational Science, Vol. 92
DOI: 10.1016/S1877-1173(10)92006-1

belonging mostly to three families, basic helix-loop-helix (bHLH), homeobox (HOX), and c-MYB. This prevalence is indicative of their importance in the T lineage, and their dominant mechanisms of transformation. For example, bHLH oncoproteins inhibit E2A and HEB, revealing their tumor suppressor function in the thymus. The induction of T-ALL, nonetheless, requires collaboration with constitutive NOTCH1 signaling and the pre-TCR, as well as loss-of-function mutations for *CDKN2A* and PTEN. Significantly, NOTCH1, the pre-TCR pathway, and E2A/HEB proteins control critical checkpoints and branchpoints in early thymocyte development whereas several oncogenic transcription factors, HOXA9, c-MYB, SCL, and LYL-1 control HSC self-renewal. Together, these genetic lesions alter key regulatory processes in the cell, favoring self-renewal and subvert the normal control of thymocyte homeostasis.

I. Introduction

Upon entering the thymus, circulating hematopoietic stem cells (HSC) and bone marrow-derived pluripotent progenitors are exposed to the inductive thymic microenvironment where these cells acquire T lineage identity and lose alternative lineage potential. Various bone marrow-derived progenitors with T lineage potential have been characterized as thymus settling cells, suggesting that mechanisms of T-lineage commitment can operate on a number of progenitors (reviewed in Ref. 1). Early thymic progenitors (ETP) reside within the $CD3^{-}CD4^{-}CD8^{-}CD44^{+}CD25^{-}Kit^{hi}$ population, a subset of double negative (DN) 1 cells (reviewed in Ref. 1) (Fig. 1). These cells acquire CD25 expression at the DN2 stage and migrate to the outer cortex where they are committed to the T lineage at the DN3 stage. At this point, some DN3 cells rearrange TCRγ/δ loci to give rise to mature γ/δ T cells whereas others rearrange TCRβ loci. In the α/β lineage which is the dominant lineage in the thymus, thymocytes progress through two critical checkpoints, the first one at the DN3 stage, controlled by the pre-T cell receptor (pre-TCR), resulting from the pairing of an invariant pTα chain with a rearranged β chain, and the second one, at the $CD4^{+}CD8^{+}$ double positive (DP) stage by the TCRα/β antigen receptor, consisting of rearranged α and β chains. Following positive and negative selection, DP cells give rise to $CD4^{+}$ or $CD8^{+}$ immunocompetent cells that migrate into the periphery.

How multipotent progenitors acquire T-lineage potential and progress through distinct developmental stages, guided by the environment and the expression of nonself-antigen receptor, has been covered in detail with new perspectives in recent reviews.[1,6] Thymic development and the TCR are discussed in two other chapters in this series. This review will address the

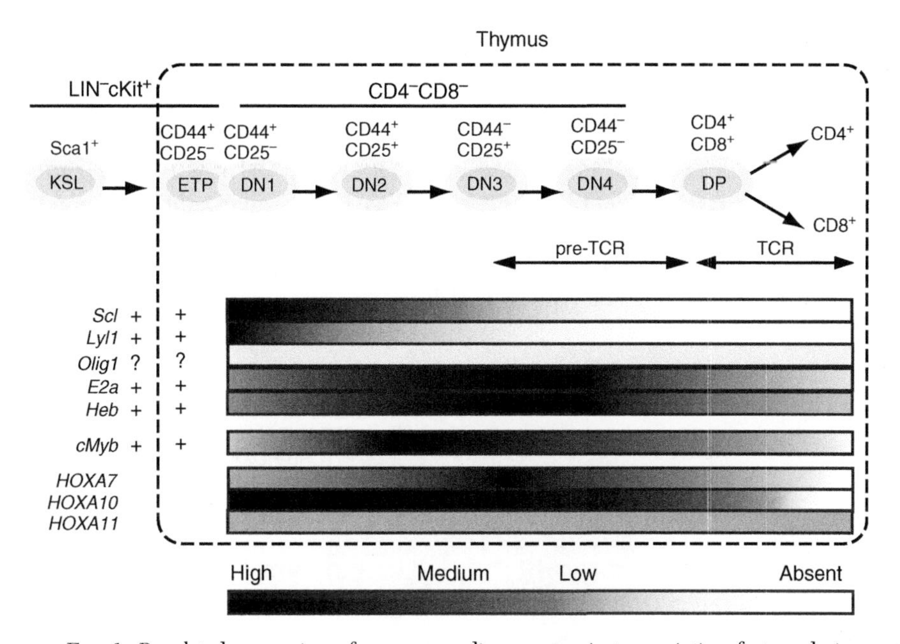

FIG. 1. Regulated expression of genes encoding oncogenic transcription factors during thymocyte differentiation. Circulating multipotent progenitors that are Kit+Sca+LIN− (KSL) enter the thymus, giving rise to early thymocyte progenitors (ETP). Depicted are the various stages of thymocyte differentiation that can be distinguished on the basis of CD4, CD8, CD44, and CD25: $CD4^{-}CD8^{-}$ double negative (DN) 1, DN2, DN3, DN4, $CD4^{+}CD8^{+}$ double positive (DP), and immunocompetent CD^{+} or $CD8^{+}$ single-positive cells. Gene expression data for bHLH factors are from Refs. 2–4 and for Hoxa genes, from Ref. 5.

molecular mechanisms underlying cell transformation in T-cell acute lymphoblastic leukemia (T-ALL) and how these oncogenic events reveal the functions of key developmental regulators and pathways in early T-cell development. We will cover information gained from the cytogenetics and molecular studies of childhood and adult T-ALL samples, as well as functional studies in the mouse. Several powerful genetic models of leukemia have emerged in recent years, but will not be included in this review for lack of space.

T-ALL is a common childhood cancer. Molecular mechanisms underlying T-ALL include aberrant oncogene activation, inactivation of tumor suppressor genes, and aberrant activation of signal transduction pathways. These genetic lesions alter key regulatory processes in the cell, favoring unlimited self-renewal and subvert the normal control of cell proliferation, cell differentiation, or apoptosis. Thus, an understanding of the molecular genetics of T-ALL will provide insight into the molecular regulation of T-cell development and by the same token, open avenues for developing novel targeted therapies.

T-ALL is characterized by recurring chromosomal anomalies, mostly translocations and aneuploidy, although 25–30% of those either harbor rare translocations or exhibit a normal karyotype (Fig. 2). Interestingly, 70% of chromosomal translocations involve transcription factors. In childhood T-ALL, the three major groups involve MYB (19%), TAL1/SCL (18%), and (homeobox) HOX11L2/TLX3 (18%) whereas in the adult, the two major groups involve HOX11/TLX1 (29%) and SCL (13%) (Fig. 2). These transcription factors are master regulators of cell fate[15] and determine the gene signature and the leukemic cell type in T-ALL.[16,17]

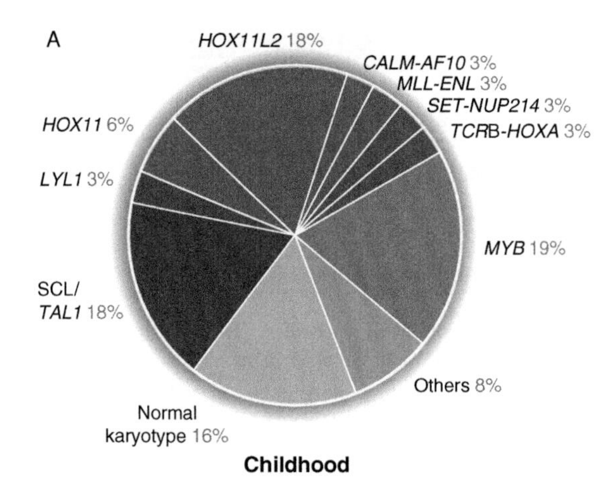

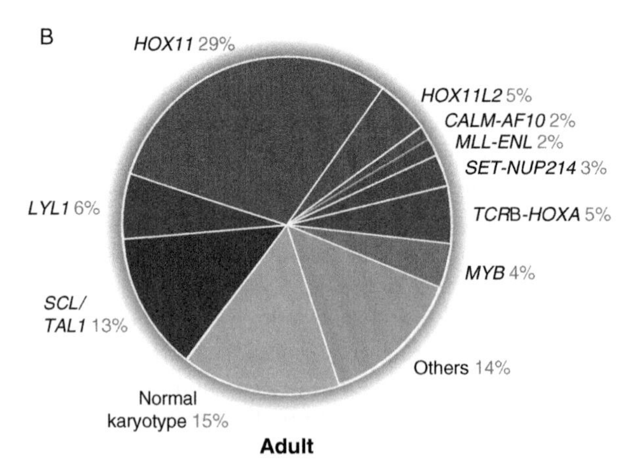

FIG. 2. Chromosomal abnormalities in T-cell acute lymphoblastic leukemia (T-ALL). The relative frequencies of chromosomal aberrations found in (A) childhood and (B) adult cases of T-ALL were calculated from the combined studies of Refs. 7–14.

Oncogenic transcription factors are not sufficient *per se* to transform primary cells as observed in monochorionic twin pairs in which the translocation was detectable in both twins but only one of the two was leukemic.[18] Similarly, transgenic mouse models in which the expression of these oncogenes is directed in the thymus rarely develop T-ALL (reviewed in Refs. 17,19). Disease induction in these experimental models requires complementation by other classes of gene products. T-ALL is therefore a multigenic disease involving additional cooperating genetic or epigenetic events that are not detected by cytogenetics. Gene specific approaches as well as several genome wide approaches have uncovered the importance of signaling pathways, for example, NOTCH1, PTEN, and (pre)TCR signaling (7; reviewed in Ref. 17) and tumor suppressors, mostly of CDKN2A/2B.[20] Remarkably, several of these oncogenic transcription factors as well as signaling pathways have been shown to control critical checkpoints in thymocyte development.[21]

II. Oncogenic Transcription Factors

A majority of chromosomal translocations in T-ALL cause ectopic expression of transcriptional regulators in thymocytes, for example, SCL, LYL1, TAL2, LMO1, LMO2, c-MYB, HOX11, HOX11L2, and HOXA, and in rarer cases, these translocations produce chimeric fusion genes, PICALM-MLLT10 (CALM-AF10), MLL-MLLT1 (MLL-ENL), and SET-NUP214.[17,22] Structurally, these transcription factors belong mostly to three families: helix-loop-helix (HLH) proteins that include SCL,[23] TAL2,[24] LYL1,[8,25] and OLIG2[26]; HOX genes, that is, HOX11,[27] HOX11L2,[28,29] and the *HOXA* cluster[9]; and c-MYB[10] which is characterized by three repeats of the helix-turn-helix (HTH) motif. The homeodomain fold also consists of an HTH structure with three alpha helices connected by short loop regions. Basic helix-loop-helix (bHLH) factors and c-MYB are overrepresented in childhood T-ALL whereas *HOX* genes, are overrepresented in the adult (Fig. 1). Together, translocations involving the three gene families bHLH, MYB-HTH, and HOX represent ~65% of all T-ALL.[30,31] Three other recurring translocations involve two highly homologous genes encoding LIM-only domain proteins, LMO1 and LMO2,[32] accounting for ~10% of T-ALL. Both genes genetically interact with *SCL* to induce T-ALL in transgenic mice[33,34] and both proteins directly associate with SCL[35–38] in the same transcription activation complex. LMO2 also associates with LYL1.[30,38] Furthermore, microarray studies indicate a common gene signature for T-ALL subgroups with *SCL*, *TAL2*, *LMO1*, and *LMO2* anomalies.[9,16,39] Thus, although LMO proteins are structurally different, these proteins functionally belong to the SCL/TAL1 group. Finally, the three fusion proteins, PICALM-MLLT10,[40] MLL-MLLT1,[41] and SET-NUP214,[42] all interact with the histone

H3K79 methyltransferase DOTL1 and activate the *HOXA* cluster.[42–44] These three fusion genes have therefore been considered as part of the HOXA group and together, represent 12% of T-ALL samples.

The prevalence of bHLH, *Myb*, and *HOX* genes in T-ALL indicates the importance of these three families of transcriptional regulators in the T lineage and their dominant mechanisms of leukemic transformation.

A. E Proteins in Thymocyte Development: Inhibition by SCL/TAL1

Converging evidence points to a critical role for HLH transcription factors in driving thymocyte development.

Class I HLH transcription factors, known as E proteins, are implicated in cell homeostasis, through the regulation of cell cycle genes, of proliferation and differentiation processes. This family of genes includes *E2A*, *HEB*, *E2-2*, and the *Drosophila* orthologue, *daughterless*. The *E2A* gene encodes for two isoforms, known as E12 and E47, generated through alternative splicing of exons encoding the DNA binding region and the dimerization domain.[45] In B-lineage cells, the predominant E-box-binding transcription complex implicates E47 homodimers, whereas in thymocytes, E47 forms heterodimers with HEB.[46,47] E proteins and their antagonists, such as Id family members, are key regulators in the commitment and differentiation to the lymphoid lineages. E proteins control the initial commitment step in the α/β lineage as E2A inhibits cell-cycle progression prior to *TCR*β gene rearrangement[48] and regulates the expression of the invariant pTα chain of the pre-TCR.[2] Furthermore, E2A prevents the progression to the $CD4^+/CD8^+$ T-cell stage in the absence of pre-TCR signaling.[49] After β-selection associated with cell survival and expansion, thymocytes that have successfully rearranged their TCRα loci undergo positive selection through a process controlled by E2A and Id3.[48,50]

Although the *E2A* and *HEB* genes are not found at chromosomal breakpoints in T-ALL, the *SCL* oncogene has been shown to inhibit E protein activity in the thymus[2,51–53] and to partially block T-cell differentiation at the DN3 to DP transition. SCL and LYL1 belong to class II bHLH transcription factors that exhibit more restricted expression patterns compared to class I factors. We previously showed that SCL and LYL1 are coexpressed in DN1 thymocytes.[3] SCL is downregulated at the DN3 stage[2] whereas LYL1 is switched off in DN2 cells[3] (Fig. 1). Chromosomal translocations place *SCL* or *LYL1* under the influence of regulatory sequences that are active at later stages of thymocyte development (for a recent review Ref. 17), causing inappropriate SCL or LYL1 expression which interferes with E protein activity and perturbs critical developmental checkpoints in the thymus.

B. Regulated *HOX* Gene Expression During Thymocyte Development

Class I *HOX* genes are located in four tightly linked physical clusters. The *HOXA* cluster is on 7p15.3. HOX-A7, HOX-A9, HOX-A10, and HOX A11 are expressed during the early stages of T-cell development, and maturation is accompanied by the sequential downregulation of the most 3′ region gene of this cluster.[5] The 5′ *HOXA* cluster genes, especially *HOXA11*, is expressed throughout all stages, whereas high *HOXA10* expression was found in the earliest human T-cell precursors and downregulated in CD4 and CD8 single-positive (SP) mature thymocytes (Fig. 1). This pattern of expression suggests that their regulated expression may be important for proper thymocyte differentiation. Indeed, several gene expression studies in human T-ALL directly or indirectly implicate dysregulated HOXA expression in this disease. Both HOXA10 and A11 are upregulated in T-ALL because of inv(7)(p15q34) that places the *HOXA* gene cluster under the influence of strong enhancers within the *TCRβ* locus (7q34).[54] HOXA9 and HOXA10 are also upregulated in MLL-rearranged, in SET-NUP T-ALL, and possibly in NUP214–ABL1 T-ALL.[9,42,55,56]

The importance of *HOXA* genes in T-cell development and in leukemia was assessed by gain-of-function and loss-of-function studies. Enforced expression of *HOXA10* and *HOXA11* impairs T-cell differentiation which is not sufficient, however, to induce T-ALL, indicating that additional genetic perturbations are required.[57] The *Hoxa* cluster, between *Hoxa6* and *Hoxa10*, was identified as a target of frequent integration sites in a small scale functional screen for genes that collaborate with E2A-PBX1 in B-ALL by retroviral infection.[58] This is associated with a dramatic increase in *Hoxa6* to *a10* expression. Transgenic studies confirm that Hoxa9 collaborates to some extent with E2A-PBX1 in T-ALL, with the provision that a single *Hoxa* gene was misexpressed in the transgenic model, compared to several genes of the *Hoxa* cluster in the retroviral screen or in human T-ALL.[59] These studies revealed redundancies, compensatory and dosage effects of *Hoxa* genes, which were corroborated by loss-of-function studies. *Hoxa9*-deficiency appears to be the single *Hoxa*-deficiency that is the most severe, as *Hoxa9*$^{-/-}$ embryos have a smaller thymus and decreased CD25^{+} thymocyte progenitors due to apoptosis.[60] Thus, unlike the HLH family in which the translocation misdirects expression of a single HLH protein, several genes within the *HOXA* cluster are upregulated either by the proximity of *TCRβ* regulatory sequences, or by the activation of upstream HOX regulators (MLL, SET-NUP214).

Unlike class I *HOX* genes, class II *HOX* genes that include *HOX11* (*TLX1*) and *HOX11L2* (*TLX3*) are dispersed in the genome. HOX11 is expressed in the thymic primordium at e13.5 in the developing mouse embryo and decreases below the limit of detection thereafter.[61] HOX11 is upregulated in T-ALL by juxtaposition of *TCRα* or β regulatory sequences due to chromosomal translocations. Ectopic expression of HOX11 in murine HSCs induces T-ALL with long latency,[62] whereas *Hox11*-deficient mice are asplenic. Therefore, the mechanism underlying cell transformation by HOX11 remains to be ascertained, despite its high frequency in adult T-ALL (Fig. 2; reviewed in Ref. 63). The *HOX11* translocation is of good prognosis whereas outcome for the *HOX11L2* translocation is debatable, as it is not favorable in some studies, but this is not confirmed in others (reviewed in Ref. 17). Global gene profiling suggests that the two types of leukemias share a common signature,[16] which merits further investigation in view of the differing disease outcomes in this study.

C. C-MYB is Essential at Several Stages of Thymocyte Development

c-MYB is predominantly expressed in immature thymocytes (Fig. 1) and is essential for the generation of the earliest thymocyte progenitors.[64] Subsequently, c-Myb regulates thymocyte proliferation following β-selection[65] and controls the survival of preselection DP thymocytes and the differentiation of CD4 thymocytes.[66] Given the nonredundant role of c-Myb at critical stages in thymocyte development, it was surprising that the *MYB* gene was not associated with chromosomal aberrations in T-ALL. However, two recent reports indicate that MYB is misexpressed either following translocation involving the *TCRβ* locus or by somatic genomic duplications at the *c-MYB* locus,[10,67] causing differentiation blockade, proliferation, and viability of the leukemic cells (reviewed in Ref. 68).

D. Toward a Mechanistic Understanding of T-ALL and Early Thymocyte Development

1. To Be or Not to Be with C-MYB

Genetic lesions involving *HOX* genes or genes of the bHLH families are mutually exclusive, suggesting that these oncogenic transcription factors potentially affect common cellular properties in the thymus. As *HOX* genes are known regulators of HSC self-renewal, one may infer that *HOX* genes may also enhance the self-renewal of preleukemic stem cells, thereby setting the stage for secondary genetic lesions. Whether or not bHLH oncogenic transcription factors control self-renewal remains to be elucidated. In contrast to the above, *MYB* rearrangements exhibit 14% overlap with *SCL*, suggesting that MYB is likely to complement SCL. In transgenic mouse models, we previously showed that the *SCL* and *LMO1* oncogenes cause increased apoptosis and decreased proliferation in preleukemic DN3/DN4 thymocytes due to decreased pre-TCR expression.[2] As c-MYB controls cell proliferation and cell survival following β selection, it is possible that c-MYB expression recues *SCL–LMO1* transgenic thymocytes from cell death and restores cell proliferation despite decreased pre-TCR expression. In addition to the above, cooccurrence of *c-MYB* translocation and MLL-MLLT1 was found in one T-ALL sample. The significance of this single observation remains to be addressed. However, MYB was also found as a downstream target of MLL-MLLT1, and of HOXA9-Meis1 by retroviral gene transfer in hematopoietic precursors.[69] Thus, the question whether c-MYB is downstream of or parallel to MLL-MLLT1 remains to be further addressed, although these two possibilities need not be mutually exclusive.

2. Illegitimate Recombination and Transcription

The complex repertoire of antigen receptors is generated through a series of site-specific somatic DNA rearrangements referred to as V(D)J recombination mediated by RAG-1 and RAG-2. These recombination events are highly regulated, showing both lineage and stage-specificity. Furthermore, the process is temporally ordered and show restriction to the G0–G1 stage of the cell cycle. Site-specificity is imparted by the RAG-recognition sequence that follows the 12/23 rule. Furthermore, RAG1–RAG2 reads marks of accessible chromatin by binding histone H3 trimethylated on the lysine at position 4 (H3K4me3) via its PHD finger (reviewed in Ref. 70). This histone mark is also associated with transcription at antigen receptor loci which therefore become poised for V(D)J recombination.

Illegitimate recombination can occur at loci other than those encoding antigen receptors, causing for example interstitial deletion of chromosome 1 that places *SIL* regulatory sequences immediately upstream of the *SCL* locus.[71] Initially, the breakpoint was found to coincide with imperfect RSS sequences.[71] Consequently, we observed that SCL, LMO2, and RAG1 are coexpressed in DN1–DN2 cells and predicted that transcriptionally active loci in early thymocyte progenitors may be more prone to illegitimate recombination.[2] The finding that RAGs read the H3K4me3 histone mark[70] concurs with these predictions and suggests that in addition to cryptic RSS sequences at the *SCL* and *LMO2* loci, transcription at these loci in ETP-DN2 cells may facilitate RAG-mediated chromosomal breaks. Additionally, an active chromatin configuration at the *LYL1*, *HOXA*, *c-MYB*, *MLL*, and *MLLT1* loci may also favor chromosomal translocations that are not necessarily dependent on RAG activity. Finally, among the thousands of genes that might be transcriptionally active in thymocyte progenitors, the selection for chromosomal translocations implicating these families of oncogenes is indicative of their functional importance during thymocyte development. This does not exclude the role of other regulators but clearly point to the dominant action of these oncogenic transcription factors in T-ALL.

III. Signal Transduction Pathways

The above genetic events, associated with chromosomal translocations, are considered as primary genetic lesions promoting transformation of normal thymocytes. Nonetheless, these genetic events are not sufficient *per se* to fully induce T-cell leukemia, which requires collaborating genetics events such as pathways that regulate T-cell homeostasis.

A. Notch1 Signaling Pathway

Notch is highly conserved throughout evolution of multicellular organisms. NOTCH family members control cell-fate determination, by regulating cell proliferation, survival, and differentiation during development.[72] The human *NOTCH1* gene was first identified in T-ALL patients with the translocation t(7;9)(q34;q34.3), causing *TCR*β-driven expression of a truncated activated form of NOTCH1 (*TAN1*).[73] However, this translocation occurs in less than 1% of T-ALL cases, whereas activating mutations of *NOTCH1* are frequent and were found in more than 50% of T-ALL cases.[7,74]

Translocation groups differ with regard to the proportion of samples with *NOTCH1* mutations (Fig. 3), the highest was identified in the LYL1 (64%) and HOX11L2 (56%) groups as well as the group with normal karyotype (58%).

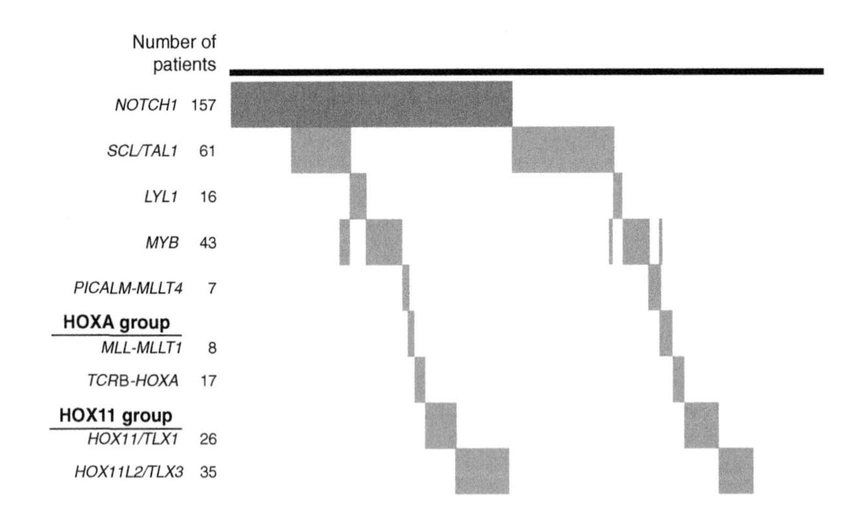

Fig. 3. Cosegregation of *NOTCH1* activating mutations and chromosomal aberrations associated with molecular subgroups of T-ALL cases. The reference number of T-ALL patients is depicted by a black line, while *NOTCH1*-associated T-ALL cases are represented by a dark grey lane and cases of each molecular subgroups of T-ALL are represented by a light grey lane. This cosegregation alignment is based on the combination of results published in Refs. 7,10,14,75–77.

The next group consists of the HOXA and MYB groups (47%). Finally, the HOX11 and SCL groups have the lowest frequencies, 38% and 30%, respectively (Fig. 3). Therefore, activating *NOTCH1* mutations potentially collaborate with all these oncogenic transcription factors to promote leukemogenesis in the T-cell lineage, which was later demonstrated for SCL/TAL1 in mouse models of T-ALL ([78,79], our unpublished results), consistent with the critical importance of NOTCH1 in the thymus.[80–82] Nonetheless, the lower frequency found in SCL and HOX11 groups suggests that additional events or pathways contribute to leukemogenesis in these groups.

NOTCH1 mutations affect the heterodimerization domain and the C-terminal PEST domain,[7] which is targeted to the proteasome following the binding of the E3-ubiquitin ligase FBXW7. Inactivating mutations of *FBXW7* could therefore represent an alternative mechanism for increased NOTCH1 signaling in T-ALL. Indeed, heterozygous *FBXW7* point mutations were identified in 8–30% of T-ALL patients. These mutations were mutually exclusive with PEST domain mutations[83–85] and mostly occurred in combination with NOTCH1 HD mutations,[85] suggesting that *NOTCH1* PEST mutations may relieve the mutational pressure on *FBXW7*. Thus, *FBXW7* inactivation could partially explain the activation of the NOTCH1 pathway in the absence of somatic activating *NOTCH1* mutations observed in some T-ALL patients. Constitutive NOTCH1 pathway activation could be an early event in leukemogenesis, as mutations have been reported *in utero*. It is, however, the combination of collaborating events which is crucial in T-ALL development, rather than the sequence of events.

Activation of the Notch pathway is initiated through interaction of the extracellular domain of NOTCH receptors (NOTCH1, 2, 3, and 4) with a DSL (Delta, Jagged, and Serrate) ligand located on the surface of an adjacent cell. This interaction allows ligand-mediated proteolytic cleavages of the receptor that result in the generation of intracellular domain of NOTCH1 (NICD). Once produced, the activated form of NOTCH translocates to the nucleus to regulate transcription through the formation of transcriptional activation complexes.[86,87] Various NOTCH1-dependent target genes include *HES1*, *HES5*, *MYC*, *PTCRA*, *DTX1*,[88–92] and members of the NF-κB pathway.[93]

Therefore, leukemogenesis associated with *NOTCH1* activating mutations likely arises from increased transcription of NOTCH1 target genes, such as *HES1*[91] and *MYC*.[89] In addition, NOTCH1 pathway activation may also modulate other signaling pathways. For example, NOTCH1 signaling causes phosphorylation of several members of the mTOR pathway,[94] resulting from the repression of *PTEN* (phosphatase and tensin homolog) by HES1.[11] NOTCH1-mediated *PTEN* repression also enhances the PI3K–Akt signaling pathway activity to promote precursor T-cell growth.[95]

In the mouse, *Notch1* gain-of-function leads to leukemogenesis[96,97] whereas *Notch1*-inactivation confirmed the indispensable role of Notch1 signaling pathway for normal T-cell lymphopoiesis.[98–102] The transforming potential of Notch1 can ultimately be viewed as a reflection of its normal function during T-cell differentiation. Notch1 signaling plays an essential role during the B- versus T-cell-fate determination in early hematopoietic precursors.[98,101,103,104] Conversely, retroviral expression of NICD in bone marrow progenitors suppresses B-cell development and induces extrathymic T-cell differentiation that leads to T-ALL.[105] The Notch1 pathway is required for commitment of the earliest T-cell progenitors in the thymus and promotes their proliferation, differentiation, and survival.[106–108] Hence, NOTCH1 signaling upregulates target genes that are critical at various stages of thymocyte development, *Il-7rα*,[109] *Tcrβ* germline transcription,[102] and, cooperatively with E2A, the expression of the *Ptcrα* gene at the DN3 stage.[99,110] T cells undergo further commitment following their *TCR* gene rearrangement into αβ- or γδ-lineage where Notch1 might play a dual role by favoring the γδ lineage development at low levels in early thymocyte populations[111] and further promoting the αβ T-cell development at later stages,[100,102] whereas a higher Notch1 signal strength favors the αβ lineage. Consistent with the latter, all leukemias harboring activating *NOTCH1* mutations arise from the αβ-lineage T-cell populations.[95,100,102,114]

Although the functional implication of Notch1 in the maturation of DP T cells and the CD4- versus CD8-lineage commitment in preleukemic mice is still controversial,[112,113] leukemias associated with NICD in transgenic mice show an increased $CD8^+$ SP population concomitantly with a decrease in $CD4^+$ SP thymocytes.[112,114] Finally, the strong oncogenic potential of NICD revealed by genetic studies and human T-ALL analysis is consistent with the essential function of Notch in T-cell commitment, along with its implication in the proliferation and survival of committed T cells.

B. (pre)TCR Signaling

The selection of T cells presenting a high affinity for foreign antigens without reactivity for self-antigens occurs through several critical checkpoints (reviewed in Ref. 115). β-Selection, which depends on signals generated through the pre-TCR complex, is essential to discriminate cells having successfully rearranged TCR-β chains in the DN3 subset. The pre-TCR, resulting from the pairing of the invariant pTα with a rearranged TCRβ chain, collaborates with Notch1 signaling to suppress the activity of E2A, leading to the proliferation of TCRβ-selected DN4 cells.[116–118] In post β-selected cells, this inhibition of E2A activity decreases Notch1 expression and signaling,[119] which is important for normal development and progression to the DP stage. Unlike the antigen receptor, the pre-TCR[120] promotes autonomous signaling due to self-oligomerization.

The pre-TCR controls cell survival, cell proliferation, and differentiation from the DN3 stage to the DP stage. Pathways controlling cell survival downstream of the pre-TCR involve NFκB, p53, FoxO3, and Bmi-1 that upregulate *Bcl2a1* and downregulate *Bid*, *Bim*, and *Cdkn2a*, respectively.[121–124] In addition, pre-TCR signaling induces ID3, that inactivates E2A allowing for cell proliferation.[125] Initiating events in pre-TCR-mediated differentiation involve CD3ε,[126] LCK, a member of the SRC family of tyrosine kinases highly expressed in T-cells,[127] Syk, and ZAP70 (reviewed in Ref. 120). Apart from frequent deletions of the *CDKN2A* locus which will be discussed later, as well as ectopic *LCK* expression driven by the t(1;7)(p34;q34) rearrangement in a few cases of T-ALL,[128] most of these genes have not been so far associated with genetic lesions in T-ALL. In contrast, *ZAP70* and *SYK* are frequently implicated in B-cell neoplasms and in chronic lymphocytic leukemias, suggesting lineage-specific mechanisms of oncogenic collaborating events. This possibility remains to be addressed but additional mutations may perhaps be unravelled in the future by genome-wide approaches in T-ALL.[19]

Signals generated through the TCR have been extensively studied (Fig. 4) and involve downstream activation of the PI3K–AKT pathway, the RAS–MAPK (mitogen-activated protein kinase) signaling cascade, and the calcineurin-nuclear factor of activated T (NFAT) pathway (reviewed in Refs. 138–140). Several of these are also implicated in pre-TCR signaling, and in T-ALL. The RAS protein plays an important role in the transmission of TCR signals from the membrane to ERK (Fig. 4), leading to MAPK cascade activation. Activating *RAS* mutations have been identified in 4–10% of T-ALL cases.[141,142] Following TCR activation, PI3K–AKT signaling and calcineurin-NFAT pathway cooperate with activated RAS–MAPK and NF-κB signaling to ultimately regulate T-cell proliferation and differentiation, along with the immune response to foreign antigens presented to naive T cells in the thymus (Fig. 4). Both NF-κB and NFAT play an essential role in negative and positive selection during T-cell development.[143] Constitutively activated calcineurin mutant was shown to induce T-ALL in transgenic mice,[144] indicating that calcineurin activation contributes to leukemogenesis. The central role of calcineurin-NFAT was confirmed by preclinical studies using calcineurin inhibitors, which induced apoptosis of leukemic cells in *TEL-JAK2*- and *NICD*-leukemic mouse models.[144] Thus, an understanding of (pre)TCR signaling pathways may reveal new targets for therapeutic strategies in T-ALL.

Although only 25% of T-ALL samples carry chromosomal translocations on the t(14q11) region, affecting the *TCRα* and δ gene locus,[145] the (pre)TCR signaling pathway plays an extended role to favor the outgrowth of transformed thymocytes in many T-ALL mouse models. For example, the *pTα* gene collaborates with Notch to induce leukemia in transgenic mice[146,147] and plays an important role in *Ikaros*-mediated leukemogenesis.[148] Recent evidence

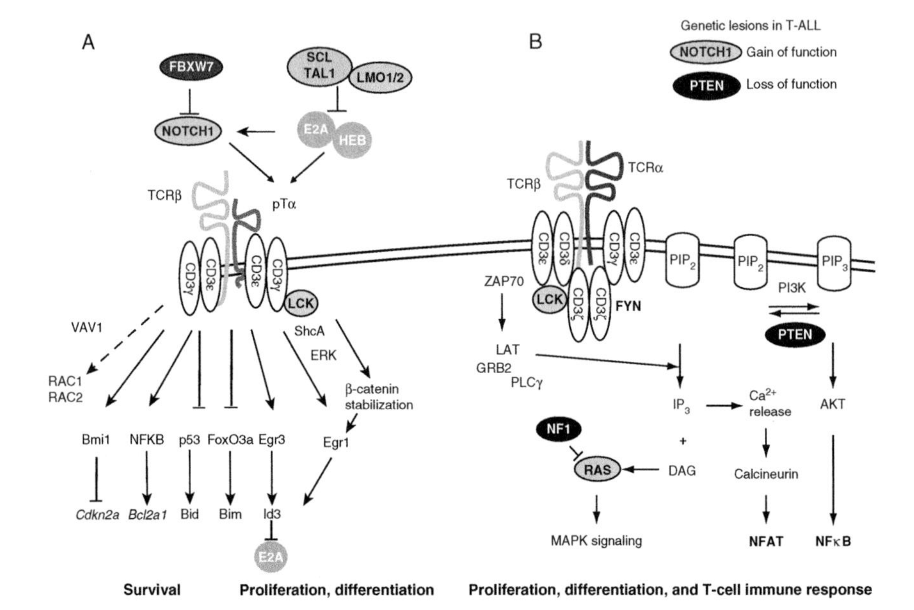

FIG. 4. Schematic representation of the (pre)TCR signaling network. (A) pre-TCR signaling. Both NOTCH1 and E2A regulate the expression of the pre-TCR alpha (*pTα*) gene. Activation pre-TCR signaling leads to Egr3 expression, Id3 induction, and inhibition of E2A activity.[125] NOTCH1 cooperates with the pre-TCR prior to and during β-selection.[95,129] However, Notch1 is an E2A target gene and therefore, Notch1 is downregulated after β-selection.[119] Pre-TCR signaling also promotes the expression of the Rho family, small GTP binding proteins RAC1 and RAC2, which is dependent on VAV1, a guanine nucleotide exchange factor. The pre-TCR ultimately controls the DN to DP transition stage, by regulating differentiation, proliferation, and TCRα rearrangements in T-cell precursors. (B) The TCR signaling network. Antigen stimulation of the TCR activates CD3-associated LCK and FYN tyrosine kinases, followed by the phosphorylation and recruitment of ZAP70 and PI3K.[130] ZAP70-mediated phosphorylation of LAT promotes the recruitment of GRB2 and PLCγ, which hydrolyzes PIP2 into IP_3 and DAG.[131,132] Ca^{2+} release induced by IP_3 activates calcineurin phosphatase activity, dephosphorylating NFAT and resulting in NFAT nuclear translocation.[133] In parallel, DAG activates RAS and MAPK signaling.[134] Furthermore, AKT activation downstream of PI3K via PIP3 phosphorylation ultimately leads to NFκB activation.[135] PTEN antagonizes PI3K activity, by dephosphorylating PIP3 into PIP2.[136] These pathways cooperate in T-cell proliferation, differentiation, or induction of a T-cell immune response (adapted from Ref. 17). Gain-of-function mutations in T-ALL are depicted in dark grey with black characters and loss-of-function mutations in light gray with white characters. The *E2A* and *HEB* genes are not mutated in T-ALL but are proposed to have a tumor suppressor function.[2,21,137]

suggests that the pre-TCR signal is also important for *TEL-JAK2*-induced leukemias,[146,149,150] as well as *SCL–LMO1*-dependent leukemias[150,151,114] but dispensable for T-ALL induced by *E47*- or *Trp53*-deficiency.[116,152] These differing requirements for pre-TCR signaling led us to investigate its importance in T-ALL.

First, we used the Survival Forest (SF) method[153–155] and multidimensional scaling[156] to perform a survival analysis of pediatric T-cell acute lymphoblatic leukemia based on gene expression data available from Ferrando *et al.*[16] The SF method is an extension of the well-known Random Forest method[157] which has been used to identify gene expression signature in pediatric T-ALL.[158–160] This analysis revealed that there are three major prognostic groups, H for *HOX11+* cases and the like, group L for *LYL1+* cases and the like, and group S for *SCL+* cases and the like and two intermediate groups M for *MLL-ENL+* cases and HL for *HOX11L2+* cases (Fig. 5A). This pattern turns out to be coherent with previous results[16] but further allows for the inclusion of all patient samples within these five groups.

Second, we took advantage of the Kyoto Encyclopedia of Genes and Genomes (Kegg) pathways and applied Goeman's global test[161] to determine whether the global expression pattern of a group of genes or a pathway is significantly related to the five prognostic groups identified above. The global test results for the top five Kegg pathways with the adjusted p-values (FWER adjusted $p < 0.01$) revealed that hematopoietic cell lineage, cytokine–cytokine receptor interaction, focal adhesion, TCR signaling, and phosphatidylinositol signaling pathways are the most significant pathways that distinguish the five prognostic groups in pediatric T-ALL (Fig. 5B). Furthermore, within the hematopoietic lineage gene list that discriminates these prognostic groups, the genes that were most upregulated in the SCL+ (T) group (Fig. 5C) were all associated with TCR signaling (Table I): CD3δ and CD3ϵ proteins, required for TCR surface expression and signaling; CD2 costimulatory molecule localized either at the immunological synapse[162] or in lipid rafts.[163] Within the cytokine–cytokine receptor interaction pathway, genes that are preferentially expressed by the SCL^+ group are also related to the activation of TCR signaling, that is, *LTB*, *IL2*, and *CXCR4* (Table I). Together, these observations indicate that TCR signaling is a distinctive feature of the SCL^+ prognostic group, contrasting to all other T-ALL prognostic groups, and suggest that the TCR signaling pathway may be relevant to the pathophysiology of *SCL*-associated T-ALL[16,146] (and reviewed by Ref. 164). The importance of pre-TCR signaling in SCL-associated T-ALL was confirmed in mouse models[114,150]. It is possible that the prevalence of the (pre)TCR in the *SCL/TAL1* group alleviates in part the requirement for activating *NOTCH1* mutations in *SCL/TAL1*-associated T-ALL.

C. PTEN and PI3K–AKT Pathway

Phosphatidylinositol 3-kinase (PI3K)–AKT signaling pathway plays an essential role in T-cell proliferation, cell-cycle progression, and apoptosis.[165] Growth factor stimulation activates PI3K, induces the production of phophatidylinositol 3,4,5 triphosphate (PIP3) from PIP2, and activates a number of

A

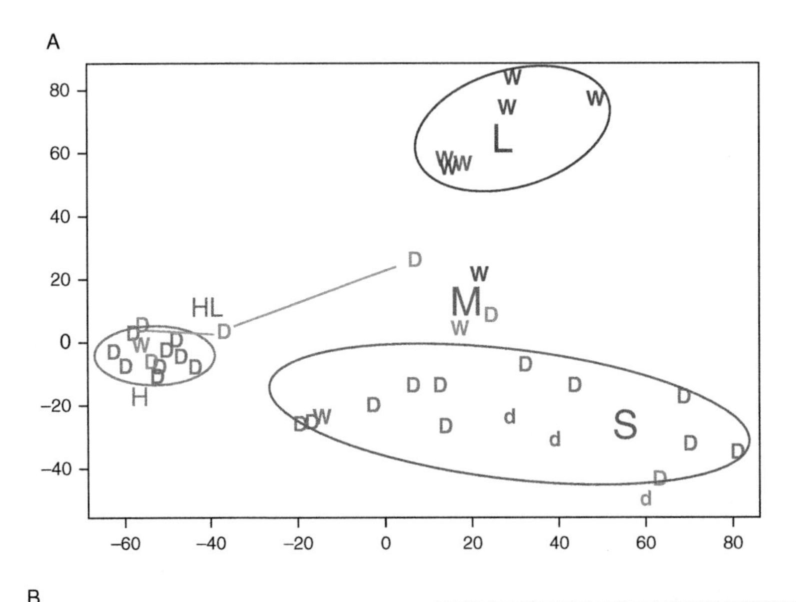

B

	Top 5 Kegg pathways		
	No. of genes	*p*-value	FWER. adjusted
Hematopoietic cell lineage	119	3.394e−06	0.00051
Cytokine–cytokine receptor interaction	243	4.510e−06	0.00068
Focal adhesion	234	2.710e−05	0.00409
T cell receptor signaling pathway	91	3.093e−05	0.00463
Phosphatidylinositol signaling system	77	6.003e−05	0.00894

C

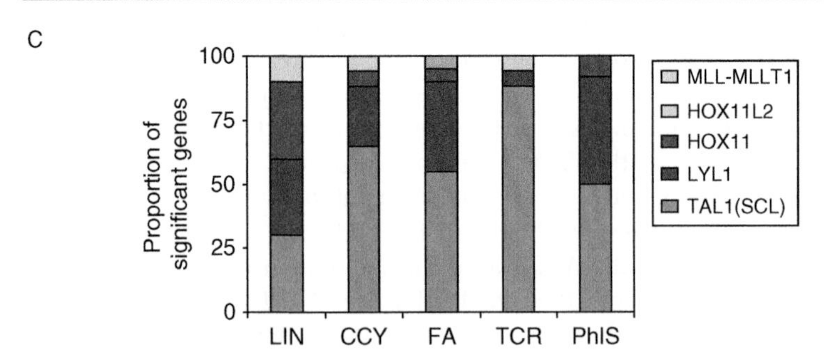

FIG. 5. Analysis of gene expression signatures identifies molecular pathways correlating with patient survival. This analysis is based on microarray results published by Ferrando *et al.*[16] (A) Five prognostic groups in T-ALL based on global gene expression and patient survival. We performed a survival analysis of pediatric T-ALL based on gene expression data available from Ferrando *et al.*[16] using the Survival Forest (SF) method. This method predicted for each patient a survival curve that in turn can serve to define prognostic groups. Multidimensional scaling provided the graphical representation of these individual survival curves in a geometric space. In this two-dimensional

downstream targets, including AKT. An important AKT target is mTOR, which supports tumor cell proliferation.[166] A major negative regulator of AKT activity is the lipid phosphatase PTEN, which antagonizes PI3K activity, thus downmodulating AKT signaling.[167,168]

In nonmalignant T cells, PTEN regulates thymic negative selection and self-tolerance. In addition, *PTEN* expression is tightly regulated by NOTCH1 signaling, which upregulates the *HES1* repressor, causing *PTEN* downregulation and increased PI3K–AKT pathway signaling.[89] As soon as the NOTCH pathway is silenced, *PTEN* expression returns to baseline, and signaling through the PI3K–AKT pathway is again modulated by PTEN. This regulatory feedback loop of PTEN and PI3K–AKT signaling by the NOTCH1 pathway is essential for cell-cycle progression and cell growth in early T-cell progenitors. On the other hand, the activation of the PI3K–AKT pathway by the (pre)TCR promotes the DN to DP transition, as well as the survival of DP thymocytes.[169]

Mutational loss of *PTEN* and activation of the PI3K–AKT pathways have been reported in all molecular subgroups of T-ALL and a very broad range of human cancers.[170–173] It is possible that PTEN–AKT favors the survival of transformed cells. Homozygous *PTEN* inactivation has been found in 8% of primary T-ALL, complete loss of the PTEN protein in 17% of patient samples (reviewed in Ref. 174), and activating mutations in *AKT* and *PI3K* genes in 11% of T-ALL cases.[175] These mutations result in constitutive AKT activation and together represent 35% of T-ALL samples, correlating with independent studies indicating that AKT is phosphorylated in more than 30% of T-ALL

---

survival dot plot, each point corresponds to a patient for whom a survival curve was predicted using SF. High similarities between survival curves correspond to small distances and clustered points on the plot and conversely. The plot shows three major prognostic groups T, H, and L (for *TAL1+* and the like, *HOX11+* and the like, and *LYL1+* and the like, respectively) at the edge of a virtual triangle and two minor groups, HL and M (for *HOX11L1+* and *MLL-ENL+*). Among the 10 samples (illustrated in green) that could not be classified previously,[16] two were found here to belong to the T prognostic group, and two to the H prognostic group, two to the M group, and three to the HL group according to survival times predicted from gene expression profiles. Furthermore, one *SCL+* case was previously clustered in the H group[16] and was found to carry the 10q24 translocation commonly associated with *HOX11* rearrangement. While the *LYL1* group was previously classified within the SCL bHLH family,[17,25] this graphical representation clearly demarcates the *LYL1+* group from the others. The status of the *CDKN2A* gene is shown for each patient sample: D, homozygous deletion; d, hemizygote deletion; W, wild type. Note that *LYL1* turns out to be the only group in T-ALL in which the *CDKN2A* locus is wild type in all samples. (B) Top five Kegg pathways derived from Goeman's global test that best discriminate the five prognostic groups. Global test gives one p-value for a pathway and adjusted p-values in multiple testing of many pathways, as well as z-scores for measuring the impact of individual genes on the test results. (C) The proportions of influent genes are represented by Kegg pathways and prognostic groups. LIN, hematopoietic cell lineage; CCY, cytokine–cytokine receptor interaction; FA, focal adhesion; TCR, T cell receptor signaling pathway; PhIS, phosphatidylinositol signaling pathway. (See Color Insert.)

TABLE I
TOP GENES ASSOCIATED WITH THE THREE PATHWAYS

Rank	Gene	Influence	Expected	S.D.	z-score	High expression
TCR pathway						
1	LCK	237	25	21	10.3	T
2	ZAP70	53	6	5	9.0	T
3	PTPN6	92	13	10	7.6	M
4	NFATC1	71	10	8	7.5	T
5	CD3E	41	7	6	5.6	T
6	CD28	33	6	5	5.3	T
7	LCP2	54	10	8	5.2	T
8	CD40LG	52	10	8	5.2	T
9	PLCG1	23	5	4	5.1	T
10	CD3D	18	4	3	5.0	T
Hematopoietic lineage						
1	CD1A	902	101	82	9.8	H
2	CD1B	1096	151	123	7.7	H
3	CD1D	249	39	32	6.5	H
4	CD34	298	47	38	6.5	L
5	FLT3	196	33	27	6.0	L
6	CD2	599	105	86	5.7	T
7	CD3E	41	7	6	5.6	T
8	CD3D	18	4	3	5.0	T
9	CD37	41	8	7	4.9	L
10	TFRC	79	16	13	4.9	L
Cytokine–cytokine receptor						
1	LTB	1406	112	91	14.1	T
2	IL2RG	71	8	7	9.1	T
3	AMH	248	31	26	8.5	H
4	FLT3	196	33	27	6.0	L
5	IL8RA	115	20	16	5.9	T
6	CXCR4	144	27	22	5.4	T
7	FLT1	18	3	3	5.4	L
8	CX3CR1	217	41	34	5.2	T
9	CD40LG	52	10	8	5.2	T
10	TNFSF4	77	15	12	5.2	L

The global test method based on the Cox proportional hazards model computes a statistic to measure the influence of a group of genes, here the TCR signaling pathway, the hematopoietic lineage, or the cytokine–cytokine receptor pathway on survival. To assess the contribution of each gene on this global statistic, the influence of the "singleton pathway" containing only that gene is estimated against the expected value. The weight of each gene is given by the z-score obtained from the standardized statistic, that is, (influence – expected influence) /S.D. (influence), and ranked among all genes in the pathway. The z-score of a gene indicates how many standard deviations its influence exceeds the reference level determined under the null hypothesis that the gene is not associated with survival.

cases.[172,175,176] In the mouse, *PTEN*-deficiency leads to T-ALL,[177,178] providing functional evidence for the importance of PTEN in preventing leukemia and in allowing normal T-cell development.

Mutational loss of *PTEN* is mutually exclusive with gain-of-function mutations of *PI3K* and *AKT* in T-ALL patient samples,[175] concurring with the view that these regulators are effectors of the same pathway. In contrast, between 10% and 50% of NOTCH1-associated T-ALL cases also carry *PTEN* mutations,[11,179,180] causing resistance to treatment using γ-secretase inhibitors (GSI), which blocks aberrant NOTCH1 signaling in leukemias.[11,181] Consistent with the presence of mutational loss of *PTEN* in a proportion of samples from each molecular subtypes of T-ALL patients, all GSI-resistant leukemic cell lines analyzed showed increased levels of AKT phosphorylation. *PTEN* mutations were also identified in relapse samples while being absent at diagnosis, suggesting that PTEN inactivation could represent a marker of tumor progression during leukemogenesis, rather than an initiating event.

D. Episomal NUP214–ABL1 Fusion Gene

Constitutive activation of ABL1 was initially found in CML, caused by the BCR-ABL translocation. More recently, episomal amplification of the NUP214–ABL1 fusion gene was found in 4–5% of T-ALL, in particular within the HOX11 and HOX11L2 molecular subgroups.[56] These observations suggest a complementation between the TLX1–TLX1L2 oncogenes and the ABL1 pathway, and that these T-ALLs may be susceptible to treatment with the tyrosine kinase inhibitor imatinib.

E. Crosstalks Between Signaling Pathways

Malignant T cells have acquired transforming properties resulting from alterations in multiple cellular signaling pathways, which are highly integrated networks (Fig. 6). Notch signaling promotes DN3 cell survival through Akt signaling activation.[95] The NOTCH1 pathway directly regulates *PTEN* expression, independently of the (pre)TCR pathway, which mediates a strong activation of the PI3K–AKT pathway.[11] PI3K activation can also be mediated by cytokine-mediated stimulation of the JAK-signal transducer and activator of transcription (STAT) pathway. STAT5 is required for bone marrow progenitor cell development and mediates DN thymocyte survival.[186] The TEL-JAK2 fusion gene, associated with t(9;12)(p24;p13) T-ALL samples,[187] as well as the activating *JAK1* mutations, observed in 18% of adult T-ALL cases, result in constitutive tyrosine kinase activity and STAT5 activation.[188] STAT3 activation could also be induced by the NOTCH1 pathway,[184] and STAT1, 3, and 5 by ERK, a member of MAPK signaling cascade.[189]

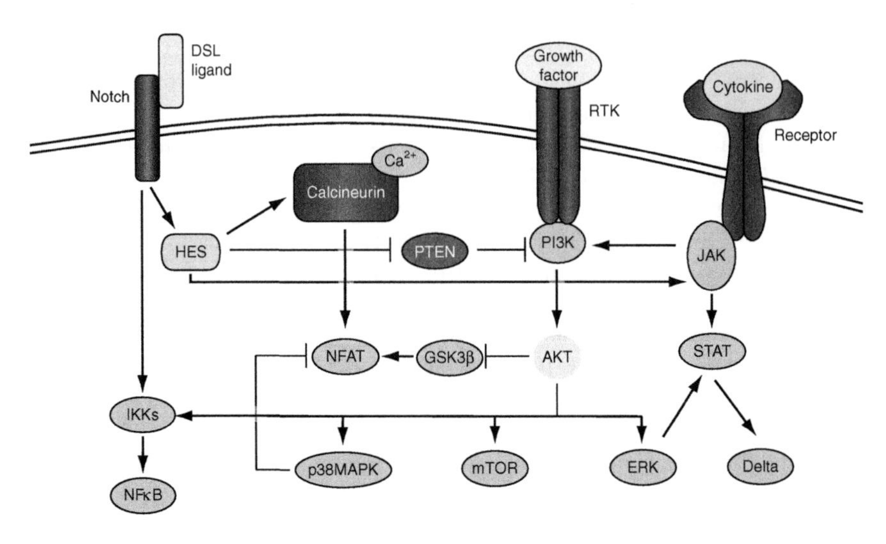

FIG. 6. Feed forward and signaling pathway crosstalks in thymocytes. NOTCH associates with IκB kinases (IKKs) and enhances (NF)-κB transcriptional activity.[182] Hairy and enhancer of Split (HES), a NOTCH1 target gene, inhibits PTEN expression, resulting in AKT pathway activation. NOTCH-induced HES1 also down regulates calcipressin, an endogenous calcineurin inhibitor, thereby stimulating the calcineurin-NFAT pathway.[183] HES also binds to STAT3 and thus mediates crosstalk between the NOTCH and JAK-STAT pathways.[184] Glycogen synthase kinase-3β (GSK3β) phosphorylates and inactivates TCR-induced NFATc1. AKT inhibits GSK3β, thereby allowing for NFATc1 nuclear retention. AKT also activates IKKs, mTOR, and p38MAPK, in the latter inhibiting NFAT, and the ERK pathway, upstream of STAT (adapted from Ref. 185).

TCR signaling modulates ERK activity to control positive selection in normal thymocytes,[190] whereas the p38-MAPK pathway regulates negative selection.[191] Moreover, p38-MAPK inhibits calcineurin-NFAT pathway activation,[192] which plays an important role in Notch1 and JAK-STAT leukemias.[193] Together, the functional interactions between NOTCH1, (pre)TCR, JAK-STAT, MAPKs, and calcineurin-NFAT pathways suggest that inappropriate regulation of only one of these pathways can deregulate this entire signaling network, ultimately favoring T-cell transformation.

IV. Tumor Suppressors

A. *CDKN2A* and *CDKN2B* Gene Deletion

A number of checkpoints distributed along the cell-cycle coordinate cell growth, cell division, and differentiation (as reviewed in Ref. 194). The *CDKN2A* and *CDKN2B* genes are located in tandem on chromosome 9p21.

Mutagenic inactivation of these genes is the most recurrent genetic abnormality identified in T-ALL[195]. In more than 90% of cases, the *CDKN2A/B* loci are inactivated through cryptic deletions, promoter hypermethylation, inactivating mutations, or posttranscriptional mutations.[196–198] The *CDKN2A* and *CDKN2B* loci encode for p16 and p15 respectively, and act as inhibitors of the cyclinD-cyclin-dependent kinase (CDK4) complex (INK4). The *CDKN2A* locus also encodes for the alternative $p19^{ARF}$ product, which is a negative regulator of MDM2, as part of the p53-DNA damage response circuitry. Therefore, deletions of the *CDKN2A* and *CDKN2B* loci promote uncontrolled cell-cycle entry, along with disabled p53-controlled cell-cycle checkpoint and apoptosis machinery. T-cell homeostasis is thus regulated through functional interactions of many different cell-cycle regulators, including the INK4 and ARF proteins. Deletions of *CDKN2A/B* loci are distributed in all molecular subgroups of T-ALL described above,[16] except for LYL1 as discussed underneath.

Our analysis of gene expression data published by Ferrando *et al.*[16] as well as graphical representation clearly demarcates the LYL1 group from the other groups, that is, HOX11, SCL, and MLL-MLLT1 (Fig. 5A). Interestingly, *CDKN2A* gene deletion is found in all SCL^{+} and $HOX11^{+}$ cases except one in each group,[16] whereas the *CDKN2A* gene was wild type in all $LYL1^{+}$ cases. Thus, the demarcation of the LYL1 group based on differential gene expression and patient survival exhibits perfect correlation with the status of the *CDKN2A* gene, strongly suggesting different oncogenic pathways for SCL and LYL1, although both genes code for bHLH transcription factors of the same subfamily.[8,12,16,23,199] Together, SF analysis of gene expression data and multidimensional scaling illustrate the biologic difference among the prognostic groups identified by Ferrando *et al.*[16] The frequent disruption of the *CDKN2A* locus is indicative of its nonredundant function in T cell homeostasis.

B. Unsuspected Tumor Suppressors, "Le Médecin Malgré Lui"[1]

1. E2A and HEB Loss-of-Function

Overexpression of Id2 protein, an inhibitor of E2A activity, blocks T-cell differentiation and leads to T-ALL development.[200] Moreover, thymocyte development is partially blocked at the DN1 stage in *E2a*-deficient mice and at the DN4 stage in *Heb*-deficient mice. E2A[48] and HEB[201] proteins enforce a proliferation checkpoint prior to and during β-selection and $E2a^{-/-}$ mice that survive to adulthood eventually succumb to T-cell leukemia,[202] which is strongly accelerated by disruption of pre-TCR signaling.[116] Together, these observations suggest that E2A collaborates with pre-TCR signaling to prevent

[1] "The Doctor despite himself" by Molière.

leukemogenesis. In addition, NICD overexpression in *E2a*$^{-/-}$ fetal thymocyte progenitors partially rescues the developmental arrest caused by *E2a* deficiency.[110] As E2A and Notch1 collaborate during T-cell development, it remains to be addressed if *E2A*-deficiency accelerates *Notch1*-mediated leukemogenesis as observed with SCL–LMO1.

V(D)J recombination takes place in G0–G1 and is regulated by E2A at several levels. Thus, E proteins regulate *Rag1/2* expression, remodel chromatin structure,[46] and prevent G_1 to S progression.[203] Conversely, *E2A*-deficiency causes increased cell cycle,[204] possibly through the regulation of *Cdkn1a*,[205] an inhibitor of CDK. Ectopic TAL1/SCL expression perturbs normal T-cell development through the repression of E2A-HEB activity,[2,52,53,137,204] leading to dysregulated T-cell homeostasis. Moreover, deletion of one allele of *E2a* or *Heb* accelerates leukemogenesis in *SCL* transgenic mice.[137] Together, these observations suggest a novel tumor suppressor function for E2A or HEB in the T lineage.[21] However, it is not clear whether the tumor suppressor activity of E2A can be accounted for by *Cdkn1a* expression or involves additional genes and pathways. Indeed, global gene profiling using an inducible E47 transgene indicates that E47 modulates the expression of genes involved in cell cycle, cytokine signaling, and T cell development.[206]

2. Rare Targeted Inactivations

The *Ikaros* gene encodes six lymphoid-restricted zinc finger proteins that are differentially expressed during lymphoid development in both human and mouse (reviewed in Ref. 207). During murine early fetal hematopoiesis, Ikaros is essential for the development of the lymphoid system,[208] propably through regulation of the expression of Notch target genes, such as *Hes1*, *Deltex*, and *pTα*.[209] Ikaros functions as a tumor suppressor in the T-lineage, as exemplified by rapid leukemia development in *Ikaros*-null mouse models.[148] Nonetheless, the role of *IKAROS* as a tumor suppressor in human T-cell lymphopoiesis is still unclear. Inactivation of *IKZF1* encoding IKAROS by deletion or mutation is a rare event, and was found in less than 5% of T-ALL cases,[210–212] which appears to differ from the high frequency of *IKZF1*, in B-ALL,[213] that is, 28.6% identified by genome-wide studies for copy number alterations. It is not clear whether this difference reflects intrinsic differences between the two types of leukemia, or might be attributed to sampling size.

Lymphoid enhancer binding factor 1 (LEF1) is downstream of several pathways, like WNT/β-Catenin, TGFβ via its interaction with SMAD4, and NOTCH1. A recent study reports mono and biallelic microdeletions of LEF1 in 11% of T-ALL, which were identified by array CGH.[214] These deletions are mutually exclusive with expression of *SCL/TAL1*, *LYL1*, *HOX11L2*, and *HOXA* cluster genes but cooccur with *NOTCH1* activating mutations as well as *CDKN2A* deletions. These observations concur with an importance of LEF1 for thymocyte differentiation.[215]

Somatic inactivation of neurofibromatosis type 1 (*NF1*) gene in hematopoietic cells results in a progressive myeloproliferative disorder in mice,[216] confirming that NF1 acts as a tumor suppressor gene.[217,218] Microdeletions in the *NF1* gene locus were found in T-ALL patient samples from HOX11L2 and CALM-AF10 molecular subgroups.[216] Moreover, activating *NOTCH1* mutations were found in two of three del(17)(p11.2)-positive T-ALL samples, associated with mutational loss of *NF1* expression. NF1 is a negative regulator of the RAS signaling pathway[219] and therefore NF1 inactivation could act as an alternative RAS activation mechanism, partially recapitulating pre-TCR signaling.

V. Concluding Remarks

Deregulation of transcription factor expression and stimuli response signaling plays a major role in leukemogenesis, as a result of perturbed T-cell homeostasis in the thymus. Thus, the molecular genetics of T-ALL revealed the cooperation of NOTCH1 signaling and of the pre-TCR with oncogenic transcription factors. These studies also revealed the tumor suppressor functions of PTEN, E2A, and HEB in the thymus. Significantly, these pathways and E proteins control critical checkpoints and branchpoints in early thymocyte development. In contrast to the above, chromosomal rearrangements implicate oncogenic transcription factors that have been shown to control HSC self-renewal. *Hoxa9* overexpression expands HSC[220] and *HOX11* immortalizes myeloid bone marrow cells[62] whereas the conditional inactivation of *Myb* in HSC leads to loss of self-renewal.[221] Both *SCL* and *LYL1* are highly expressed in HSCs and control long-term HSC functions.[4,222] In particular, *SCL* gene dosage determines the integrity of HSCs under conditions requiring extensive proliferation.[4] We therefore propose that dysregulated expression of these transcription factors driven by chromosomal rearrangements may confer extended self-renewal properties to preleukemic stem cells, thereby predisposing these cells to accumulate secondary genetic anomalies.

T-cell development in the thymus cannot be sustained by thymocyte progenitors and requires input from circulating bone marrow-derived multipotent cells (reviewed in Ref. 223). Indeed, a hallmark of stem cell populations is their capacity for sustained self-renewal, whereas progenitors are devoid of self-renewal capacity. Under exceptional circumstances, however, these progenitors may acquire self-renewal capacity.[224] The expression of a set of "stem cell" genes, that is, *Scl*, *Lyl1*, *Lmo2*, as well as genes of the *Hoxa* cluster in DN1 and to some extent, in DN2 thymocytes, is consistent with the view that these cells originate from "multipotent" progenitors that settle in the thymus and have retained some degree of multi- or oligopotency. DN1–DN3 cells have the highest proliferative potential in the thymus but appear however, to have lost sustainability.

Whether some of the oncogenic transcription factors in T-ALL may confer self-renewal properties to thymic progenitors remains an unexplored question. In support of this possibility, a recent study provides compelling evidence that ectopic LMO2 expression confers long-term thymic engraftment capacity to DN progenitors[225] (reviewed in Ref. 226). Pathways to cell transformation minimally involve "enhanced self-renewal capacity, increase in autonomous growth and, ultimately, escape from terminal differentiation".[227] In principle, two genes with the same biological outcome are unlikely to collaborate with each other. As *Notch1* mutations are found in LMO2-transgenic mouse models[78] and NOTCH1 activation collaborates with LMO2 to induce T-ALL (our unpublished results), NOTCH1, and LMO2 would be predicted to confer different biological outcomes. Consistent with these predictions, the *Notch3* oncogene did not confer engraftment capacity to thymocyte progenitors, suggesting a different mechanism of cell transformation.[225]

Finally, how the interactions between thymic settling cells and the thymic stroma shape cell fate during early T-cell development is a question of considerable interest.[228] Nonetheless, the thymic microenvironment may not be as critical in T-ALL, as these cells can acquire autonomy from stromal ligands during the process of multistep leukemogenesis, as exemplified by the acquisition of activating NOTCH1 mutations. A significant proportion of T-ALL develops in the periphery with minimal thymic involvement, with or without activating *NOTCH1* mutations. Perhaps there is still plenty to learn from the molecular genetics of T-ALL.

Acknowledgments

The work was funded by grants from the Canadian Cancer Society Research Institute, the Canadian Institutes of Health Research and the Canada Research Chair program (Trang Hoang), and in part by an FRSQ group grant (IRIC infrastructure). C. S. Tremblay received a postdoctoral Research Fellowship Award of the Terry Fox Foundation (#700153). We wish to thank Dr Jean-Sébastien Delisle for his comments on the manuscript. We apologize to our colleagues for not referencing their work due to space limitation.

References

1. Bhandoola A, von Boehmer H, Petrie HT, Zuniga-Pflucker JC. Commitment and developmental potential of extrathymic and intrathymic T cell precursors: plenty to choose from. *Immunity* 2007;**26**:678–89.

2. Herblot S, Steff AM, Hugo P, Aplan PD, Hoang T. SCL and LMO1 alter thymocyte differentiation: inhibition of E2A-HEB function and pre-T alpha chain expression. *Nat Immunol* 2000;**1**:138–44.
3. Ferrando AA, Herblot S, Palomero T, Hansen M, Hoang T, Fox EA, et al. Biallelic transcriptional activation of oncogenic transcription factors in T-cell acute lymphoblastic leukemia. *Blood* 2004;**103**:1909–11.
4. Lacombe J, Herblot S, Rojas-Sutterlin S, Haman A, Barakat S, Iscove NN, et al. Scl regulates the quiescence and the long-term competence of hematopoietic stem cells. *Blood* 2009;**115**(4):792–803.
5. Taghon T, Thys K, De Smedt M, Weerkamp F, Staal FJ, Plum J, et al. Homeobox gene expression profile in human hematopoietic multipotent stem cells and T-cell progenitors: implications for human T-cell development. *Leukemia* 2003;**17**:1157–63.
6. Rothenberg EV, Moore JE, Yui MA. Launching the T-cell-lineage developmental programme. *Nat Rev Immunol* 2008;**8**:9–21.
7. Weng AP, Ferrando AA, Lee W, Morris JPt, Silverman LB, Sanchez-Irizarry C, et al. Activating mutations of NOTCH1 in human T cell acute lymphoblastic leukemia. *Science* 2004;**306**:269–71.
8. Mellentin JD, Smith SD, Cleary ML. lyl-1, a novel gene altered by chromosomal translocation in T cell leukemia, codes for a protein with a helix-loop-helix DNA binding motif. *Cell* 1989;**58**:77–83.
9. Soulier J, Clappier E, Cayuela JM, Regnault A, Garcia-Peydro M, Dombret H, et al. HOXA genes are included in genetic and biologic networks defining human acute T-cell leukemia (T-ALL). *Blood* 2005;**106**:274–86.
10. Clappier E, Cuccuini W, Kalota A, Crinquette A, Cayuela JM, Dik WA, et al. The C-MYB locus is involved in chromosomal translocation and genomic duplications in human T-cell acute leukemia (T-ALL), the translocation defining a new T-ALL subtype in very young children. *Blood* 2007;**110**:1251–61.
11. Palomero T, Sulis ML, Cortina M, Real PJ, Barnes K, Ciofani M, et al. Mutational loss of PTEN induces resistance to NOTCH1 inhibition in T-cell leukemia. *Nat Med* 2007;**13**:1203–10.
12. van Grotel M, Meijerink JP, Beverloo HB, Langerak AW, Buys-Gladdines JG, Schneider P, et al. The outcome of molecular-cytogenetic subgroups in pediatric T-cell acute lymphoblastic leukemia: a retrospective study of patients treated according to DCOG or COALL protocols. *Haematologica* 2006;**91**:1212–21.
13. Armstrong SA, Look AT. Molecular genetics of acute lymphoblastic leukemia. *J Clin Oncol* 2005;**23**:6306–15.
14. Mansur MB, Emerenciano M, Splendore A, Brewer L, Hassan R, Pombo-de-Oliveira MS. T-cell lymphoblastic leukemia in early childhood presents NOTCH1 mutations and MLL rearrangements. *Leuk Res* 2010;**34**(4):483–6.
15. Rabbitts TH. Translocations, master genes, and differences between the origins of acute and chronic leukemias. *Cell* 1991;**67**:641–4.
16. Ferrando AA, Neuberg DS, Staunton J, Loh ML, Huard C, Raimondi SC, et al. Gene expression signatures define novel oncogenic pathways in T cell acute lymphoblastic leukemia. *Cancer Cell* 2002;**1**:75–87.
17. Van Vlierberghe P, Pieters R, Beverloo HB, Meijerink JP. Molecular-genetic insights in paediatric T-cell acute lymphoblastic leukaemia. *Br J Haematol* 2008;**143**:153–68.
18. Hong D, Gupta R, Ancliff P, Atzberger A, Brown J, Soneji S, et al. Initiating and cancer-propagating cells in TEL-AML1-associated childhood leukemia. *Science* 2008;**319**:336–9.
19. Mullighan CG, Downing JR. Genome-wide profiling of genetic alterations in acute lymphoblastic leukemia: recent insights and future directions. *Leukemia* 2009;**23**:1209–18.

20. Hebert J, Cayuela JM, Berkeley J, Sigaux F. Candidate tumor-suppressor genes MTS1 (p16INK4A) and MTS2 (p15INK4B) display frequent homozygous deletions in primary cells from T- but not from B-cell lineage acute lymphoblastic leukemias. *Blood* 1994;**84**:4038–44.
21. Murre C. Intertwining proteins in thymocyte development and cancer. *Nat Immunol* 2000;**1**:97–8.
22. Look AT. Oncogenic transcription factors in the human acute leukemias. *Science* 1997;**278**:1059–64.
23. Begley CG, Aplan PD, Davey MP, Nakahara K, Tchorz K, Kurtzberg J, et al. Chromosomal translocation in a human leukemic stem-cell line disrupts the T-cell antigen receptor delta-chain diversity region and results in a previously unreported fusion transcript. *Proc Natl Acad Sci USA* 1989;**86**:2031–5.
24. Xia Y, Brown L, Yang CY, Tsan JT, Siciliano MJ, Espinosa 3rd R, et al. TAL2, a helix-loop-helix gene activated by the (7;9)(q34;q32) translocation in human T-cell leukemia. *Proc Natl Acad Sci USA* 1991;**88**:11416–20.
25. Green AR, Begley CG. SCL and related hemopoietic helix-loop-helix transcription factors. *Int J Cell Cloning* 1992;**10**:269–76.
26. Wang J, Jani-Sait SN, Escalon EA, Carroll AJ, de Jong PJ, Kirsch IR, et al. The t(14;21)(q11.2; q22) chromosomal translocation associated with T-cell acute lymphoblastic leukemia activates the BHLHB1 gene. *Proc Natl Acad Sci USA* 2000;**97**:3497–502.
27. Hatano M, Roberts CW, Minden M, Crist WM, Korsmeyer SJ. Deregulation of a homeobox gene, HOX11, by the t(10;14) in T cell leukemia. *Science* 1991;**253**:79–82.
28. Mauvieux L, Leymarie V, Helias C, Perrusson N, Falkenrodt A, Lioure B, et al. High incidence of Hox11L2 expression in children with T-ALL. *Leukemia* 2002;**16**:2417–22.
29. Hansen-Hagge TE, Schafer M, Kiyoi H, Morris SW, Whitlock JA, Koch P, et al. Disruption of the RanBP17/Hox11L2 region by recombination with the TCRdelta locus in acute lymphoblastic leukemias with t(5;14) (q34;q11). *Leukemia* 2002;**16**:2205–12.
30. Lecuyer E, Lariviere S, Sincennes MC, Haman A, Lahlil R, Todorova M, et al. Protein stability and transcription factor complex assembly determined by the SCL–LMO2 interaction. *J Biol Chem* 2007;**282**:33649–58.
31. Xu Z, Huang S, Chang LS, Agulnick AD, Brandt SJ. Identification of a TAL1 target gene reveals a positive role for the LIM domain-binding protein Ldb1 in erythroid gene expression and differentiation. *Mol Cell Biol* 2003;**23**:7585–99.
32. Boehm T, Foroni L, Kaneko Y, Perutz MF, Rabbitts TH. The rhombotin family of cysteine-rich LIM-domain oncogenes: distinct members are involved in T-cell translocations to human chromosomes 11p15 and 11p13. *Proc Natl Acad Sci USA* 1991;**88**:4367–71.
33. Larson RC, Lavenir I, Larson TA, Baer R, Warren AJ, Wadman I, et al. Protein dimerization between Lmo2 (Rbtn2) and Tal1 alters thymocyte development and potentiates T cell tumorigenesis in transgenic mice. *EMBO J* 1996;**15**:1021–7.
34. Aplan PD, Jones CA, Chervinsky DS, Zhao X, Ellsworth M, Wu C, et al. An scl gene product lacking the transactivation domain induces bony abnormalities and cooperates with LMO1 to generate T-cell malignancies in transgenic mice. *EMBO J* 1997;**16**:2408–19.
35. Wadman IA, Osada H, Grutz GG, Agulnick AD, Westphal H, Forster A, et al. The LIM-only protein Lmo2 is a bridging molecule assembling an erythroid, DNA-binding complex which includes the TAL1, E47, GATA-1 and Ldb1/NLI proteins. *EMBO J* 1997;**16**:3145–57.
36. Lecuyer E, Herblot S, Saint-Denis M, Martin R, Begley CG, Porcher C, et al. The SCL complex regulates c-kit expression in hematopoietic cells through functional interaction with Sp1. *Blood* 2002;**100**:2430–40.
37. Lahlil R, Lecuyer E, Herblot S, Hoang T. SCL assembles a multifactorial complex that determines glycophorin A expression. *Mol Cell Biol* 2004;**24**:1439–52.

38. Schlaeger TM, Schuh A, Flitter S, Fisher A, Mikkola H, Orkin SH, et al. Decoding hematopoietic specificity in the helix-loop-helix domain of the transcription factor SCL/Tal-1. *Mol Cell Biol* 2004;**24**:7491–502.
39. Van Vlierberghe P, Beverloo HB, Buijs-Gladdines J, van Wering ER, Horstmann M, Pieters R, et al. Monoallelic or biallelic LMO2 expression in relation to the LMO2 rearrangement status in pediatric T-cell acute lymphoblastic leukemia. *Leukemia* 2008;**22**:1131–7.
40. Bohlander SK, Muschinsky V, Schrader K, Siebert R, Schlegelberger B, Harder L, et al. Molecular analysis of the CALM/AF10 fusion: identical rearrangements in acute myeloid leukemia, acute lymphoblastic leukemia and malignant lymphoma patients. *Leukemia* 2000;**14**:93–9.
41. Rubnitz JE, Behm FG, Curcio-Brint AM, Pinheiro RP, Carroll AJ, Raimondi SC, et al. Molecular analysis of t(11;19) breakpoints in childhood acute leukemias. *Blood* 1996;**87**:4804–8.
42. Van Vlierberghe P, van Grotel M, Tchinda J, Lee C, Beverloo HB, van der Spek PJ, et al. The recurrent SET-NUP214 fusion as a new HOXA activation mechanism in pediatric T-cell acute lymphoblastic leukemia. *Blood* 2008;**111**:4668–80.
43. Okada Y, Feng Q, Lin Y, Jiang Q, Li Y, Coffield VM, et al. hDOT1L links histone methylation to leukemogenesis. *Cell* 2005;**121**:167–78.
44. Okada Y, Jiang Q, Lemieux M, Jeannotte L, Su L, Zhang Y. Leukaemic transformation by CALM-AF10 involves upregulation of Hoxa5 by hDOT1L. *Nat Cell Biol* 2006;**8**:1017–24.
45. Massari ME, Murre C. Helix-loop-helix proteins: regulators of transcription in eucaryotic organisms. *Mol Cell Biol* 2000;**20**:429–40.
46. Murre C. Helix-loop-helix proteins and lymphocyte development. *Nat Immunol* 2005;**6**:1079–86.
47. Sawada S, Suzuki G, Kitamura K, Takaku F. Irreversible suppression of CD8 expression in CD4-CD8+ thymocytes upon in vitro stimulation. *Immunol Invest* 1993;**22**:301–18.
48. Engel I, Murre C. E2A proteins enforce a proliferation checkpoint in developing thymocytes. *EMBO J* 2004;**23**:202–11.
49. Engel I, Johns C, Bain G, Rivera RR, Murre C. Early thymocyte development is regulated by modulation of E2A protein activity. *J Exp Med* 2001;**194**:733–45.
50. Rivera RR, Johns CP, Quan J, Johnson RS, Murre C. Thymocyte selection is regulated by the helix-loop-helix inhibitor protein, Id3. *Immunity* 2000;**12**:17–26.
51. Hsu HL, Wadman I, Tsan JT, Baer R. Positive and negative transcriptional control by the TAL1 helix-loop-helix protein. *Proc Natl Acad Sci USA* 1994;**91**:5947–51.
52. Chervinsky DS, Zhao XF, Lam DH, Ellsworth M, Gross KW, Aplan PD. Disordered T-cell development and T-cell malignancies in SCL LMO1 double-transgenic mice: parallels with E2A-deficient mice. *Mol Cell Biol* 1999;**19**:5025–35.
53. Tremblay M, Herblot S, Lecuyer E, Hoang T. Regulation of pT alpha gene expression by a dosage of E2A, HEB, and SCL. *J Biol Chem* 2003;**278**:12680–7.
54. Cauwelier B, Speleman F. *HOXA11 (homeobox A11)*. In: *Atlas Genet Cytogenet Oncol Haematol*. 2006, URL: http://AtlasGeneticsOncology.org/Genes/HOXA11ID40847ch7p15.html.
55. Dik WA, Brahim W, Braun C, Asnafi V, Dastugue N, Bernard OA, et al. CALM-AF10+ T-ALL expression profiles are characterized by overexpression of HOXA and BMI1 oncogenes. *Leukemia* 2005;**19**:1948–57.
56. Graux C, Cools J, Melotte C, Quentmeier H, Ferrando A, Levine R, et al. Fusion of NUP214 to ABL1 on amplified episomes in T-cell acute lymphoblastic leukemia. *Nat Genet* 2004;**36**:1084–9.
57. Argiropoulos B, Humphries RK. Hox genes in hematopoiesis and leukemogenesis. *Oncogene* 2007;**26**:6766–76.

58. Bijl J, Sauvageau M, Thompson A, Sauvageau G. High incidence of proviral integrations in the Hoxa locus in a new model of E2a-PBX1-induced B-cell leukemia. *Genes Dev* 2005;**19**:224–33.
59. Bijl J, Krosl J, Lebert-Ghali CE, Vacher J, Mayotte N, Sauvageau G. Evidence for Hox and E2A-PBX1 collaboration in mouse T-cell leukemia. *Oncogene* 2008;**27**:6356–64.
60. Izon DJ, Rozenfeld S, Fong ST, Komuves L, Largman C, Lawrence HJ. Loss of function of the homeobox gene Hoxa-9 perturbs early T-cell development and induces apoptosis in primitive thymocytes. *Blood* 1998;**92**:383–93.
61. Bult CJ, Eppig JT, Kadin JA, Richardson JE, Blake JA. The Mouse Genome Database (MGD): mouse biology and model systems. *Nucleic Acids Res* 2008;**36**:D724–8.
62. Hawley RG, Fong AZ, Reis MD, Zhang N, Lu M, Hawley TS. Transforming function of the HOX11/TCL3 homeobox gene. *Cancer Res* 1997;**57**:337–45.
63. Graux C, Cools J, Michaux L, Vandenberghe P, Hagemeijer A. Cytogenetics and molecular genetics of T-cell acute lymphoblastic leukemia: from thymocyte to lymphoblast. *Leukemia* 2006;**20**:1496–510.
64. Allen 3rd RD, Bender TP, Siu G. c-Myb is essential for early T cell development. *Genes Dev* 1999;**13**:1073–8.
65. Pearson R, Weston K. c-Myb regulates the proliferation of immature thymocytes following beta-selection. *EMBO J* 2000;**19**:6112–20.
66. Bender TP, Kremer CS, Kraus M, Buch T, Rajewsky K. Critical functions for c-Myb at three checkpoints during thymocyte development. *Nat Immunol* 2004;**5**:721–9.
67. Lahortiga I, De Keersmaecker K, Van Vlierberghe P, Graux C, Cauwelier B, Lambert F, et al. Duplication of the MYB oncogene in T cell acute lymphoblastic leukemia. *Nat Genet* 2007;**39**:593–5.
68. Ramsay RG, Gonda TJ. MYB function in normal and cancer cells. *Nat Rev Cancer* 2008;**8**:523–34.
69. Hess JL, Bittner CB, Zeisig DT, Bach C, Fuchs U, Borkhardt A, et al. c-Myb is an essential downstream target for homeobox-mediated transformation of hematopoietic cells. *Blood* 2006;**108**:297–304.
70. Matthews AG, Oettinger MA. RAG: a recombinase diversified. *Nat Immunol* 2009;**10**:817–21.
71. Aplan PD, Lombardi DP, Ginsberg AM, Cossman J, Bertness VL, Kirsch IR. Disruption of the human SCL locus by "illegitimate" V-(D)-J recombinase activity. *Science* 1990;**250**:1426–9.
72. Lai EC. Notch signaling: control of cell communication and cell fate. *Development* 2004;**131**:965–73.
73. Ellisen LW, Bird J, West DC, Soreng AL, Reynolds TC, Smith SD, et al. TAN-1, the human homolog of the *Drosophila* notch gene, is broken by chromosomal translocations in T lymphoblastic neoplasms. *Cell* 1991;**66**:649–61.
74. O'Neil J, Look AT. Mechanisms of transcription factor deregulation in lymphoid cell transformation. *Oncogene* 2007;**26**:6838–49.
75. Cauwelier B, Cave H, Gervais C, Lessard M, Barin C, Perot C, et al. Clinical, cytogenetic and molecular characteristics of 14 T-ALL patients carrying the TCRbeta-HOXA rearrangement: a study of the Groupe Francophone de Cytogenetique Hematologique. *Leukemia* 2007;**21**:121–8.
76. Mansur MB, Emerenciano M, Brewer L, Sant'Ana M, Mendonca N, Thuler LC, et al. SIL-TAL1 fusion gene negative impact in T-cell acute lymphoblastic leukemia outcome. *Leuk Lymphoma* 2009;**50**:1318–25.
77. Zhu YM, Zhao WL, Fu JF, Shi JY, Pan Q, Hu J, et al. NOTCH1 mutations in T-cell acute lymphoblastic leukemia: prognostic significance and implication in multifactorial leukemogenesis. *Clin Cancer Res* 2006;**12**:3043–9.

78. Lin YW, Nichols RA, Letterio JJ, Aplan PD. Notch1 mutations are important for leukemic transformation in murine models of precursor-T leukemia/lymphoma. *Blood* 2006;**107**:2540–3.
79. O'Neil J, Calvo J, McKenna K, Krishnamoorthy V, Aster JC, Bassing CH, et al. Activating Notch1 mutations in mouse models of T-ALL. *Blood* 2006;**107**:781–5.
80. Schmitt TM, Zuniga-Pflucker JC. Induction of T cell development from hematopoietic progenitor cells by delta-like-1 in vitro. *Immunity* 2002;**17**:749–56.
81. Schmitt TM, Ciofani M, Petrie HT, Zuniga-Pflucker JC. Maintenance of T cell specification and differentiation requires recurrent notch receptor–ligand interactions. *J Exp Med* 2004;**200**:469–79.
82. Tan JB, Visan I, Yuan JS, Guidos CJ. Requirement for Notch1 signals at sequential early stages of intrathymic T cell development. *Nat Immunol* 2005;**6**:671–9.
83. Malyukova A, Dohda T, von der Lehr N, Akhoondi S, Corcoran M, Heyman M, et al. The tumor suppressor gene hCDC4 is frequently mutated in human T-cell acute lymphoblastic leukemia with functional consequences for Notch signaling. *Cancer Res* 2007;**67**:5611–6.
84. O'Neil J, Grim J, Strack P, Rao S, Tibbitts D, Winter C, et al. FBW7 mutations in leukemic cells mediate NOTCH pathway activation and resistance to gamma-secretase inhibitors. *J Exp Med* 2007;**204**:1813–24.
85. Thompson BJ, Buonamici S, Sulis ML, Palomero T, Vilimas T, Basso G, et al. The SCFFBW7 ubiquitin ligase complex as a tumor suppressor in T cell leukemia. *J Exp Med* 2007;**204**:1825–35.
86. Artavanis-Tsakonas S, Rand MD, Lake RJ. Notch signaling: cell fate control and signal integration in development. *Science* 1999;**284**:770–6.
87. Aster JC, Pear WS, Blacklow SC. Notch signaling in leukemia. *Annu Rev Pathol* 2008;**3**:587–613.
88. Grabher C, von Boehmer H, Look AT. Notch1 activation in the molecular pathogenesis of T-cell acute lymphoblastic leukaemia. *Nat Rev Cancer* 2006;**6**:347–59.
89. Palomero T, Lim WK, Odom DT, Sulis ML, Real PJ, Margolin A, et al. NOTCH1 directly regulates c-MYC and activates a feed-forward-loop transcriptional network promoting leukemic cell growth. *Proc Natl Acad Sci USA* 2006;**103**:18261–6.
90. Sharma VM, Calvo JA, Draheim KM, Cunningham LA, Hermance N, Beverly L, et al. Notch1 contributes to mouse T-cell leukemia by directly inducing the expression of c-myc. *Mol Cell Biol* 2006;**26**:8022–31.
91. Tomita K, Hattori M, Nakamura E, Nakanishi S, Minato N, Kageyama R. The bHLH gene Hes1 is essential for expansion of early T cell precursors. *Genes Dev* 1999;**13**:1203–10.
92. Weng AP, Millholland JM, Yashiro-Ohtani Y, Arcangeli ML, Lau A, Wai C, et al. c-Myc is an important direct target of Notch1 in T-cell acute lymphoblastic leukemia/lymphoma. *Genes Dev* 2006;**20**:2096–109.
93. Vilimas T, Mascarenhas J, Palomero T, Mandal M, Buonamici S, Meng F, et al. Targeting the NF-kappaB signaling pathway in Notch1-induced T-cell leukemia. *Nat Med* 2007;**13**:70–7.
94. Chan SM, Weng AP, Tibshirani R, Aster JC, Utz PJ. Notch signals positively regulate activity of the mTOR pathway in T-cell acute lymphoblastic leukemia. *Blood* 2007;**110**:278–86.
95. Ciofani M, Zuniga-Pflucker JC. Notch promotes survival of pre-T cells at the beta-selection checkpoint by regulating cellular metabolism. *Nat Immunol* 2005;**6**:881–8.
96. Pear WS, Aster JC, Scott ML, Hasserjian RP, Soffer B, Sklar J, et al. Exclusive development of T cell neoplasms in mice transplanted with bone marrow expressing activated Notch alleles. *J Exp Med* 1996;**183**:2283–91.
97. Pear WS, Aster JC. T cell acute lymphoblastic leukemia/lymphoma: a human cancer commonly associated with aberrant NOTCH1 signaling. *Curr Opin Hematol* 2004;**11**:426–33.

98. Radtke F, Wilson A, Stark G, Bauer M, van Meerwijk J, MacDonald HR, et al. Deficient T cell fate specification in mice with an induced inactivation of Notch1. *Immunity* 1999;**10**:547–58.
99. Reizis B, Leder P. Direct induction of T lymphocyte-specific gene expression by the mammalian Notch signaling pathway. *Genes Dev* 2002;**16**:295–300.
100. Tanigaki K, Tsuji M, Yamamoto N, Han H, Tsukada J, Inoue H, et al. Regulation of alphabeta/gammadelta T cell lineage commitment and peripheral T cell responses by Notch/RBP-J signaling. *Immunity* 2004;**20**:611–22.
101. Wilson A, MacDonald HR, Radtke F. Notch 1-deficient common lymphoid precursors adopt a B cell fate in the thymus. *J Exp Med* 2001;**194**:1003–12.
102. Wolfer A, Wilson A, Nemir M, MacDonald HR, Radtke F. Inactivation of Notch1 impairs VDJbeta rearrangement and allows pre-TCR-independent survival of early alpha beta Lineage Thymocytes. *Immunity* 2002;**16**:869–79.
103. Feyerabend TB, Terszowski G, Tietz A, Blum C, Luche H, Gossler A, et al. Deletion of Notch1 converts pro-T cells to dendritic cells and promotes thymic B cells by cell-extrinsic and cell-intrinsic mechanisms. *Immunity* 2009;**30**:67–79.
104. Han H, Tanigaki K, Yamamoto N, Kuroda K, Yoshimoto M, Nakahata T, et al. Inducible gene knockout of transcription factor recombination signal binding protein-J reveals its essential role in T versus B lineage decision. *Int Immunol* 2002;**14**:637–45.
105. Pui JC, Allman D, Xu L, DeRocco S, Karnell FG, Bakkour S, et al. Notch1 expression in early lymphopoiesis influences B versus T lineage determination. *Immunity* 1999;**11**:299–308.
106. Lehar SM, Dooley J, Farr AG, Bevan MJ. Notch ligands Delta 1 and Jagged1 transmit distinct signals to T-cell precursors. *Blood* 2005;**105**:1440–7.
107. Sambandam A, Maillard I, Zediak VP, Xu L, Gerstein RM, Aster JC, et al. Notch signaling controls the generation and differentiation of early T lineage progenitors. *Nat Immunol* 2005;**6**:663–70.
108. Taghon T, Yui MA, Pant R, Diamond RA, Rothenberg EV. Developmental and molecular characterization of emerging beta- and gammadelta-selected pre-T cells in the adult mouse thymus. *Immunity* 2006;**24**:53–64.
109. Gonzalez-Garcia S, Garcia-Peydro M, Martin-Gayo E, Ballestar E, Esteller M, Bornstein R, et al. CSL-MAML-dependent Notch1 signaling controls T lineage-specific IL-7R{alpha} gene expression in early human thymopoiesis and leukemia. *J Exp Med* 2009;**206**:779–91.
110. Ikawa T, Kawamoto H, Goldrath AW, Murre C. E proteins and Notch signaling cooperate to promote T cell lineage specification and commitment. *J Exp Med* 2006;**203**:1329–42.
111. Washburn T, Schweighoffer E, Gridley T, Chang D, Fowlkes BJ, Cado D, et al. Notch activity influences the alphabeta versus gammadelta T cell lineage decision. *Cell* 1997;**88**:833–43.
112. Fowlkes BJ, Robey EA. A reassessment of the effect of activated Notch1 on CD4 and CD8 T cell development. *J Immunol* 2002;**169**:1817–21.
113. Robey E, Chang D, Itano A, Cado D, Alexander H, Lans D, et al. An activated form of Notch influences the choice between CD4 and CD8 T cell lineages. *Cell* 1996;**87**:483–92.
114. Tremblay M, Tremblay CS, Herblot S, Aplan PD, Hébert J, Perreault C, et al. Modeling T-cell acute lymphoblastic leukemia induced by the *SCL* and *LMO1* oncogenes. *Genes Dev.* 2010;**24**:1093–105.
115. Kruisbeek AM, Haks MC, Carleton M, Michie AM, Zuniga-Pflucker JC, Wiest DL. Branching out to gain control: how the pre-TCR is linked to multiple functions. *Immunol Today* 2000;**21**:637–44.
116. Engel I, Murre C. Disruption of pre-TCR expression accelerates lymphomagenesis in E2A-deficient mice. *Proc Natl Acad Sci USA* 2002;**99**:11322–7.
117. Nie L, Xu M, Vladimirova A, Sun XH. Notch-induced E2A ubiquitination and degradation are controlled by MAP kinase activities. *EMBO J* 2003;**22**:5780–92.

118. Talora C, Campese AF, Bellavia D, Pascucci M, Checquolo S, Groppioni M, et al. Pre-TCR-triggered ERK signalling-dependent downregulation of E2A activity in Notch3-induced T-cell lymphoma. *EMBO Rep* 2003;**4**:1067–72.
119. Yashiro-Ohtani Y, He Y, Ohtani T, Jones ME, Shestova O, Xu L, et al. Pre-TCR signaling inactivates Notch1 transcription by antagonizing E2A. *Genes Dev* 2009;**23**:1665–76.
120. Yamasaki S, Saito T. Molecular basis for pre-TCR-mediated autonomous signaling. *Trends Immunol* 2007;**28**:39–43.
121. Voll RE, Jimi E, Phillips RJ, Barber DF, Rincon M, Hayday AC, et al. NF-kappa B activation by the pre-T cell receptor serves as a selective survival signal in T lymphocyte development. *Immunity* 2000;**13**:677–89.
122. Mandal M, Borowski C, Palomero T, Ferrando AA, Oberdoerffer P, Meng F, et al. The BCL2A1 gene as a pre-T cell receptor-induced regulator of thymocyte survival. *J Exp Med* 2005;**201**:603–14.
123. Mandal M, Crusio KM, Meng F, Liu S, Kinsella M, Clark MR, et al. Regulation of lymphocyte progenitor survival by the proapoptotic activities of Bim and Bid. *Proc Natl Acad Sci USA* 2008;**105**:20840–5.
124. Miyazaki M, Miyazaki K, Itoi M, Katoh Y, Guo Y, Kanno R, et al. Thymocyte proliferation induced by pre-T cell receptor signaling is maintained through polycomb gene product Bmi-1-mediated Cdkn2a repression. *Immunity* 2008;**28**:231–45.
125. Xi H, Schwartz R, Engel I, Murre C, Kersh GJ. Interplay between RORgammat, Egr3, and E proteins controls proliferation in response to pre-TCR signals. *Immunity* 2006;**24**:813–26.
126. Malissen M, Gillet A, Ardouin L, Bouvier G, Trucy J, Ferrier P, et al. Altered T cell development in mice with a targeted mutation of the CD3-epsilon gene. *EMBO J* 1995;**14**:4641–53.
127. Palacios EH, Weiss A. Function of the Src-family kinases, Lck and Fyn, in T-cell development and activation. *Oncogene* 2004;**23**:7990–8000.
128. Tycko B, Smith SD, Sklar J. Chromosomal translocations joining LCK and TCRB loci in human T cell leukemia. *J Exp Med* 1991;**174**:867–73.
129. Ciofani M, Schmitt TM, Ciofani A, Michie AM, Cuburu N, Aublin A, et al. Obligatory role for cooperative signaling by pre-TCR and Notch during thymocyte differentiation. *J Immunol* 2004;**172**:5230–9.
130. Latour S, Veillette A. Proximal protein tyrosine kinases in immunoreceptor signaling. *Curr Opin Immunol* 2001;**13**:299–306.
131. Zhang W, Sloan-Lancaster J, Kitchen J, Trible RP, Samelson LE. LAT: the ZAP-70 tyrosine kinase substrate that links T cell receptor to cellular activation. *Cell* 1998;**92**:83–92.
132. Wange RL. LAT, the linker for activation of T cells: a bridge between T cell-specific and general signaling pathways. *Sci STKE* 2000;re1.
133. Wang CR, Hashimoto K, Kubo S, Yokochi T, Kubo M, Suzuki M, et al. T cell receptor-mediated signaling events in CD4+CD8+ thymocytes undergoing thymic selection: requirement of calcineurin activation for thymic positive selection but not negative selection. *J Exp Med* 1995;**181**:927–41.
134. Teixeira C, Stang SL, Zheng Y, Beswick NS, Stone JC. Integration of DAG signaling systems mediated by PKC-dependent phosphorylation of RasGRP3. *Blood* 2003;**102**:1414–20.
135. Carrera AC, Rodriguez-Borlado L, Martinez-Alonso C, Merida I. T cell receptor-associated alpha-phosphatidylinositol 3-kinase becomes activated by T cell receptor cross-linking and requires pp 56lck. *J Biol Chem* 1994;**269**:19435–40.
136. Wang X, Gjorloff-Wingren A, Saxena M, Pathan N, Reed JC, Mustelin T. The tumor suppressor PTEN regulates T cell survival and antigen receptor signaling by acting as a phosphatidylinositol 3-phosphatase. *J Immunol* 2000;**164**:1934–9.
137. O'Neil J, Shank J, Cusson N, Murre C, Kelliher M. TAL1/SCL induces leukemia by inhibiting the transcriptional activity of E47/HEB. *Cancer Cell* 2004;**5**:587–96.

138. Germain RN, Stefanova I. The dynamics of T cell receptor signaling: complex orchestration and the key roles of tempo and cooperation. *Annu Rev Immunol* 1999;**17**:467–522.
139. Samelson LE, Donovan JA, Isakov N, Ota Y, Wange RL. Signal transduction mediated by the T-cell antigen receptor. *Ann NY Acad Sci* 1995;**766**:157–72.
140. Stefanova I, Dorfman JR, Tsukamoto M, Germain RN. On the role of self-recognition in T cell responses to foreign antigen. *Immunol Rev* 2003;**191**:97–106.
141. Kawamura M, Ohnishi H, Guo SX, Sheng XM, Minegishi M, Hanada R, et al. Alterations of the p53, p21, p16, p15 and RAS genes in childhood T-cell acute lymphoblastic leukemia. *Leuk Res* 1999;**23**:115–26.
142. Yokota S, Nakao M, Horiike S, Seriu T, Iwai T, Kaneko H, et al. Mutational analysis of the N-ras gene in acute lymphoblastic leukemia: a study of 125 Japanese pediatric cases. *Int J Hematol* 1998;**67**:379–87.
143. Mora AL, Stanley S, Armistead W, Chan AC, Boothby M. Inefficient ZAP-70 phosphorylation and decreased thymic selection in vivo result from inhibition of NF-kappaB/Rel. *J Immunol* 2001;**167**:5628–35.
144. Medyouf H, Ghysdael J. The calcineurin/NFAT signaling pathway: a novel therapeutic target in leukemia and solid tumors. *Cell Cycle* 2008;**7**:297–303.
145. Douet-Guilbert N, Morel F, Le Bris MJ, Herry A, Le Calvez G, Marion V, et al. Cytogenetic studies in T-cell acute lymphoblastic leukemia (1981–2002). *Leuk Lymphoma* 2004;**45**:287–90.
146. Bellavia D, Campese AF, Checquolo S, Balestri A, Biondi A, Cazzaniga G, et al. Combined expression of pTalpha and Notch3 in T cell leukemia identifies the requirement of preTCR for leukemogenesis. *Proc Natl Acad Sci USA* 2002;**99**:3788–93.
147. Campese AF, Garbe AI, Zhang F, Grassi F, Screpanti I, von Boehmer H. Notch1-dependent lymphomagenesis is assisted by but does not essentially require pre-TCR signaling. *Blood* 2006;**108**:305–10.
148. Winandy S, Wu P, Georgopoulos K. A dominant mutation in the Ikaros gene leads to rapid development of leukemia and lymphoma. *Cell* 1995;**83**:289–99.
149. dos Santos NR, Rickman DS, de Reynies A, Cormier F, Williame M, Blanchard C, et al. Pre-TCR expression cooperates with TEL-JAK2 to transform immature thymocytes and induce T-cell leukemia. *Blood* 2007;**109**:3972–81.
150. Fasseu M, Aplan PD, Chopin M, Boissel N, Bories JC, Soulier J, et al. p16INK4A tumor suppressor gene expression and CD3epsilon deficiency but not pre-TCR deficiency inhibit TAL1-linked T-lineage leukemogenesis. *Blood* 2007;**110**:2610–9.
151. Chervinsky DS, Lam DH, Melman MP, Gross KW, Aplan PD. scid Thymocytes with TCRbeta gene rearrangements are targets for the oncogenic effect of SCL and LMO1 transgenes. *Cancer Res* 2001;**61**:6382–7.
152. Liao MJ, Zhang XX, Hill R, Gao J, Qumsiyeh MB, Nichols W, et al. No requirement for V(D)J recombination in p53-deficient thymic lymphoma. *Mol Cell Biol* 1998;**18**:3495–501.
153. Breiman L. Software for the masses. In: Lecture W, editor. *277th meeting of the Institute of Mathematical Statistics. Banff, Alberta, Canda*; 2002.
154. Hoang T, Parsons VL. Bagging survival trees for prognosis based on gene profiles. In: Antoch J, editor. *Compstat*. Berlin: Physica Verlag; 2004. p. 1201–8.
155. Ishwaran H, Kogalur U. Random survival forests. *Rnews* 2007;**7**:25–31.
156. Hastie T, Friedman J, Tibshirani R. *Elements of Statistical Learning: Data Mining, Inference and Prediction.* Berlin, Germany: Springer-Verlag; 2001.
157. Breiman L. Random forests. In: Schapire RE, editor. *Machine Learning*. The Netherlands: Kluwer Academic Publishers; 2001. p. 5–32.
158. Beesley AH, Cummings AJ, Freitas JR, Hoffmann K, Firth MJ, Ford J, et al. The gene expression signature of relapse in paediatric acute lymphoblastic leukaemia: implications for mechanisms of therapy failure. *Br J Haematol* 2005;**131**:447–56.

159. Hoffmann K, Firth MJ, Beesley AH, de Klerk NH, Kees UR. Translating microarray data for diagnostic testing in childhood leukaemia. *BMC Cancer* 2006;**6**:229.
160. Gottardo NG, Hoffmann K, Beesley AH, Freitas JR, Firth MJ, Perera KU, et al. Identification of novel molecular prognostic markers for paediatric T-cell acute lymphoblastic leukaemia. *Br J Haematol* 2007;**137**:319–28.
161. Goeman JJ, van de Geer SA, de Kort F, van Houwelingen HC. A global test for groups of genes: testing association with a clinical outcome. *Bioinformatics* 2004;**20**:93–9.
162. Reinherz EL, Chang HC, Clayton LK, Gardner P, Howard FD, Koyasu S, et al. The biology of human CD2. *Cold Spring Harb Symp Quant Biol* 1989;**54**(Pt 2):611–25.
163. Tibaldi EV, Salgia R, Reinherz EL. CD2 molecules redistribute to the uropod during T cell scanning: implications for cellular activation and immune surveillance. *Proc Natl Acad Sci USA* 2002;**99**:7582–7.
164. Aifantis I, Raetz E, Buonamici S. Molecular pathogenesis of T-cell leukaemia and lymphoma. *Nat Rev Immunol* 2008;**8**:380–90.
165. Vivanco I, Sawyers CL. The phosphatidylinositol 3-Kinase AKT pathway in human cancer. *Nat Rev Cancer* 2002;**2**:489–501.
166. Brown VI, Seif AE, Reid GS, Teachey DT, Grupp SA. Novel molecular and cellular therapeutic targets in acute lymphoblastic leukemia and lymphoproliferative disease. *Immunol Res* 2008;**42**:84–105.
167. McCubrey JA, Steelman LS, Abrams SL, Bertrand FE, Ludwig DE, Basecke J, et al. Targeting survival cascades induced by activation of Ras/Raf/MEK/ERK, PI3K/PTEN/Akt/mTOR and Jak/STAT pathways for effective leukemia therapy. *Leukemia* 2008;**22**:708–22.
168. Steelman LS, Navolanic PM, Sokolosky ML, Taylor JR, Lehmann BD, Chappell WH, et al. Suppression of PTEN function increases breast cancer chemotherapeutic drug resistance while conferring sensitivity to mTOR inhibitors. *Oncogene* 2008;**27**:4086–95.
169. Juntilla MM, Koretzky GA. Critical roles of the PI3K/Akt signaling pathway in T cell development. *Immunol Lett* 2008;**116**:104–10.
170. Bussaglia E, del Rio E, Matias-Guiu X, Prat J. PTEN mutations in endometrial carcinomas: a molecular and clinicopathologic analysis of 38 cases. *Hum Pathol* 2000;**31**:312–7.
171. Celebi JT, Wanner M, Ping XL, Zhang H, Peacocke M. Association of splicing defects in PTEN leading to exon skipping or partial intron retention in Cowden syndrome. *Hum Genet* 2000;**107**:234–8.
172. Silva A, Yunes JA, Cardoso BA, Martins LR, Jotta PY, Abecasis M, et al. PTEN posttranslational inactivation and hyperactivation of the PI3K/Akt pathway sustain primary T cell leukemia viability. *J Clin Invest* 2008;**118**:3762–74.
173. Wang SI, Parsons R, Ittmann M. Homozygous deletion of the PTEN tumor suppressor gene in a subset of prostate adenocarcinomas. *Clin Cancer Res* 1998;**4**:811–5.
174. Palomero T, Dominguez M, Ferrando AA. The role of the PTEN/AKT Pathway in NOTCH1-induced leukemia. *Cell Cycle* 2008;**7**:965–70.
175. Gutierrez A, Sanda T, Grebliunaite R, Carracedo A, Salmena L, Ahn Y, et al. High frequency of PTEN, PI3K, and AKT abnormalities in T-cell acute lymphoblastic leukemia. *Blood* 2009;**114**:647–50.
176. Remke M, Pfister S, Kox C, Toedt G, Becker N, Benner A, et al. High-resolution genomic profiling of childhood T-ALL reveals frequent copy-number alterations affecting the TGF-beta and PI3K–AKT pathways and deletions at 6q15-16.1 as a genomic marker for unfavorable early treatment response. *Blood* 2009;**114**:1053–62.
177. Zhang J, Grindley JC, Yin T, Jayasinghe S, He XC, Ross JT, et al. PTEN maintains haematopoietic stem cells and acts in lineage choice and leukaemia prevention. *Nature* 2006;**441**:518–22.

178. Yilmaz OH, Valdez R, Theisen BK, Guo W, Ferguson DO, Wu H, et al. Pten dependence distinguishes haematopoietic stem cells from leukaemia-initiating cells. *Nature* 2006;**441**:475–82.
179. Medyouf H, Gao X, Armstrong F, Gusscott S, Liu Q, Larson Gedman A, et al. Acute T-cell leukemias remain dependent on notch signaling despite PTEN and INK4A/ARF loss. *Blood* 2009.
180. Silva A, Jotta PY, Silveira AB, Ribeiro D, Brandalise SR, Yunes JA, et al. Regulation of PTEN by CK2 and Notch1 in primary T-cell acute lymphoblastic leukemia: rationale for combined use of CK2- and gamma-secretase inhibitors. *Haematologica* 2010;**95**:674–8.
181. Larson Gedman A, Chen Q, Kugel Desmoulin S, Ge Y, LaFiura K, Haska CL, et al. The impact of NOTCH1, FBW7 and PTEN mutations on prognosis and downstream signaling in pediatric T-cell acute lymphoblastic leukemia: a report from the Children's Oncology Group. *Leukemia* 2009;**23**:1417–25.
182. Song LL, Peng Y, Yun J, Rizzo P, Chaturvedi V, Weijzen S, et al. Notch-1 associates with IKKalpha and regulates IKK activity in cervical cancer cells. *Oncogene* 2008;**27**:5833–44.
183. Mammucari C, Tommasi di Vignano A, Sharov AA, Neilson J, Havrda MC, Roop DR, et al. Integration of Notch 1 and calcineurin/NFAT signaling pathways in keratinocyte growth and differentiation control. *Dev Cell* 2005;**8**:665–76.
184. Kamakura S, Oishi K, Yoshimatsu T, Nakafuku M, Masuyama N, Gotoh Y. Hes binding to STAT3 mediates crosstalk between Notch and JAK-STAT signalling. *Nat Cell Biol* 2004;**6**:547–54.
185. Zhao WL. Targeted therapy in T-cell malignancies: dysregulation of the cellular signaling pathways. *Leukemia* 2010;**24**:13–21.
186. Constantinescu SN, Girardot M, Pecquet C. Mining for JAK-STAT mutations in cancer. *Trends Biochem Sci* 2008;**33**:122–31.
187. Lacronique V, Boureux A, Valle VD, Poirel H, Quang CT, Mauchauffe M, et al. A TEL-JAK2 fusion protein with constitutive kinase activity in human leukemia. *Science* 1997;**278**:1309–12.
188. Schwaller J, Frantsve J, Aster J, Williams IR, Tomasson MH, Ross TS, et al. Transformation of hematopoietic cell lines to growth-factor independence and induction of a fatal myelo- and lymphoproliferative disease in mice by retrovirally transduced TEL/JAK2 fusion genes. *EMBO J* 1998;**17**:5321–33.
189. Pircher TJ, Petersen H, Gustafsson JA, Haldosen LA. Extracellular signal-regulated kinase (ERK) interacts with signal transducer and activator of transcription (STAT) 5a. *Mol Endocrinol* 1999;**13**:555–65.
190. McNeil LK, Starr TK, Hogquist KA. A requirement for sustained ERK signaling during thymocyte positive selection in vivo. *Proc Natl Acad Sci USA* 2005;**102**:13574–9.
191. Sugawara T, Moriguchi T, Nishida E, Takahama Y. Differential roles of ERK and p38 MAP kinase pathways in positive and negative selection of T lymphocytes. *Immunity* 1998;**9**:565–74.
192. Liu Q, Wilkins BJ, Lee YJ, Ichijo H, Molkentin JD. Direct interaction and reciprocal regulation between ASK1 and calcineurin-NFAT control cardiomyocyte death and growth. *Mol Cell Biol* 2006;**26**:3785–97.
193. Medyouf H, Alcalde H, Berthier C, Guillemin MC, dos Santos NR, Janin A, et al. Targeting calcineurin activation as a therapeutic strategy for T-cell acute lymphoblastic leukemia. *Nat Med* 2007;**13**:736–41.
194. Nurse P. A long twentieth century of the cell cycle and beyond. *Cell* 2000;**100**:71–8.
195. Cayuela JM, Madani A, Sanhes L, Stern MH, Sigaux F. Multiple tumor-suppressor gene 1 inactivation is the most frequent genetic alteration in T-cell acute lymphoblastic leukemia. *Blood* 1996;**87**:2180–6.

196. Herman JG, Jen J, Merlo A, Baylin SB. Hypermethylation-associated inactivation indicates a tumor suppressor role for p15INK4B. *Cancer Res* 1996;**56**:722–7.
197. Merlo A, Herman JG, Mao L, Lee DJ, Gabrielson E, Burger PC, et al. 5′ CpG island methylation is associated with transcriptional silencing of the tumour suppressor p16/CDKN2/MTS1 in human cancers. *Nat Med* 1995;**1**:686–92.
198. Okamoto A, Demetrick DJ, Spillare EA, Hagiwara K, Hussain SP, Bennett WP, et al. p16INK4 mutations and altered expression in human tumors and cell lines. *Cold Spring Harb Symp Quant Biol* 1994;**59**:49–57.
199. Begley CG, Aplan PD, Denning SM, Haynes BF, Waldmann TA, Kirsch IR. The gene SCL is expressed during early hematopoiesis and encodes a differentiation-related DNA-binding motif. *Proc Natl Acad Sci USA* 1989;**86**:10128–32.
200. Morrow MA, Mayer EW, Perez CA, Adlam M, Siu G. Overexpression of the Helix-Loop-Helix protein Id2 blocks T cell development at multiple stages. *Mol Immunol* 1999;**36**:491–503.
201. Wojciechowski J, Lai A, Kondo M, Zhuang Y. E2A and HEB are required to block thymocyte proliferation prior to pre-TCR expression. *J Immunol* 2007;**178**:5717–26.
202. Bain G, Engel I, Robanus Maandag EC, te Riele HP, Voland JR, Sharp LL, et al. E2A deficiency leads to abnormalities in alphabeta T-cell development and to rapid development of T-cell lymphomas. *Mol Cell Biol* 1997;**17**:4782–91.
203. Peverali FA, Ramqvist T, Saffrich R, Pepperkok R, Barone MV, Philipson L. Regulation of G1 progression by E2A and Id helix-loop-helix proteins. *EMBO J* 1994;**13**:4291–301.
204. Herblot S, Aplan PD, Hoang T. Gradient of E2A activity in B-cell development. *Mol Cell Biol* 2002;**22**:886–900.
205. Prabhu S, Ignatova A, Park ST, Sun XH. Regulation of the expression of cyclin-dependent kinase inhibitor p21 by E2A and Id proteins. *Mol Cell Biol* 1997;**17**:5888–96.
206. Schwartz R, Engel I, Fallahi-Sichani M, Petrie HT, Murre C. Gene expression patterns define novel roles for E47 in cell cycle progression, cytokine-mediated signaling, and T lineage development. *Proc Natl Acad Sci USA* 2006;**103**:9976–81.
207. Georgopoulos K, Winandy S, Avitahl N. The role of the Ikaros gene in lymphocyte development and homeostasis. *Annu Rev Immunol* 1997;**15**:155–76.
208. Georgopoulos K, Bigby M, Wang JH, Molnar A, Wu P, Winandy S, et al. The Ikaros gene is required for the development of all lymphoid lineages. *Cell* 1994;**79**:143–56.
209. Chari S, Winandy S. Ikaros regulates Notch target gene expression in developing thymocytes. *J Immunol* 2008;**181**:6265–74.
210. Kuiper RP, Schoenmakers EF, van Reijmersdal SV, Hehir-Kwa JY, van Kessel AG, van Leeuwen F, et al. High-resolution genomic profiling of childhood ALL reveals novel recurrent genetic lesions affecting pathways involved in lymphocyte differentiation and cell cycle progression. *Leukemia* 2007;**21**:1258–66.
211. Marcais A, Jeannet R, Hernandez L, Soulier J, Sigaux F, Chan S, et al. Genetic inactivation of Ikaros is a rare event in human T-ALL. *Leuk Res* 2010;**34**:426–9.
212. Maser RS, Choudhury B, Campbell PJ, Feng B, Wong KK, Protopopov A, et al. Chromosomally unstable mouse tumours have genomic alterations similar to diverse human cancers. *Nature* 2007;**447**:966–71.
213. Mullighan CG, Su X, Zhang J, Radtke I, Phillips LA, Miller CB, et al. Deletion of IKZF1 and prognosis in acute lymphoblastic leukemia. *N Engl J Med* 2009;**360**:470–80.
214. Gutierrez A, Sanda T, Ma W, Zhang J, Grebliunaite R, Dahlberg S, et al. Inactivation of LEF1 in T-cell acute lymphoblastic leukemia. *Blood* 2010;**115**:2845–51.
215. Okamura RM, Sigvardsson M, Galceran J, Verbeek S, Clevers H, Grosschedl R. Redundant regulation of T cell differentiation and TCRalpha gene expression by the transcription factors LEF-1 and TCF-1. *Immunity* 1998;**8**:11–20.

216. Balgobind BV, Van Vlierberghe P, van den Ouweland AM, Beverloo HB, Terlouw-Kromosoeto ER, van Wering ER, et al. Leukemia-associated NF1 inactivation in patients with pediatric T-ALL and AML lacking evidence for neurofibromatosis. *Blood* 2008;**111**:4322–8.
217. Le DT, Kong N, Zhu Y, Lauchle JO, Aiyigari A, Braun BS, et al. Somatic inactivation of Nf1 in hematopoietic cells results in a progressive myeloproliferative disorder. *Blood* 2004;**103**:4243–50.
218. Side L, Taylor B, Cayouette M, Conner E, Thompson P, Luce M, et al. Homozygous inactivation of the NF1 gene in bone marrow cells from children with neurofibromatosis type 1 and malignant myeloid disorders. *N Engl J Med* 1997;**336**:1713–20.
219. McCormick F. Ras signaling and NF1. *Curr Opin Genet Dev* 1995;**5**:51–5.
220. Thorsteinsdottir U, Mamo A, Kroon E, Jerome L, Bijl J, Lawrence HJ, et al. Overexpression of the myeloid leukemia-associated Hoxa9 gene in bone marrow cells induces stem cell expansion. *Blood* 2002;**99**:121–9.
221. Lieu YK, Reddy EP. Conditional c-myb knockout in adult hematopoietic stem cells leads to loss of self-renewal due to impaired proliferation and accelerated differentiation. *Proc Natl Acad Sci USA* 2009;**106**:21689–94.
222. Souroullas GP, Salmon JM, Sablitzky F, Curtis DJ, Goodell MA. Adult hematopoietic stem and progenitor cells require either Lyl1 or Scl for survival. *Cell Stem Cell* 2009;**4**:180–6.
223. Bhandoola A, Sambandam A. From stem cell to T cell: one route or many? *Nat Rev Immunol* 2006;**6**:117–26.
224. Krivtsov AV, Twomey D, Feng Z, Stubbs MC, Wang Y, Faber J, et al. Transformation from committed progenitor to leukaemia stem cell initiated by MLL-AF9. *Nature* 2006;**442**:818–22.
225. McCormack MP, Young LF, Vasudevan S, de Graaf CA, Codrington R, Rabbitts TH, et al. The Lmo2 oncogene initiates leukemia in mice by inducing thymocyte self-renewal. *Science* 2010;**327**:879–83.
226. Hoang T. Of mice and men: how an oncogene transgresses the limits and predisposes to acute leukemia. *Sci Transl Med* 2010;**2**: 21ps10.
227. Jonkers J, Berns A. Retroviral insertional mutagenesis as a strategy to identify cancer genes. *Biochim Biophys Acta* 1996;**1287**:29–57.
228. von Boehmer H, Aifantis I, Gounari F, Azogui O, Haughn L, Apostolou I, et al. Thymic selection revisited: how essential is it? *Immunol Rev* 2003;**191**:62–78.

Section III

T Cell Immunity in the Periphery

Lymphoid Tissue Inducer Cells and the Evolution of CD4 Dependent High-Affinity Antibody Responses

Peter J.L. Lane, Fiona M. McConnell, David Withers, Fabrina Gaspal, Manoj Saini, and Graham Anderson

MRC Centre for Immune Regulation, Institute for Biomedical Research, Birmingham Medical School, Birmingham, United Kingdom

Phylogeny indicates that in mammals memory CD4-dependent antibody responses evolved after monotremes split from the common ancestor of marsupial and eutherian mammals. This was strongly associated with the development of segregated B and T cell areas and the development of a linked lymph node network. The evolution of the lymphotoxin beta receptor in these higher mammals was key to the development of these new functions. Here, we argue that lymphoid tissue inducer cells played a pivotal role not only in the development of organized lymphoid structures but also in the subsequent genesis of the CD4-dependent class-switched memory antibody responses that depend on an organized infrastructure to work. In this review, we concentrate on the role of this cell type in the making of a tolerant CD4 T cell repertoire and in the sustenance of CD4 T cell responses for protective immunity.

Progress in Molecular Biology
and Translational Science, Vol. 92
DOI: 10.1016/S1877-1173(10)92007-3

I. Introduction

A key feature of protective immunity in mammals is the capacity to generate over time high-affinity antibody responses that are then remembered. In Burnet's remarkably insightful original treatise on clonal selection,[1] he speculated that affinity maturation of the antibody response was due to the selection of somatic variants bearing high-affinity receptors, and that retention of these high-value lymphocytes could explain antibody memory. Burnet also realized that comparing immune responses in different classes of vertebrate was a powerful tool to understand their evolution, and had noted the discrepancy between cellular (rejection) and humoral (antibody) immunity. Whereas all classes of vertebrate could reject allogeneic skin grafts, with accelerated "memory" rejection by regrafted animals, antibody responses in lower vertebrates were feeble, and did not show evidence of memory.

We now have a much better cellular and molecular understanding of the basis for these differences, informed by comparison of the immunological genomes and immune structures in different classes of vertebrate. What is clear is that evolution of mammalian CD4-dependent memory antibody responses is linked closely with the development of a lymph node network, extant only in mammals.[2] In contrast, CD8 memory, which probably underpins the capacity of all animals to reject second skin grafts with accelerated kinetics, evolved much earlier[3,4] and does not depend on conventional secondary lymphoid tissues in mice.[5]

The link between development of organized lymphoid structures and CD4 memory antibody responses becomes obvious when you consider how they are generated and maintained. Although CD4 memory cells can reside outside secondary lymphoid tissues, for example, in the lamina propria of the gut[6] and bone marrow,[7] B follicles are required for the CD4 dependent selection of somatically mutating B cells in germinal centers (GCs),[8] and secondary lymphoid tissues are required for memory B and T cell collaboration for memory antibody responses.[9] With the exception of spleen, the development of organized lymph nodes depends on lymphoid tissue inducer cells (LTi) and the evidence that we have accumulated indicates that these cells are crucial to the development and maintenance of CD4 but not CD8 memory.

II. LTi and the Development of Organized Lymphoid Structures

LTi were first described in detail in developing lymph node anlage.[10] The development of these cells depends on the splice variant of the transcription factor, retinoic acid orphan receptor gamma (RORγt).[11–13] They are required for not only the induction of both conventional lymph nodes[14,15] but also the formation of isolated lymphoid follicles (ILFs) mainly located in the large intestine but also at other mucosal sites.[16] RORγt LTi are not essential for the development of some lymphoid tissues: spleen, nasal associated lymphoid tissues,[17] and primitive aggregates of lymphocytes in the omentum.[18]

In mice, classical RORγt$^+$ LTi are characterized by their expression of the interleukin 7 receptor alpha (IL7Rα); most express CD4, but lack expression of CD3 (T cell), B220 (B and plasmacytoid dendritic cell (DC)), or CD11c (DC) markers (CD4$^+$IL7Rα$^+$CD11c$^-$CD3$^-$B220$^-$),[14] but in humans they lack expression of CD4.[19]

Although conventional RORγt$^+$ LTi are not essential for the formation of white pulp areas of the spleen,[11,20] conventional LTi are present in the fetal spleen[21,22] where they attach to fixed stromal cells, which subsequently express homeostatic chemokines that guide the recruitment of cells (T cells, DCs, and B cells) that form the white pulp microenvironments for CD4 T cell responses.

Although originally only thought to be functional in the embryo, LTi are also present in adult life. Adult LTi have been implicated in the induction of *de novo* ILFs in the gut[16] and in the repair of secondary lymphoid tissues,[23] a function analogous to the one they perform in fetal life.

Both murine fetal and adult splenic LTi are heterogeneous with regard to their expression of homeostatic chemokine receptors, CXCR5 and CCR7,[24] consistent with the three locations where they are found in adult murine spleen: some are associated with B follicle stromal cells (CXCR5$^+$CCR7$^-$) and a few are in the central T zone (CXCR5$^+$CCR7$^+$)[25] but the majority are at the boundary between the B and T cell areas (CXCR5$^+$CCR7$^+$).[26]

In the T zone, interactions between LTi and the underlying stroma, as well as with DCs and T cells, can readily be identified,[25] and the associations between primed CD4 T cells and LTi in B follicles, particularly at the B/T interface, are easy to observe.[26] The latter is consistent with provision by LTi of signals to CD4 T cells.

III. LTi Express Recently Evolved Tumor Necrosis Family Member Ligands

The Tumor Necrosis Family (TNF) of ligands and receptors plays a pivotal role in the organization of the immune system and development of immune responses. For example, CD4 T cell CD40-ligand (TNFSF5) through its interaction with B cell CD40 (TNFRSF5) plays the pivotal role in T cell help for B cells[27] and this function is conserved in all vertebrates with adaptive T and B cell responses.[28] In addition, other TNF-ligands, BAFF and APRIL, implicated in B[29] and plasma cell[30,31] survival, respectively, are found in fish genomes.[32] However, the development of lymph nodes and CD4 memory evolved much later, as these functions are lacking in lower vertebrates.[33]

Local gene duplication is a key source of variation between individuals.[34] Iterated over millions of generations as species diverge, diversification of duplicated genes provides the genetic fuel for the acquisition of new genes that control new functions. A simple way to identify genes responsible for lymph node development and organization and CD4 memory therefore is to identify new gene family members present in mammals but not in lower vertebrates. Because of the link between TNF members and fundamental immune function, the TNF family is a good place to look for new genes directing the new functions of organization and memory in mammals. Comparison of fish and mammalian TNF members identifies six new genes.[32] These are the lymphotoxin genes (LTα and LTβ), OX40-ligand (OX40L)(TNFSF4), GITRL (TNFSF18), CD30L(TNFSF8), and CD27L(TNFSF7).

The receptors for these genes are located in two TNF clusters at the tip of human chromosomes 1 and 12 (Table I). In addition to TNFR family members, the two gene clusters contain closely related serine proteases, MASP2 (chromosome 1) and C1rs (chromosome 12), which catalyze the activation of complement component 4 via the mannan binding and the classical pathways, respectively. It seems likely that the two TNF clusters may have arisen from an ancestral chromosomal duplication that predated the evolution of jawed

TABLE I
HUMAN TNF RECEPTORS ON CHROMOSOMES 1 AND 12

	Chromosome 1		Chromosome 12
TNF cluster 1	Position (MB)	TNF cluster 2	Position (MB)
GITR	1.13	TNFR1*	12:6.3
OX40**	1.14	**LTβR****	12:6.36
HVEM***	2.5	CD27	12:6.4
DR3*	6.4	CD4	12:6.7
4-1BB	7.9	C1rs$	12:7.1
MASP2$	11	AID	12:8.6
CD30	12		
TNFR2***	12.1		

Genes marked in black letters show TNF receptors present in all vertebrates. Genes marked in bold show TNF receptors whose ligands are only present in higher vertebrates. Genes marked in with $ sign show the two serine proteases that catalytically cleave complement component, C4. Asterisked (*) TNF receptors are close paralogues. Data from www.ensembl.org.

vertebrates, as originally proposed by Ohno[35] and for which there is direct evidence.[36] In support of this, although teleosts lack some TNF members, both gene clusters are present in fish genomes (www.ensembl.org).

LTi are characterized by their expression of many different TNF ligands. However, they do not express any of the TNF members linked with B cell activation and survival (CD40, BAFF, and APRIL).[37] The expressed molecules include TNFL shared with primitive vertebrates (TNFα (TNFSF2), LIGHT (TNFSF14), and TRANCE (TNFSF11A)) and also the more recently evolved TNFL. Adult and embryonic LTi both express high levels of the lymphotoxins, LTα (TNFSF1) and LTβ (TNFSF3), and adult LTi express high levels of OX40L (TNFSF4) and CD30L (TNFSF8).[37]

IV. High-Affinity Class Switched Antibodies Depend on LTα and LTβ Induced Organization

LTα and LTβ, by binding to the TNF receptor (TNFRSF1A) and the lymphotoxin beta receptor (LTβR) expressed in fixed stromal cells, provide the key signals for the development of organized secondary lymphoid tissues,[38] although we have found an additional role for CD30 in lymphoid tissue organization[39] particularly in the absence of LT signals.[40] During ontogeny, LTi are the first cells to provide these LT signals to stroma in both lymph node[14] and spleen,[21] but B cell LTβR signals are crucial for the induction and maintenance of CXCL13 and B follicle formation.[41]

The formation of discrete B follicles is crucial for CD4 T cells to orchestrate the CD40-dependent proliferation, somatic mutation, and selection of B cells within GCs, with the consequent production of high-affinity class switched antibody responses. Consequently, LT deficient mice, which lack these structures, produce neither high-affinity IgG in the systemic circulation[42] nor IgA at mucosal surfaces.[43] In contrast, T-dependent primary low affinity nonswitched IgM responses do not depend on organized lymphoid structures and B follicles.

Antibody responses in disorganized LT deficient mice mimic the antibody immune system found in lower vertebrates, which lack these genes, despite having the activation induced cytidine deaminase (AID)[44] required for class switching and somatic mutation,[45] intact T cell help through CD40,[28] and somatic mutation of immunoglobulin genes.[46] This demonstrates the key importance of the LT dependent B follicle as the structure for generation of high-affinity antibodies, consistent with the idea that the driving force for the evolution of organized lymphoid structures was the acquisition of a mechanism to select efficiently the B cell precursors of high-affinity class switched antibodies.

Lymphoid tissue organization is also required for effective collaboration of memory B and T cells. Under normal circumstances, memory B and T cells collaborate with each other in the outer T zone of secondary lymphoid structures.[47] In the absence of this organization, memory B and T cells fail to find each other.[48] This implies that lymphotoxin-dependent lymphoid tissue organization was a prerequisite for the development of the generation of memory B and T cells and also for their recall function, that is, memory.

V. AID-Dependent Class Switching Anticipated Affinity Maturation and CD4 Memory Antibody Responses

The capacity to make high-affinity antibody responses developed after therian mammals split from monotremes ~166 mya.[49] Marsupials have evidence of memory[50,51] whereas monotremes do not.[52,53] The development of CD4 memory antibody responses is directly associated with the development of organized lymphoid structures and lymph nodes. Unlike marsupials and eutherian mammals, monotremes lack proper encapsulated macroscopic lymph nodes, but have instead numerous ILFs at sites where lymph nodes form in higher animals.[52] Furthermore, these ILFs have GC like structures that are traditionally thought to be sites of affinity maturation of the B cell responses.[8] This suggests that GCs in ILFs were ancestral to GC in lymph nodes, but were inefficient at providing an environment for the generation and recall of memory antibodies.

What was the function of GCs in ILFs in primitive mammals if not to make high-affinity antibodies? Recently, an intriguing insight into the evolution of CD4 dependent memory class switched antibody responses was revealed by the observation that LTi in ILFs present in the gut of all mammals and probably lower animals[33] were required for AID-dependent but T-independent IgA class switching.[54] This IgA class switching depends on the TNF-ligands, BAFF and APRIL, but not CD40-ligand signals, and was provided by CD11c$^+$ dendritic-like cells and stromal cells but not CD4 T cells.[54] Although LTi do not directly provide the BAFF and APRIL signals, by recruiting B cells and CD11c$^+$ cells, they are essential for forming the cellular environment for IgA class switching.

Birds and mammals share IgA but not IgG homologs,[55] suggesting that IgA class switching evolved first in their common reptilian ancestor. In contrast, monotremes do have the genes encoding IgG and IgE,[56] indicating that these major immunoglobulin classes were present in the common mammalian ancestor, which had centrally located ILFs at sites of lymph node development in higher mammals. In contrast to IgA, the production of IgG and IgE is much more T cell CD40-dependent.[57,58] It seems plausible therefore to suggest that CD4 T cell-driven IgG and IgE class switching but not affinity maturation in monotremes occurs in centrally located ILFs.

How might have ILFs present in the lamina propria of the gut moved to centrally located sites? An intriguing possibility is that they might have been "metastasized" by local lymphatics draining the gut. ILFs depend on LTi, and the proximity of LTi in ILFs to gut draining lymphatics is consistent with this view.[52]

The key difference between ILFs and lymph nodes is that the latter are macroscopic encapsulated structures where B and T cells are segregated, and unlike ILFs, $CD11c^+$ cells are excluded from B follicles. This key difference allows division of labor in AID-dependent function: iterative selection of AID-dependent B cell somatic mutants by T cells in GCs for affinity maturation of the B cell response, versus AID-dependent class switching at the B/T interface associated with $CD11c^+$ cells.

Figure 1 is a schematic for the putative evolution of ILFs from structures in which LTi orchestrated the accumulation of BAFF and APRIL bearing $CD11c^+$ cell dependent but T-independent class switching, into structures in which T cells were involved in B cell selection. Subsequently, CD4 T cell dependent class switching was segregated into B follicles, primarily for the iterative CD4 T cell selection of high-affinity B cells, and $CD11c^+$ cell induced class switching and formation of plasma cells settled in the outer T zone.

VI. The Genes for Organization Show Linkage with CD4 and AID

TNFRSF1A and the LTβR are the key genes expressed in stromal cells that regulate the expression of homeostatic chemokines that orchestrate both lymph node development[59] and B/T segregation. In therian mammals, they are genetically closely linked to the genes for CD4 and AID. Although these genes are structurally unrelated, functionally they are all required for generating high-affinity class switched antibodies. CD4 is required for the selection of the T cells that select B cells in GCs, AID is required for somatic hypermutation and immunoglobulin class switching in B cells, and TNFRSF1A and LTβR enable the formation of B follicles within which CD4 T cells select AID-expressing GC B cells. Although monotremes lack true lymph nodes and memory, they have a good homolog of TNFRSF1A but lack a convincing homolog of LTβR (www.ensembl.org), suggesting that the evolution by gene duplication of this gene was instrumental in facilitating the development of structures where B and T cells were segregated.

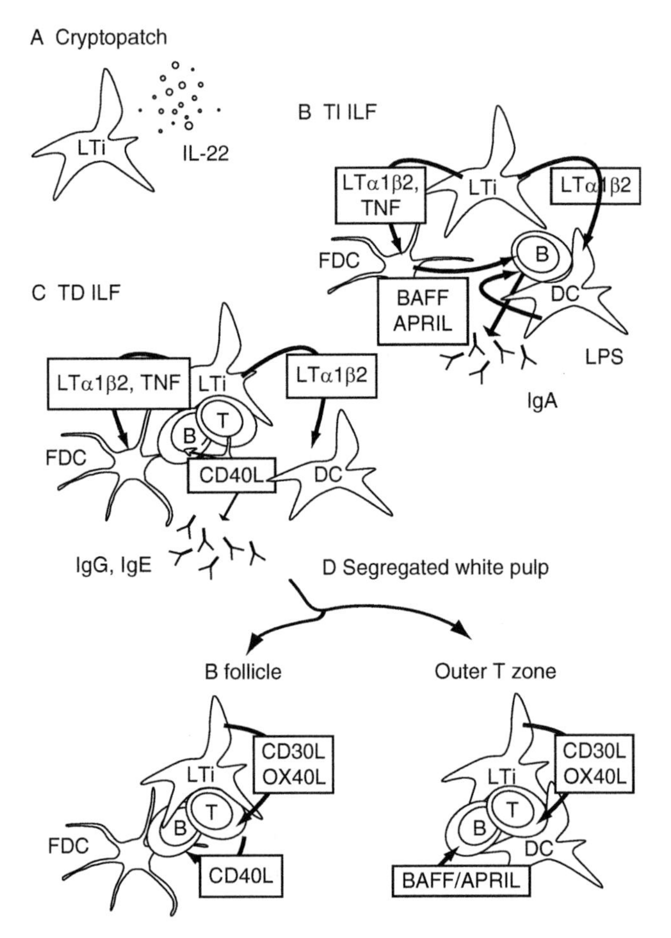

Fig. 1. A model for the evolution of classical lymphoid structures. (A) IL22 LTi like cells promote epithelial integrity at mucosal surfaces. (B) LTi, adapted through expression of lymphotoxins, promote T-independent class switching to IgA mediated by BAFF and APRIL expressing $CD11c^{+}$ cells, again to promote mucosal integrity. (C) Evolution of CD40 dependent T cell selection within B follicles, with acquisition of new TNF ligands (CD30 and OX40 promoting CD4 memory. (D) Classical TD responses in lymph node and spleen. In B follicles, CD40L expressing CD4 T cells, LTi, and FDCs foster AID-dependent germinal center affinity maturation. In the outer T zone, T (CD40L) and $CD11c^{+}$ (BAFF and APRIL) foster AID-dependent class switching in B cells that become plasma cells.

In lower vertebrates, CD4 and AID are also linked genetically (www.ensembl.org), indicating that this configuration was established in the ancestors of jawed vertebrates which have homologs of TNFRSF1A, but homologs of

LTβR are exclusive to mammals. We think that proximity of genes favors selection of related functions, that is, the function of the LTβR in context of B/T segregation and lymph node development arose in the context of the respective functions of CD4 and AID in the generation and selection of high-affinity antibody responses. If it is correct that monotremes lack a true homolog of the LTβR whereas marsupial mammals clearly have it (www.ensembl.org), then this supports the idea that new genes map to the evolution of new functions, in this case lymph nodes and the capacity to make memory antibody responses.

VII. OX40L and CD30L are Expressed by LTi

In addition to the LT genes required for organization, adult but not embryonic LTi express high levels of OX40L and CD30L.[26,60] These TNF ligands are not exclusive to LTi, but LTi isolated from T cell deficient adult mice express high levels of these molecules, particularly OX40L.[26] In contrast to T cell deficient mice, LTi in mice with a normal repertoire of T cell have lower but still readily detectable OX40L levels, unlike embryonic LTi which lack OX40L expression. Adult and embryonic LTi OX40L is upregulated by the proinflammatory TNF ligand, TL1A,[37] produced by activated DCs,[61] consistent with the idea that LTi OX40L expression is regulated.[37]

VIII. Lack of CD4 Memory in Mice Deficient in OX40 and CD30

The receptors for OX40 and CD30 are primarily expressed on activated T cells.[62,63] Because of our evidence that primed CD4 T cells could interact directly with LTi[26] which could express high levels of the ligands, we made mice singly deficient in OX40 or CD30, or deficient in both molecules. Individual contributions of OX40 and CD30 signals to CD4 T cell memory were identified,[26,64] but the striking result was that when both signaling pathways were removed, CD4 memory generation was abrogated.

There was no effect on primary antibody responses, and *in vitro* and *in vivo*, CD4 T cells deficient in OX40 and CD30 proliferated quite normally initially.[64] Therefore, there does not seem to be a requirement for OX40 and CD30 for priming CD4 T cells. The impaired memory responses were not due to lack of appropriate positive selection in the thymus as T cell receptor transgenic CD4 T cells deficient in OX40 and CD30 were positively selected, as they are normally. The key defect caused by the double deficit is in the capacity of primed CD4 T cells to survive, and OX40 and CD30 have synergistic effects. The premature death of primed CD4 T cells was reflected in the

CD4 T cell driven GC response. GCs were initiated normally but terminated prematurely with the consequence that affinity maturation of the antibody response was grossly impaired. However, the most dramatic effect was on recall memory immune responses which were vestigial in mice deficient in OX40 and CD30.

IX. CD30-Signals Evolved Before OX40 and the LTβR, and Share Functions in Both Organization and CD4 Memory

Birds and mammals share a common reptilian ancestor. The chicken genome contains homologs of both CD30 and OX40, and CD30L, but not LTβR. There is a possible but not very convincing chicken homolog of OX40L (www.ensembl.org). CD30 signals not only have effects on CD4 T cell survival[64] but also have a modest effect on the stromal expression of the T zone chemokines, CCL21, in the absence of mammalian LT signals.[40] This has led us to suggest that the dual functions of CD30 in organization (recruitment of T cells and DC) and CD4 T cell survival evolved first to recruit and sustain CD4 T cells. This was further refined in mammals via OX40 (dominant role in CD4 memory) and LTβR (B/T segregation and lymph node organization).

X. Direct Evidence that LTI are Required for CD4 Memory

Although OX40L and CD30L are expressed at high levels on LTi, many different cell types can express these molecules, particularly when activated.[62,63] We now have, however, unpublished (Withers and Lane) but direct evidence that LTi are essential for both the generation and maintenance of CD4 memory. We first looked at $ROR\gamma^{KO}$ mice[11] and found that they had grossly reduced numbers of CD4 but not CD8 memory cells in spleen and bone marrow. However, because T cell development is abnormal in $ROR\gamma^{KO}$ mice,[11,20] we have gone on to examine the fate of $ROR\gamma^{+}$ CD4 memory T cells in $ROR\gamma^{KO}$ mice, and found that the generation and persistence of CD4 memory is virtually completely LTi dependent. These data provide strong evidence that LTi, through their expression of OX40L and CD30L, are pivotal to both forming the microenvironments within which CD4 T cells are activated to help B cells, and maintaining primed memory CD4 T cells, functions essential for the generation of CD4 memory antibody responses.[2,48]

XI. LTi and the Induction and Maintenance of Tolerance

LTi are also present in murine thymus, located adjacent to thymic epithelium.[65] We have linked LTi expression of the TNF ligand, TRANCE (TNFSF11), with the generation of the first cohort of Autoimmune Regulator (AIRE)-expressing medullary thymic epithelial cells (mTECs) in the embryonic thymus which enables intrathymic expression of tissue restricted antigens for self-tolerance. Recently, in the adult thymus, positively selected thymocytes have also been shown to induce AIRE-expressing mTECs.[66] In this context, it is interesting to note that it has been shown that AIRE expression in the perinatal period appears both essential and sufficient to prevent autoimmunity,[67] suggesting that the contribution made by LTi in the development of AIRE+ mTECs is an important component of self-tolerance.

Thus, LTi are linked with the induction of tolerance to self as well as to the generation of adaptive memory immune responses. Self-tolerance is a cornerstone of adaptive immunity,[1] and is particularly relevant in the context of CD4 memory where autoimmunity becomes a significant risk. The presence of both TRANCE and AIRE in early vertebrate genomes (www.ensembl.org) is consistent with AIRE-dependent mechanisms evolving early. However, although the late evolving LTβR is not required for AIRE-expression,[68] expression of it does affect the turnover of AIRE-expressing cells of thymic mTECs, and $LT\beta R^{KO}$ mice have evidence of autoantibody production[69] which is remarkable considering the impairment in production of high-affinity antibodies in $LT\beta R^{KO}$ mice.[59]

We think that the evolution of more sophisticated antibody immune responses in secondary lymphoid tissues demanded more stringent negative selection of potentially autoreactive T cells, so that the evolution of lymph nodes and high-affinity antibodies dependent on LTβR was paralleled by LTβR-dependent improved deletion in the thymus.

XII. Putative Origin of LTi and Summary

We have speculated elsewhere how LTi function evolved.[70,71] This is summarized in Fig. 1. We think that the role of LTi in the development of lymphotoxin dependent organization of lymph nodes and in OX40 and CD30 dependent memory evolved late, and that the high expression of interleukin 22 (IL22),[72] which is implicated strongly in mucosal epithelial immunity[73,74] might indicate an earlier innate role for LTi located in crypopatches.

We have speculated that LTi evolved first as innate cells maintaining epithelial integrity through expression of IL22 (Fig. 1A). Recent data show extensive synteny between the mammalian and the fish IL22 genetic loci, suggesting that this cytokine cluster was established in a common ancestor of mammals and fish,[75] so such an origin for LTi is possible. This is supported by evidence that the IL22 transcription factor, the aryl hydrocarbon receptor (AHR) is present in jawless lamprey,[76] and expressed in the lamprey T lymphocyte equivalent.[77] Cryptopatches, as their name suggests, are located between crypts at the base of the villi, and RORγt dependent LTi are required for their formation,[13] so they are well placed to provide IL22 signals to the local intestinal epithelium in vertebrates (Fig. 1A).

Both IL22 and IgA promote mucosal defenses, but the former is innate whereas the latter is part of the adaptive immune system. The evidence from Fagarasan's group that cryptopatches became a focus for T cell independent B cell selection and class switching to IgA via recruitment of CD11c+ cells expressing BAFF and APRIL[54] defines a transition in LTi function, linking them with both innate and adaptive immunity (Fig. 1B). These new functions were dependent on the acquisition by LTi of the new TNF genes encoding the lymphotoxins. They would enable the development of LTi and lymphotoxin dependent ILFs into structures that supported CD4 and CD40 dependent B cell selection (Fig. 1C), with the capacity to substantially increase the affinity of these antibodies. Given the advantage against pathogens conferred by high-affinity memory antibody responses, it is not too difficult to imagine that LTi, already keeping other cell types alive through TNF ligand expression, were the right niche for the expression of the late evolving TNF ligands that supported survival of primed CD4 T cells.[70] Finally, we would point out that the development of mechanisms to make CD4 memory responses significantly raised the risk of developing CD4 driven autoimmune diseases, and speculate that this selected for more efficient deletion of self-reactive cells in thymus.

Acknowledgments

This work was supported by a Wellcome Trust Programme Grant to P. L. and G. A.

References

1. Burnet FM. A modification of Jerne's theory of antibody production using the concept of clonal selection. *Aust J Sci* 1957;**20**:67–9.
2. Lane PJL, Gaspal MC, Kim M-Y. Two sides of a cellular coin: CD4+CD3− cells orchestrate memory antibody responses and lymph node organisation. *Nat Rev Immunol* 2005;**5**:655–60.
3. Haas R. Transplantation reactions in the African lungfish, *Protopterus amphibius*. *Transplantation* 1982;**33**:249–53.
4. Morales HD, Robert J. Characterization of primary and memory CD8 T-cell responses against ranavirus (FV3) in *Xenopus laevis*. *J Virol* 2007;**81**:2240–8.

5. Moyron-Quiroz JE, Rangel-Moreno J, Hartson L, Kusser K, Tighe MP, Klonowski KD, et al. Persistence and responsiveness of immunologic memory in the absence of secondary lymphoid organs. *Immunity* 2006;**25**:643–54.
6. Reinhardt RL, Khoruts A, Merica R, Zell T, Jenkins MK. Visualizing the generation of memory CD4 T cells in the whole body. *Nature* 2001;**410**:101–5.
7. Tokoyoda K, Zehentmeier S, Hegazy AN, Albrecht I, Grun JR, Lohning M, et al. Professional memory CD4+ T lymphocytes preferentially reside and rest in the bone marrow. *Immunity* 2009;**30**:721–30.
8. MacLennan ICM. Germinal centers. *Annu Rev Immunol* 1994;**12**:117–39.
9. MacLennan ICM, GulbransonJudge A, Toellner KM, CasamayorPalleja M, Chan E, Sze DMY, et al. The changing preference of T and B cells for partners as T-dependent antibody responses develop. *Immunol Rev* 1997;**156**:53–66.
10. Mebius RE, Rennert P, Weissman IL. Developing lymph nodes collect CD4+CD3− LTbeta+ cells that can differentiate to APC, NK cells, and follicular cells but not T or B cells. *Immunity* 1997;**7**:493–504.
11. Sun Z, Unutmaz D, Zou YR, Sunshine MJ, Pierani A, Brenner-Morton S, et al. Requirement for RORgamma in thymocyte survival and lymphoid organ development. *Science* 2000;**288**:2369–73.
12. Eberl G, Marmon S, Sunshine MJ, Rennert PD, Choi Y, Littman DR. An essential function for the nuclear receptor RORgamma(t) in the generation of fetal lymphoid tissue inducer cells. *Nat Immunol* 2004;**5**:64–73.
13. Eberl G, Littman DR. Thymic origin of intestinal alphabeta T cells revealed by fate mapping of RORgammat+ cells. *Science* 2004;**305**:248–51.
14. Mebius RE. Organogenesis of lymphoid tissues. *Nat Rev Immunol* 2003;**3**:292–303.
15. Eberl G, Littman DR. The role of the nuclear hormone receptor RORgammat in the development of lymph nodes and Peyer's patches. *Immunol Rev* 2003;**195**:81–90.
16. Eberl G. Opinion: inducible lymphoid tissues in the adult gut: recapitulation of a fetal developmental pathway? *Nat Rev Immunol* 2005;**5**:413–20.
17. Harmsen A, Kusser K, Hartson L, Tighe M, Sunshine MJ, Sedgwick JD, et al. Cutting edge: organogenesis of nasal-associated lymphoid tissue (NALT) occurs independently of lymphotoxin-alpha (LT alpha) and retinoic acid receptor-related orphan receptor-gamma, but the organization of NALT is LT alpha dependent. *J Immunol* 2002;**168**:986–90.
18. Rangel-Moreno J, Moyron-Quiroz JE, Carragher DM, Kusser K, Hartson L, Moquin A, et al. Omental milky spots develop in the absence of lymphoid tissue-inducer cells and support B and T cell responses to peritoneal antigens. *Immunity* 2009;**30**:731–43.
19. Cupedo T, Crellin NK, Papazian N, Rombouts EJ, Weijer K, Grogan JL, et al. Human fetal lymphoid tissue-inducer cells are interleukin 17-producing precursors to RORC+ CD127+ natural killer-like cells. *Nat Immunol* 2009;**10**:66–74.
20. Kurebayashi S, Ueda E, Sakaue M, Patel DD, Medvedev A, Zhang F, et al. Retinoid-related orphan receptor gamma (RORgamma) is essential for lymphoid organogenesis and controls apoptosis during thymopoiesis. *Proc Natl Acad Sci USA* 2000;**97**:10132–7.
21. Withers DR, Kim MY, Bekiaris V, Rossi SW, Jenkinson WE, Gaspal F, et al. The role of lymphoid tissue inducer cells in splenic white pulp development. *Eur J Immunol* 2007;**37**:3240–5.
22. Vondenhoff MF, Desanti GE, Cupedo T, Bertrand JY, Cumano A, Kraal G, et al. Separation of splenic red and white pulp occurs before birth in a LTalphabeta-independent manner. *J Leukoc Biol* 2008;**84**:152–61.
23. Scandella E, Bolinger B, Lattmann E, Miller S, Favre S, Littman DR, et al. Restoration of lymphoid organ integrity through the interaction of lymphoid tissue-inducer cells with stroma of the T cell zone. *Nat Immunol* 2008;**9**:667–75.

24. Kim MY, Rossi S, Withers D, McConnell F, Toellner KM, Gaspal F, et al. Heterogeneity of lymphoid tissue inducer cell populations present in embryonic and adult mouse lymphoid tissues. *Immunology* 2008;**124**:166–74.
25. Kim MY, McConnell FM, Gaspal FM, White A, Glanville SH, Bekiaris V. Function of CD4+ CD3− cells in relation to B- and T-zone stroma in spleen. *Blood* 2007;**109**:1602–10.
26. Kim MY, Gaspal FM, Wiggett HE, McConnell FM, Gulbranson-Judge A, Raykundalia C, et al. CD4(+)CD3(−) accessory cells costimulate primed CD4 T cells through OX40 and CD30 at sites where T cells collaborate with B cells. *Immunity* 2003;**18**:643–54.
27. Banchereau J, Bazan F, Blanchard D, Briere F, Galizzi JP, van Kooten C, et al. The CD40 antigen and its ligand. *Annu Rev Immunol* 1994;**12**:881–922.
28. Gong YF, Xiang LX, Shao JZ. CD154–CD40 interactions are essential for thymus-dependent antibody production in zebrafish: insights into the origin of costimulatory pathway in helper T cell-regulated adaptive immunity in early vertebrates. *J Immunol* 2009;**182**:7749–62.
29. Mackay F, Browning JL. BAFF: a fundamental survival factor for B cells. *Nat Rev Immunol* 2002;**2**:465–75.
30. Benson MJ, Dillon SR, Castigli E, Geha RS, Xu S, Lam KP, et al. Cutting edge: the dependence of plasma cells and independence of memory B cells on BAFF and APRIL. *J Immunol* 2008;**180**:3655–9.
31. Belnoue E, Pihlgren M, McGaha TL, Tougne C, Rochat AF, Bossen C, et al. APRIL is critical for plasmablast survival in the bone marrow and poorly expressed by early-life bone marrow stromal cells. *Blood* 2008;**111**:2755–64.
32. Glenney GW, Wiens GD. Early diversification of the TNF superfamily in teleosts: genomic characterization and expression analysis. *J Immunol* 2007;**178**:7955–73.
33. Zapata A, Ameimiya CT. Phylogeny of lower vertebrates and their immunological structures. In: du Pasquier L, Litman GW, editors. *Origin and evolution of the vertebrate immune system.* Berlin: Springer; 2000. p. 67–110.
34. Levy S, Sutton G, Ng PC, Feuk L, Halpern AL, Walenz BP, et al. The diploid genome sequence of an individual human. *PLoS Biol* 2007;**5**:e254.
35. Ohno S. *Evolution by Gene Duplication.* New York: Springer-Verlag; 1970.
36. Putnam NH, Butts T, Ferrier DE, Furlong RF, Hellsten U, Kawashima T, et al. The amphioxus genome and the evolution of the chordate karyotype. *Nature* 2008;**453**:1064–71.
37. Kim MY, Toellner KM, White A, McConnell FM, Gaspal FM, Parnell SM, et al. Neonatal and adult CD4+CD3− cells share similar gene expression profile, and neonatal cells up-regulate OX40 ligand in response to TL1A (TNFSF15). *J Immunol* 2006;**177**:3074–81.
38. Fu YX, Chaplin DD. Development and maturation of secondary lymphoid tissues. *Annu Rev Immunol* 1999;**17**:399–433.
39. Bekiaris V, Withers D, Glanville SH, McConnell FM, Parnell SM, Kim MY, et al. Role of CD30 in B/T segregation in the spleen. *J Immunol* 2007;**179**:7535–43.
40. Bekiaris V, Gaspal F, Kim MY, Withers DR, McConnell FM, Anderson G, et al. CD30 is required for CCL21 expression and CD4 T cell recruitment in the absence of lymphotoxin signals. *J Immunol* 2009;**182**:4771–5.
41. Ansel KM, Ngo VN, Hyman PL, Luther SA, Forster R, Sedgwick JD, et al. A chemokine-driven positive feedback loop organizes lymphoid follicles. *Nature* 2000;**406**:309–14.
42. Fu YX, Molina H, Matsumoto M, Huang G, Min J, Chaplin DD. Lymphotoxin-alpha (LTalpha) supports development of splenic follicular structure that is required for IgG responses. *J Exp Med* 1997;**185**:2111–20.
43. Kang HS, Chin RK, Wang Y, Yu P, Wang J, Newell KA, et al. Signaling via LTbetaR on the lamina propria stromal cells of the gut is required for IgA production. *Nat Immunol* 2002;**3**:576–82.

44. Rogozin IB, Iyer LM, Liang L, Glazko GV, Liston VG, Pavlov YI, et al. Evolution and diversification of lamprey antigen receptors: evidence for involvement of an AID-APOBEC family cytosine deaminase. *Nat Immunol* 2007;**8**:647–56.
45. Muramatsu M, Kinoshita K, Fagarasan S, Yamada S, Shinkai Y, Honjo T. Class switch recombination and hypermutation require activation-induced cytidine deaminase (AID), a potential RNA editing enzyme [see comments]. *Cell* 2000;**102**:553–63.
46. Wilson M, Hsu E, Marcuz A, Courtet M, Du Pasquier L, Steinberg C. What limits affinity maturation of antibodies in *Xenopus*—the rate of somatic mutation or the ability to select mutants? *EMBO J* 1992;**11**:4337–47.
47. Liu YJ, Zhang J, Lane PJ, Chan EY, MacLennan IC. Sites of specific B cell activation in primary and secondary responses to T cell-dependent and T cell-independent antigens. *Eur J Immunol* 1991;**21**:2951–62.
48. Fu YX, Huang G, Wang Y, Chaplin DD. Lymphotoxin-alpha-dependent spleen microenvironment supports the generation of memory B cells and is required for their subsequent antigen-induced activation. *J Immunol* 2000;**164**:2508–14.
49. Warren WC, Hillier LW, Marshall Graves JA, Birney E, Ponting CP, Grutzner F, et al. Genome analysis of the platypus reveals unique signatures of evolution. *Nature* 2008;**453**:175–83.
50. Shearer MH, Robinson ES, VandeBerg JL, Kennedy RC. Humoral immune response in a marsupial Monodelphis domestica: anti-isotypic and anti-idiotypic responses detected by species-specific monoclonal anti-immunoglobulin reagents. *Dev Comp Immunol* 1995;**19**: 237–46.
51. Kreiss A, Wells B, Woods GM. The humoral immune response of the Tasmanian devil (*Sarcophilus harrisii*) against horse red blood cells. *Vet Immunol Immunopathol* 2009;**130**: 135–7.
52. Diener E, Ealey EH. Immune system in a monotreme: studies on the Australian echidna (*Tachyglossus aculeatus*). *Nature* 1965;**208**:950–3.
53. Wronski EV, Woods GM, Munday BL. Antibody response to sheep red blood cells in platypus and echidna. *Comp Biochem Physiol A Mol Integr Physiol* 2003;**136**:957–63.
54. Tsuji M, Suzuki K, Kitamura H, Maruya M, Kinoshita K, Ivanov II, et al. Requirement for lymphoid tissue-inducer cells in isolated follicle formation and T cell-independent immunoglobulin A generation in the gut. *Immunity* 2008;**29**:261–71.
55. Bengten E, Wilson M, Miller N, Clem LW, Pilstrom L, Warr GW. Immunoglobulin isotypes: structure, function and genetics. In: du Pasquier L, Litman GW, editors. *Origin and Evolution of the Vertebrate Immune System*. Berlin: Springer; 2000. p. 189–219.
56. Vernersson M, Aveskogh M, Munday B, Hellman L. Evidence for an early appearance of modern post-switch immunoglobulin isotypes in mammalian evolution (II); cloning of IgE, IgG1 and IgG2 from a monotreme, the duck-billed platypus, *Ornithorhynchus anatinus*. *Eur J Immunol* 2002;**32**:2145–55.
57. DiSanto JP, Bonnefoy JY, Gauchat JF, Fischer A, de Saint Basile G. CD40 ligand mutations in x-linked immunodeficiency with hyper-IgM. *Nature* 1993;**361**:541–3.
58. Korthauer U, Graf D, Mages HW, Briere F, Padayachee M, Malcolm S, et al. Defective expression of T-cell CD40 ligand causes X-linked immunodeficiency with hyper-IgM. *Nature* 1993;**361**:539–41.
59. Futterer A, Mink K, Luz A, Kosco-Vilbois MH, Pfeffer K. The lymphotoxin beta receptor controls organogenesis and affinity maturation in peripheral lymphoid tissues. *Immunity* 1998;**9**:59–70.
60. Kim M-Y, Anderson G, Martensson I-L, Erlandsson L, Arlt W, White A, et al. OX40-ligand and CD30-ligand are expressed on adult but not neonatal CD4+CD3− inducer cells: evidence that IL7 signals regulate CD30-ligand but not OX40-ligand expression. *J Immunol* 2005;**174**:6686–91.

61. Meylan F, Davidson TS, Kahle E, Kinder M, Acharya K, Jankovic D, et al. The TNF-family receptor DR3 is essential for diverse T cell-mediated inflammatory diseases. *Immunity* 2008;**29**:79–89.
62. Croft M. Co-stimulatory members of the TNFR family: keys to effective T-cell immunity? *Nat Rev Immunol* 2003;**3**:609–20.
63. Watts TH. TNF/TNFR family members in costimulation of T cell responses. *Annu Rev Immunol* 2005;**23**:23–68.
64. Gaspal FM, Kim MY, McConnell FM, Raykundalia C, Bekiaris V, Lane PJ. Mice deficient in OX40 and CD30 signals lack memory antibody responses because of deficient CD4 T cell memory. *J Immunol* 2005;**174**:3891–6.
65. Rossi SW, Kim MY, Leibbrandt A, Parnell SM, Jenkinson WE, Glanville SH, et al. RANK signals from CD4(+)3(−) inducer cells regulate development of AIRE-expressing epithelial cells in the thymic medulla. *J Exp Med* 2007;**204**:1267–72.
66. Hikosaka Y, Nitta T, Ohigashi I, Yano K, Ishimaru N, Hayashi Y, et al. The cytokine RANKL produced by positively selected thymocytes fosters medullary thymic epithelial cells that express autoimmune regulator. *Immunity* 2008;**29**:438–50.
67. Guerau-de-Arellano M, Martinic M, Benoist C, Mathis D. Neonatal tolerance revisited: a perinatal window for AIRE control of autoimmunity. *J Exp Med* 2009;**206**:1245–52.
68. Martins VC, Boehm T, Bleul CC. Ltbetar signaling does not regulate AIRE-dependent transcripts in medullary thymic epithelial cells. *J Immunol* 2008;**181**:400–7.
69. Boehm T, Scheu S, Pfeffer K, Bleul CC. Thymic medullary epithelial cell differentiation, thymocyte emigration, and the control of autoimmunity require lympho-epithelial cross talk via LTbetaR. *J Exp Med* 2003;**198**:757–69.
70. Lane PJ. The architects of B and T cell immune responses. *Immunity* 2008;**29**:171–2.
71. Lane PJ, McConnell FM, Withers D, Gaspal F, Saini M, Anderson G. Lymphoid tissue inducer cells: bridges between the ancient innate and the modern adaptive immune systems. *Mucosal Immunol* 2009;**2**:472–7.
72. Takatori H, Kanno Y, Watford WT, Tato CM, Weiss G, Ivanov II, et al. Lymphoid tissue inducer-like cells are an innate source of IL-17 and IL-22. *J Exp Med* 2009;**206**:35–41.
73. Aujla SJ, Chan YR, Zheng M, Fei M, Askew DJ, Pociask DA, et al. IL-22 mediates mucosal host defense against Gram-negative bacterial pneumonia. *Nat Med* 2008;**14**:275–81.
74. Zheng Y, Valdez PA, Danilenko DM, Hu Y, Sa SM, Gong Q, et al. Interleukin-22 mediates early host defense against attaching and effacing bacterial pathogens. *Nat Med* 2008;**14**:282–9.
75. Igawa D, Sakai M, Savan R. An unexpected discovery of two interferon gamma-like genes along with interleukin (IL)-22 and -26 from teleost: IL-22 and -26 genes have been described for the first time outside mammals. *Mol Immunol* 2006;**43**:999–1009.
76. Hahn ME, Karchner SI, Shapiro MA, Perera SA. Molecular evolution of two vertebrate aryl hydrocarbon (dioxin) receptors (AHR1 and AHR2) and the PAS family. *Proc Natl Acad Sci USA* 1997;**94**:13743–8.
77. Guo P, Hirano M, Herrin BR, Li J, Yu C, Sadlonova A, et al. Dual nature of the adaptive immune system in lampreys. *Nature* 2009;**459**:796–801.

Commentary on "Lymphoid Tissue Inducer Cells and the Evolution of CD4 Dependent High-Affinity Antibody Responses"

Mark Coles,* Dimitris Kioussis,† and Henrique Veiga-Fernandes‡

*Centre for Immunology and Infection, Department of Biology and HYMS, University of York, York, United Kingdom

†Division of Molecular Immunology, MRC National Institute for Medical Research, The Ridgeway, London, United Kingdom

‡Immunobiology Unit, Instituto de Medicina Molecular, Faculdade de Medicina de Lisboa. Av. Prof. Egas Moniz. Edifício Egas Moniz, Lisboa, Portugal

The current chapter by Peter Lane and colleagues provides thoughtful insights into the question of how lymph nodes (LNs) evolved allowing for the establishment of efficient immune responses.

The conventional paradigm has defined advanced mammals by a number of characteristics such as the presence of mammary glands, a thoracic cavity, a four chambered heart, hair covered body, and ability to bear live young. However, an essential characteristic of mammals is the presence of LNs, organized lymphoid structures that first evolved in marsupials. Thus, it is possible that LNs are among the key evolutionary advantages that have made mammals so successful, allowing the establishment of protective memory T cell responses and long-term immunity in times of crisis. In our understanding, the correlation shown in this chapter between the evolution of lymphotoxin, LNs, and T cell memory provides a compelling potential connection of how structure and function coevolved in the immune system.

Lymphoid tissue inducer (LTi) cells are crucial for secondary lymphoid organ development in a lymphotoxin dependent manner. Interestingly, a series of recent published papers have shown that in adulthood LTi cells mediate key functions in innate responses to enteric pathogens, suggesting an important role for these cells in the maintenance of epithelial barriers. In the adult, LTi cells accumulate in enteric cryptal patches that develop into isolated lymphoid

Progress in Molecular Biology
and Translational Science, Vol. 92
DOI: 10.1016/S1877-1173(10)92016-4

1877-1173/10 $35.00

follicles (ILF) in response to the commensal flora. Thus, in evolutionary terms, LTi innate immune functions may have preceded their LTi potential. This is in line with Lane's hypothesis suggesting that LNs arose from primitive ILF derived LTi cells which "metastased" from the intestines to the mesenterium giving rise to organized secondary lymphoid tissues.

In adult mice, Lane and colleagues have shown key roles for LTi cells in the development of protective $CD4^+$ T cell memory. This occurs in an OX40/CD30 dependent manner implicating direct interactions between $CD4^+$ T cells and LTi cells in the generation of protective secondary immune responses. Although this is a key mechanism of how adult LTi cells modulate T cell immune responses, LTi cells may also affect the outcome of these responses indirectly by maintaining stroma microenvironments, which contribute for the homeostasis of lymphocytes and facilitate B–T cell and T—dendritic cell interactions. It has been shown that LTi cells contribute to the regeneration of stroma cell environments during infection, and therefore in the absence of LTi cells it is possible that modifications to LN architecture may affect the outcome of immune responses, potentially complicating our understanding of how T cell memory is regulated.

Through the work of Graham Anderson, an additional role for LTi cells has been found in the thymus. The evolution of memory $CD4^+$ T cells required the coevolution of mechanisms inhibiting inappropriate activation of self-reactive T cells. Although regulatory T cells evolved in fish, the formation of central tolerance requires the efficient deletion of self reactive T cells by medullar epithelial cells in the thymus. During the development of the thymus, Anderson and colleagues have shown a key role for LTi cells in the induction of functional medullar epithelial cells.

Thus, through interactions with stroma cells, lymphocytes, and epithelial cells in the thymus and intestines, LTi cells have a key role in the establishment of protective secondary T cell immune responses which have evolved along with mammals.

Cellular and Molecular Requirements in Lymph Node and Peyer's Patch Development

Mark Coles,* Dimitris Kioussis,† and Henrique Veiga-Fernandes‡

**Centre for Immunology and Infection, Department of Biology and HYMS, University of York, York, United Kingdom*

†Division of Molecular Immunology, MRC National Institute for Medical Research, The Ridgeway, London, United Kingdom

‡Immunobiology Unit, Instituto de Medicina Molecular, Faculdade de Medicina de Lisboa, Av. Prof. Egas Moniz, Edifício Egas Moniz, Lisboa, Portugal

Progress in Molecular Biology
and Translational Science, Vol. 92
DOI: 10.1016/S1877-1173(10)92008-5

1877-1173/10 $35.00

Lymphoid tissues have a unique role in the organization and function of the adaptive immune system. Mechanisms driving the development of these tissues have fascinated immunologists for the last 175 years. In this review, we will initially focus on historical literature describing lymph node (LN) anlage development and then on the contemporary understanding of the molecular mechanisms driving LN and Peyer's patch (PP) formation. Utilizing transgenic reporters and gene knockout mice, the interplay between hematopoietic inducer cells and stromal organizer cells has been shown to have a key role in the development and organization of the lymphoid tissues. Although PPs and LNs share many similarities in their development, key differences in the molecular requirements for their development have recently emerged.

I. Foreword

Secondary lymphoid organs (SLOs) are essential structures for efficient immune-surveillance, rapid immune responses, and maintenance of protective immunity. The development of these structures occurs during embryonic life and is tightly regulated by multiple and complex interactions between cells from different germ layer origins, mainly, between hematopoietic cells and their mesenchymal counterparts.[1,2] Molecules involved in these interactions include homeostatic cytokines and chemokines, which attract hematopoietic cells to prospective lymphoid organ locations and ensure their survival. Conversely, hematopoietic cells induce the differentiation and maturation of stroma cells, which in turn form the lymphoid organ scaffold. These processes ultimately lead to the aggregation and retention of more hematopoietic cells at the sites of interactions and ensure the onset of a lymphoid organ developmental program.[1,2]

In September 2009, the lymphoid tissue community held their second meeting in Birmingham (UK) titled "Microanatomy of the Immune System in Health and Disease," where new advances utilizing an array of fluorescent imaging technologies were presented. Back in the late 1800s, immunologists did not have the advantages of modern instrumentation and their analysis of immune function and development depended on highly skilled three-dimensional (3D) microanatomic and microscopic analysis of healthy and diseased tissues. Just over a century later, researchers have revisited this area of investigation with all the advantages of multicolored reagents and advanced microscopy techniques to discover that these early scientists had made remarkably insightful observations many of which proved to be accurate. Great credit, therefore, is due to these early pioneers for their patience, histological skills, anatomical knowledge, and artistic capabilities.

II. Historical Perspective: Lymphoid Tissue Development, 1836–1996

> *"Un ganglion lymphatique est un angiome caverneux lymphatique qui a été d'abord angiome simple"*
>
> Ranvier, 1896.

In this section we will highlight early work on embryos from a large variety of different species that led to the first recognition and characterization of seminal steps in lymph node (LN) anlagen development. Although some of the initial concepts, such as mesenchymal transition into lymphocytes in the anlagen, were later recognized to be incorrect, key observations on infiltration of hematopoietic cells, mesenchymal condensation, and subsequent proliferation and vascularization of the developing anlagen had very accurately described the steps in LN formation.

Our earliest understanding of lymphoid tissue development arises from the work of Gilbert Breschet, a famous French anatomist who in 1836 first linked the origin of LNs with that of lymph sacs.[3] Further work by Gullard in 1894 examined the stages in fetal life, through which human lymphatic glands pass, to "ascertain how they become fitted for the protective role which they assume in the adult," a central principle that continues to drive research on LN anlagen development.[4] Subsequent work from Saxer (1896) observed the presence of "wandering lymphocytes" (probably corresponding to what later became known as lymphoid tissue inducer cells, LTi) that colonize the earliest LN anlagen in ovine embryos.[5] Seminal observations were made in these early studies. These include the key findings that LN development occurs during fetal life, the development of the cervical LNs begins before other anlagen, LN development is associated with lymph sacs and the cell type responsible for LN development is a "wandering lymphocyte."[6] Only with application of fluorescent imaging technology have we been able to show directly that these fetal hematopoietic cells are highly motile lymphocytes that mediate the process of LN formation.[7] Summarizing LN development, Ranvier in 1897 wrote "Un ganglion lymphatique est un angiome caverneux lymphatique qui a été d'abord angiome simple," a rather simple observation that LN development resembles formation of a cavernous lymphangioma, a benign tumor of lymph vessel that contains aggregates of hematopoietic and inflammatory cells and loose connective tissue.[8] The developing LNs were thus hypothesized to represent a form of controlled embryonic inflammation, a central premise of the modern understanding of lymphoid tissue development.

Subsequent work by Kling,[9] Sabin,[10] and Lewis[11] very elegantly analyzed the development of lymph glands in both pig and human embryos. Initial work focused on mechanisms of lymphatic network formation and on identifying the sites of lymphocyte development in developing embryos. Most of the early work on LN development focused on ovine, bovine, swine, and human embryos due to the relative large size and ready availability of embryonic tissue. These studies were later extended to mice through the work of Higgins, who focused on development of the jugular lymph sac in albino mice.[12] Utilizing serial sectioning and 3D wax models (Fig. 1), Higgins resolved multiple stages of LN development in mice. Analysis of these early papers indicates that the timing and cellular interactions involved in the development of LN anlagen are conserved in mammalian species from opossums to humans (Fig. 2).

In Tables I and II we summarize studies on LN anlagen development from 1836 to 1970.

The issue of lymphocytes originating in the developing anlagen has been a matter of debate in the literature, with initial work being inspired by the concept that lymphocytes might arise from LN anlagen itself, while later work suggested that leucocytes arise from the yolk sac. In 1909, Lewis tried to address this question utilizing thin and specially stained sections.[11] Utilizing 10 μm sections, he found lymphocytes in the developing lymph glands (LN anlagen), but found no evidence of lymphocyte-like cells in the developing blood. Retrospectively, we know that the frequency of lymphocyte-like cells in circulation during these stages of development is very low, and was very likely below the detection limits of sectioning and staining technologies of that time period. In contrast, Maximow, who extensively studied the development of the first lymphocytes, thought these arose in the yolk sac from endothelial progenitors, and concluded that the origin of lymphocytes in the embryo has nothing to do with lymphoid tissues.[22] However, in a later review in 1924 Maximow reaches an alternative conclusion: "The primordium of the LN arise relatively late in the embryo, in the neighborhood of lymph-sacs or lymph-vessels. Small, fairly well outlined areas of mesenchyme become discernible, consisting of numerous small, proliferating, undifferentiated mesenchyme cells, displaying unrestricted hematopoietic potencies."[21] Indeed, this later hypothesis predominated in the literature for 70 years, with lymphocytes found in the developing anlagen thought to arise *in situ* from mesenchymal cells within the developing anlagen. This popular theory was applied not only to LNs but also to gut-associated lymphoid tissues and to T cell development in the thymus, where developing T cells were thought to arise from thymic epithelial cells in the thymus.[19]

Over the years, however, it became clear that mesenchyme in developing anlagen does not give rise to hematopoietic cells. First, seminal work in the late 1960s established that blood-derived lymphocyte precursors populate and colonize developing lymphoid organs, clearly showing that T cells do not arise from mesenchymal or epithelial cells, but rather from bloodborne precursors.[23–25] Although this work did not conclusively disprove the earlier

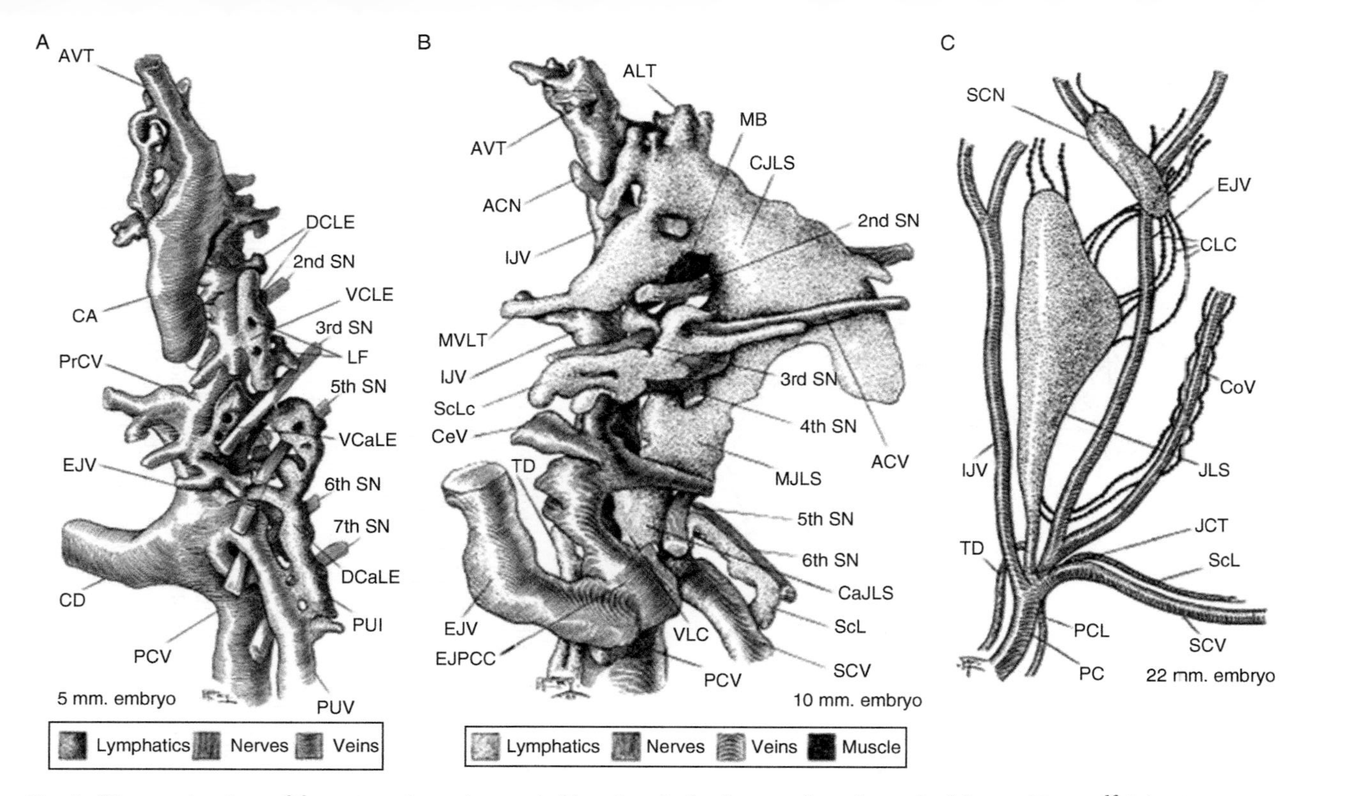

FIG. 1. 3D reconstructions of three stages in murine cervical lymph node development from the work of George Higgins.[12] (A) A 3D construction ventral–lateral view of a reconstruction of the vascular channels in a mouse embryo 5 mm long. Position of the primary lymphatic elements is shown. (B) A ventral–lateral view of a reconstruction of the jugular lymph sac and the related channels in a mouse embryo 10 mm long 100× magnification. (C) Ventral view of the jugular lymph sac and the lymph channels of a mouse embryo 22 mm long, taken from the gravid uterus and injected with India ink 20×. Abbreviations in images: ACN, accessories nerve; ACV, anterior branch cephalic vein; ALT, anterior venous tributary; CA, cephalic arch;

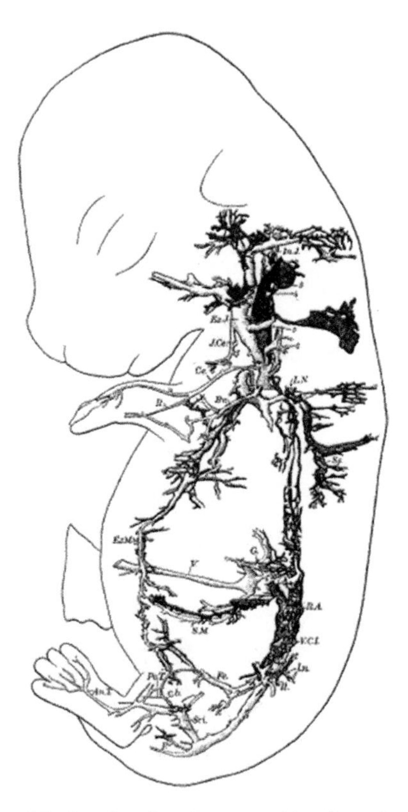

FIG. 2. Development of the lymph node anlagen in rabbits from the work of Frederic Lewis.[13] Rabbit, 20 days, 29 mm, H.E.C., Series 170, × 6.9 diams. The first lymph nodes develop at L. N., along the subscapular vein, Ss.; and at l.n., along the ilio-lumbare vein, Il. The veins of the arm are: Br., brachial; Ce., cephalic; J.Ce., jugulo-cephalic; R., radial. Those of the legs are: An. T., anterior tibial; Sci, sciatic; Po. T., posterior tibial; Fe., femoral; c. B., connecting branch between femoral and sciatic. P. Marks the pudic vein. Images from Ref. 13, pp. 108. Copyright 1905, John Willey & Sons.

CD, cuvierian duct; CaJLS, caudal portion jugular lymph sac; CeV, cephalic vein; CJA, common jugular approach; CJLS, cephalic portion jugular lymph sac; CLC, cutaneous lymph channel; DCLE, dorsal caudal lymphatic element; DVT, dorsal venous tributary; EJPCC, external jugular-precardinal confluence; EJV, external jugular vein; HG, hibernating gland; ICA, internal carotid artery; IJV, internal jugular vein; JCT, jugular confluence tap; JLS, jugular lymph sac; LT, lymphatic foramina; MB, muscle bundle; MDVT, median dorsal venous tributary; MJLS, median portion jugular lymph sac; MVLT, median ventral lymphatic tributary; NG, ganglion nodosum; PC, percava; PCL, precava lymph channel; PCV, post cardinal vein; PrCV, precardinal vein; PUL, primitive ulnar lymphatic; PUV, primitive ulnar vein; ScL, subclavian lymph channel; ScLC, subcutaneous lymph channel; SCA, subclavian vein; SG, sympathetic ganglion; SN, spinal nerve; SNT, sympathetic nerve trunk; TCA, thyro-cervical artery; TD, thoracic duct; VCLE, ventral cranial lymphatic element; VCaLE, ventral caudal lymphatic element; VLC, veno-lymphatic communication; VN, vagus nerve. All images from Ref. 12. Copyright 1926, John Willey & Sons.

TABLE I
EARLY STUDIES ON LYMPH NODE ANLAGEN DEVELOPMENT

Reference	Author	Year	Organism	Key findings
3	Breschet	1836	Human	Earliest description of LN anlagen development
4	Gulland	1894	Human	Further characteriztion of steps in human LN development
5	Saxer	1896	Ovine	Identification of "Wandering lymphocytes"
8	Ranvier	1897	Porcine	Developing anlagen is lined by endothelium
9	Kling	1904	Human	Lymphatic glands arise from a plexus of lymphatic ducts
10	Sabin	1905	Swine	Characterization of early stages in swine anlagen development
11	Lewis	1909	Human & Rabbit	Determination of first lymph glands
12	Higgins	1926	Mice	Defined stages in murine anlagen development
14	Rothermel	1929	Bovine	Characterization of mesenchymal condensation and proliferation
15	Block	1964	Opossum	–[a]
16	Archer	1964	Rabbit	–[a]
17	Miller	1965	Opossum	LNs develop despite embryonic thymectomy
18	Hostetler & Ackerman	1966	Rabbit	Developing stroma expresses alkaline phosphatase
19	Ackerman	1967	Feline	Lymphocytes are present in anlagen prior to thymus development
20	Hostetler & Ackerman	1969	Rabbit	Identification of rare "lymphocytes" in circulation prior to anlagen development. Multiple waves of "lymphocytes" in anlagen development

[a]These papers were unavailable for further analysis.

TABLE II
EARLY COMPREHENSIVE REVIEWS OF LN ANLAGEN DEVELOPMENT

Reference	Author	Year	Title
6	Sabin	1913	The origin and development of lymphatic system
21	Maximow	1927	Relation of blood cells to connective tissues and endothelium

hypothesis that lymphocytes originate *in situ* from local LN mesenchyme in LN anlagen, it prompted more careful analysis of the concept that lymphocytes in developing anlagen arise from bloodborne precursors rather than local mesenchyme.[26] Second, to readdress the origins of lymphoid cells in the developing LNs, the relative timing of lymphocyte appearance in the developing anlagen was compared to their appearance in the thymus of feline embryos. This analysis showed that lymphoid cells were present in the developing anlagen prior to thymic lymphopoiesis, demonstrating that lymphocytes found in the anlagen probably did not originate from the thymus.[19] These studies were later confirmed in rabbits, using electron microscopy, looking for the presence of very rare lymphocytes in early fetal blood. This analysis identified small numbers of lymphocyte-like cells in blood vessels of developing rabbit embryos identical to those lymphocytes found in the developing anlagen, thus identifying a putative lymphocyte or lymphocytic precursor cell that could enter into the LN anlagen from the blood.[20] These results were the first to give strong evidence that lymphocytic cells present in the primordia arise in the fetal liver and migrate through the blood to sites of anlagen development.[19,20] Finally, the thymus was shown not to be the origin of hematopoietic cells in the anlagen, from the analysis of embryonic thymectomy experiments in opossums.[17] Opossums are marsupials that allow direct access to developing embryos post birth prior to their entry in the pouch where they undergo further development. In thymectomized opossum embryos, LN development occurred normally, proving that lymphocyte output from the embryonic thymus was not critical for LN development.[17] Subsequent analysis of LNs in another marsupial, *Stonix brachyurus* (Wallaby), showed that lymphocytes were first observed in the thymus 2–4 days post birth and that LNs first appeared as aggregates of lymphocytes around lymphatic vessels at 5 days post birth, 9 days prior to the anlagen being colonized by lymphocytes from the thymus.[27] These observations provided additional evidence that lymphocytes found in the developing LN anlagen do not originate thymus.

These different anatomical studies helped identifying the following key steps in anlagen development using a number of different species[10–12,14]:

1. The joining of lymphatic vessels in an anastomosing plexus (lymph sac) is seen at places where LN anlage forms.
2. The surrounding mesenchyme is arranged into "connective tissue bridges" or trabeculae.
3. The lymphatic vessels increase in size and a mesenchymal condensation is triggered.
4. Capillaries enter the primitive anlagen.
5. Lymphocytes enter the condensing LN mesenchyme.

6. The primitive LN analge enlarges and its stroma increases in size and density due to the proliferation of mesenchymal cells.
7. Lymphocyte numbers increase, but the density of lymphocytes remains constant in the tissue for several days.
8. A rapid increase in lymphocytes in the LNs takes place.

Interestingly, recent observation of early anlage development has now been reanalyzed with the added advantage of fluorescent microscopy and an array of antibodies by a number of groups with a similar set of conclusions.

An interesting observation by Hostetler and Ackerman was that the developing mesenchymal stroma (later termed lymphoid tissue organizer, LTo) cells express alkaline phosphatase activity as they mature. This is a unique property of LN stroma not found in surrounding mesenchyme or epithelial cells.[18] Although endogenous alkaline phosphatase activity is often associated with BMP-2 expression in the bone, it has been shown previously that retinoic acid and tumor necrosis factor alpha (TNFα) can synergistically induce expression of alkaline phosphatase.[28,29] Interestingly, both retinoic acid and TNFα/TNFR1 signaling have recently been shown to have important roles in lymphoid tissue development.[30,31]

Although the development of the LN anlagen had been described, the phenotype of lymphocytes present in these structures prior to T cell egress from the thymus and their entry into peripheral lymphoid tissues was still unknown in the early 1990s. In 1992, Kelly and Scollay analyzed lymphocytes in neonatal LNs, tracking the arrival of thymic emigrants.[32] Analysis of the newborn LNs identified a previously unknown CD3$^-$ CD4$^+$ population of cells that were present *in situ* prior to infiltration of CD4$^+$ and CD8$^+$ lymphocytes from the thymus.[32] Utilizing fluorescence activated cell scanning (FACS) analysis, the authors showed that these cells did not express B220, MAC-1, or HSA, but expressed low levels of Thy-1 and high levels of CD44.[32] Giemsa-stained smears of purified cells showed the cells to be small lymphocytes, larger than naïve CD4$^+$ CD3$^+$ cells but smaller than activated lymphocytes.[32] Functional assays showed no phagocytic ability in these cells. Moreover, transfer of these cells into CD45 congenic recipients indicated no stem cell or progenitor cell activity that could be measured.[32] At this stage, the role of these cells was unclear and, indeed, the authors conclude, "We have no idea of the function or life history of these cells, nor do we know whether they are thymus derived."[32] The functional meaning of these cells only became clear 5 years later with the work of Mebius and Rennert. These seminal contributions characterized the function of CD4$^+$ CD3$^-$ LTβ^+ cells and lymphotoxin signaling in the development of LNs.[33–35] It is upon these key results that our current understanding of lymphoid tissue development is based.

III. Tools and Technologies Transforming Analysis of Lymphoid Tissue Development

A number of different technological advancements have permitted a new generation of scientists to revisit questions surrounding lymphoid tissue organogenesis. These included development of panels of monoclonal antibodies, multicolor flow cytometry, confocal microscopy, and, most importantly, genetically modified mice. It was through the discovery that LNs and Peyer's patches (PPs) fail to develop in lymphotoxin β-deficient mice that the field of lymphoid organogenesis re-emerged.[35] In fact, it was now possible to apply the power of genetics to further dissect the molecular pathways underlying lymphoid tissue formation. Further development of fluorescent proteins and *Cre* recombinase transgenic and knock-in mice has helped in providing additional insight into this developmental process. In this section we will review the role of these key technologies and discuss two emerging tools that are primed to provide new insightful perspectives on lymphoid organogenesis.

A. Gene Targeting in Lymphoid Tissue Development

The development of gene-specific knock-out and knock-in mice has transformed the understanding of immune development and function. Analysis of the key role of signaling pathways in lymphoid tissue development has been revealed using mice with specific deficiency in proteins required for LTi cell development; in molecules involved in the LTβ, RANK, and IL-7 signaling pathways, which are required for interactions between LTi/LTin cells and developing stroma cells; in chemokines required for the recruitment and retention of LTi cells; and in molecules required for LTin–stroma cell interactions.[1] Perhaps the three most important knockouts that have transformed our understanding of lymphoid tissues have been the $LT\beta^{-/-}$ mice showing a key role for LT signaling in lymphoid tissue development, $Rorc^{-/-}$ mice demonstrating a requirement for LTi cells in LN and PP organogenesis, and $Ret^{-/-}$ which revealed the role of LTin cells in the initiation of PP formation.[7,35,36] These genetic approaches have, however, been limited, since many proteins involved in lymphoid organ formation are also crucial for the early embryo development, and thus the generation of tissue-specific gene knock-out mice for some of these genes would be very welcome in the field.

B. Development of Tissue-Specific Fluorescent Protein Mice

A number of groups have utilized tissue-specific fluorescent protein expression to track the development of the LN anlagen and PPs. The development of green fluorescent protein (GFP) knock-in into the *Rorc* (RoRγ) locus provided a unique tool to analyze both early stages of LN anlagen development and the role of RoRc in LTi development.[36] The human CD2 locus controlling region coupled to GFP has also been utilized to analyze the development of LN anlagen.[37] The human CD2 LCR/promoter drives high levels of GFP expression on all T cells, on immature B cells, and on LTi and LTin cells.[7,39] During fetal development, expression of GFP in peripheral tissues is restricted to lymphoid-tissue-inducing cells, permitting real-time observation of lymphoid tissue development. Utilizing the ability to image and isolate developing LN anlagen and PPs has provided key insights into signaling events involved in both the initial steps in PP development and the role of IL-7 signals in the maturation of the LN stroma network.[7,37]

C. Generation of Tissue-Specific Cre Recombinase Mice

Tissue-specific Cre lines have been invaluable in the generation of tissue-specific knock-out mice and lineage tracing in tissue organogenesis. In lymphoid tissues, a number of different Cre transgenic and knock-in lines have been developed, permitting Cre-mediated recombination in LTi cells including the BAC transgenic RORγ Cre[38] and the human CD2 Cre.[39] Recently, a stroma-specific IL-7Cre BAC transgenic mouse line that leads to Cre activity in lymphoid stroma cells has been generated.[40] Utilization of these lines will permit lineage-specific deletion of genes required in other developmental pathways and lineage-specific tracing of LTi, LTin, and stroma cell lineages.

D. Use of Lineage-Tracing Tools

The combined use of tissue-specific *Cre* recombinase mice and Rosa26 reporter mice provides a tool to determine the lineage relationship between cell populations in lymphoid tissues. A number of different fluorescent, bioluminescent, and lacZ encoded lox-stop-lox reporter mouse strains have been generated. The most popular strains currently utilized are the Rosa26 eYFP and Rosa26 ACTB-tdTomato/EGFP reporter strains.[41,42] The Rosa26 eYFP cells are not fluorescent, and only after Cre-mediated excision of the transcriptional stop can they start to express eYFP. Conversely, Rosa26 ACTB-tdTomato/EGFP reporter cells express membrane-bound dtTomato fluorescent protein at high levels in all tissues and then express a membrane-bound eGFP after Cre-mediated recombination. The advantage of this second strain of mice is that the ACTB promoter element drives stronger expression of fluorescent proteins in comparison to native Rosa26 locus. In our hands, expression analysis of mGFP in lymphoid tissues from Rosa26 ACTB-tdTomato/EGFP reporter revealed a 10-fold increase in fluorescence when compared to Rosa26 eYFP mice both by flow cytometry and multiphoton imaging.

Because Cre-mediated deletion of lox flanked transcriptional stop signal in genetic reporters is permanent, all the progeny of cells that express or have expressed Cre are genetically marked. In the near future, we believe that stroma-specific Cre mouse lines will provide a powerful technology to trace the lineage relationship between different stroma cell populations.

E. Application of Fluorescent Stereomicroscopy

Taking advantage of fluorescent protein transgenic mice through direct fluorescent protein expression or from tissue-specific Cre bred to Rosa26 reporter mice, fluorescent stereomicroscopy permits both imaging and isolation of developing lymphoid organ primordia throughout different stages of their development. As an example, human CD2-GFP transgenic mice allow the observation and dissection of developing fetal anlagen from day E12.5 onward (Fig. 3). Utilizing fluorescent stereomicroscopy on hCD2-GFP mice, it is possible to observe single LTi cells in the developing embryo. In fact, clustering of tens of these GFP-labeled cells is sufficient for easy identification and microdissection of developing anlagen using fluorescent stereomicroscopy.

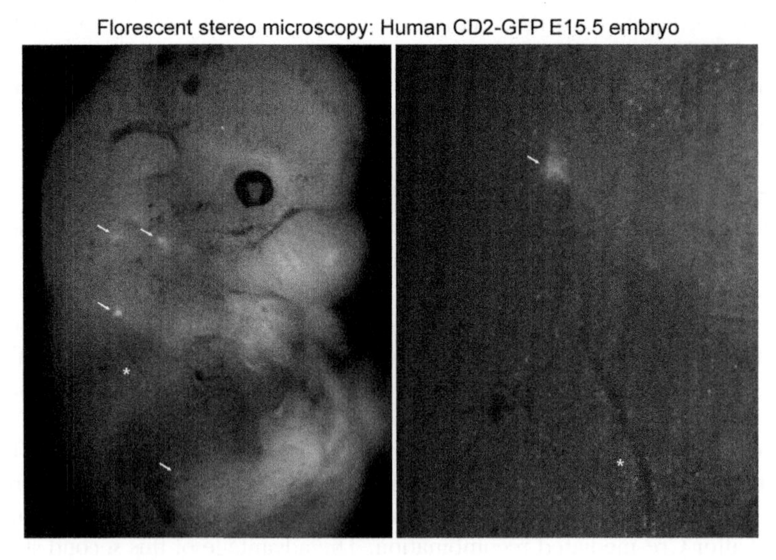

FIG. 3. Use of fluorescent stereo microscopy to visualize LN analagen development. Day 15.5 embryos from human CD2 LCR—GFP transgenic mice were analyzed using a Zeiss M2bio fluorescent stereo microscope. The LN anlagen are highlighted by the arrows in the embryo. Higher magnification of the area surrounding the axillary anlagen shows LTi cells in the vasculature (*) prior to entry into the developing anlagen. Images from Ref. 37. Copyright 2006 National Academy of Sciences, USA.

Throughout development, stereomicroscopy also allows observation of the steady condensation of the developing anlagen. Using this technology, it was observed that, in mice deficient in LTβ receptor signaling (*Aly*/*Aly* mice), hematopoietic cells accumulate in the developing anlagen, but subsequently dissipate into surrounding tissues rather than condensing.[37] Fluorescent stereomicroscopy combined with explant organ cultures and inline heating can also be utilized to create time-lapse images of developing PPs. Using real-time imaging of fetal intestines, it was shown that enteric hematopoietic cells have a similar velocity and random walk as observed by B and T lymphocytes in adult LNs.[7]

F. Applications of Confocal and Multiphoton Microscopy

Development of multicolor confocal microscopy has greatly enhanced our understandings on the cellular organization in developing LN and PPs. The utilization of whole-mount staining, clearing agents and confocal microscopy has now made it possible to image entire developing LN anlagen and PPs and creating three dimensional (3D) software reconstructions of these developing organs. This technology provides additional spatial context to the developing lymphoid tissues. The development of mice that express genetically encoded fluorescent proteins under control of tissue-specific promoters permits the application of 4D multiphoton confocal microscopy to image the process of lymphoid tissue formation in real time. Analysis of developing LN anlagen will permit direct quantification of hematopoietic cell behavior during LN anlagen formation. The development of multicolored fluorescent reporter mice allowing specific expression of different fluorescent proteins on hematopoietic and stroma will permit direct observation of cell–cell interactions in lymphoid tissue primordia.

G. Optical Projection Tomography

Optical projection tomography (OPT) is a 3D imaging technology developed by James Sharp, which is a particularly powerful methodology to image the process of organogenesis in developing embryos and structural modifications in lymphoid tissues.[43] This technology has now been applied to understanding mechanisms of tissue remodeling resulting from disease pathology.[44] Utilizing lymphocytic choriomeningitis virus (LCMV) infection, Kumar *et al.* utilized OPT to quantify changes to the structural organization of LNs, showing rapid and substantial changes to LN microarchitecture and then analyze the molecular and cellular factors required for lymphoid tissue remodeling. Using

OPT imaging, they showed a key role for LTα1β2 expressing B cells in modulating lymphoid tissue microarchitecture. This technology has the potential to provide key new insights into lymphoid tissue development.

H. Applications of Systems Biology to Lymphoid Tissue Development

Systems biology involves combining biological experimentation with computation modeling to develop a holistic understanding of the complex interactions involved in a biological process. The use of computer simulation serves two main purposes: to develop a scientific understanding of a complex biological system, and as a method to explore alternative hypothesis in lymphoid tissue development, some of which are very difficult to examine using experimental models. This is particularly powerful in the case of lymphoid tissue development that occurs in the relatively inaccessible developing embryo and where the modulation of physiological environments is very difficult. Utilizing measurements of cell velocities and behaviors with data from gene knock-outs, the formation of PPs can be described in unified modeling language (UML) (Fig. 4) and then used to construct an *in silico* simulation of PP formation. Utilizing these simulations, it is possible to test the specific role of physiology, cellular interactions, and biological factors in lymphoid tissue development.

IV. 1996 Onward: Requirements for Lymph Node and Peyer's Patch Development

Below is an account of contemporary views on how LNs develop.

A. Early Steps in Lymph Node Development

LNs develop at predetermined fixed positions in the body, which is in great contrast to PPs that form in variable numbers and not in fixed positions along the gut. This hallmark of LN development implies a developmental program possibly relying in nonmotile cells present at presumptive sites of anlagen formation. The most likely candidate for this role was for many decades the lymph sac formed by lymphatic endothelial cells.[45,46] As lymph sac formation is one of the earliest events during LN development and the clustering of hematopoietic LTi cells in those sites occurs right after their appearance, lymph sacs were thought to be a prerequisite for LN formation.[20,32–34,46] Thus, initially venous endothelial cells start to bud off to form the primitive lymph sacs at defined positions in the body, a processes that is dependent on *Prox1* expression.[43,47,48] Subsequently, lymph sac endothelial cells start to

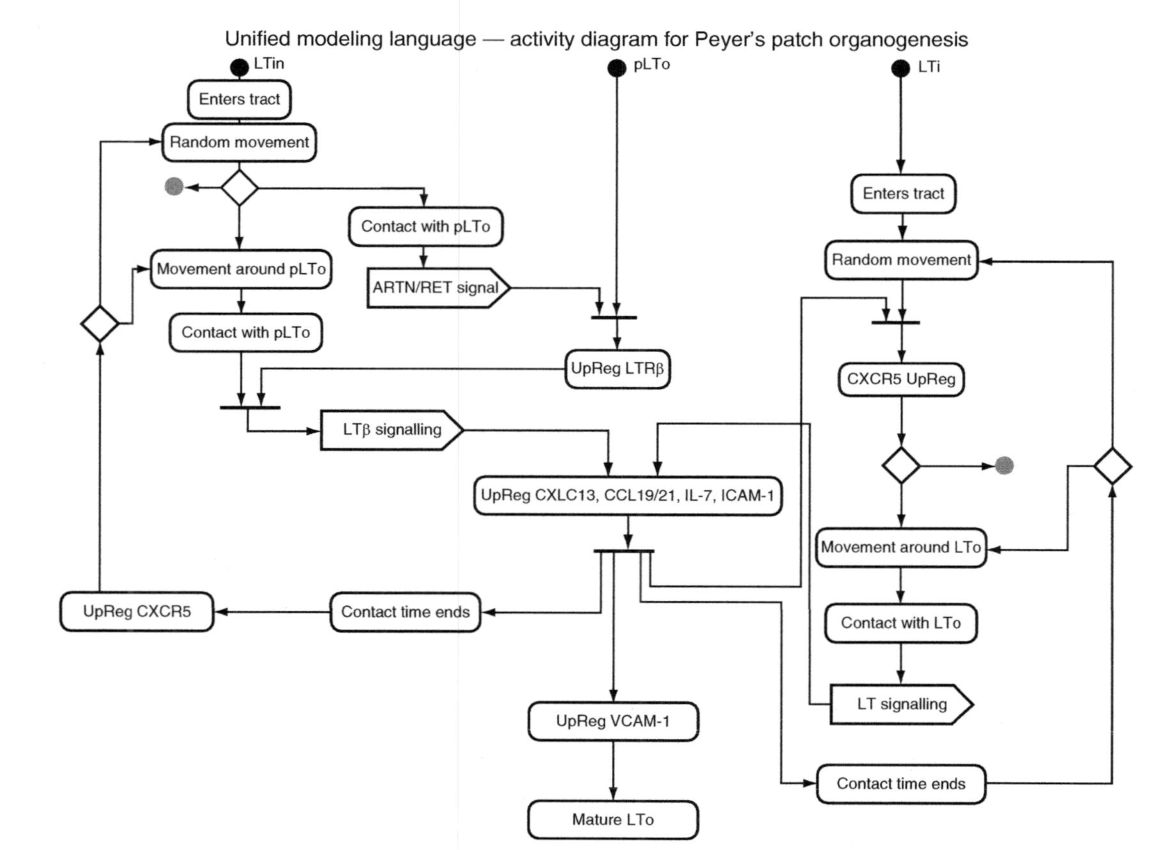

Fig. 4. Utilization of systems biology approach to understanding PP organogenesis. UML was utilized to describe the process involved in PP formation integrating information from imaging data and gene knock-out to create an activity diagram describing the interactions between three different cell types LTin, LTo, and LTin and the process involved in patch formation. This UML diagram has been utilized as the basis for a simulation model of the PP.

sprout, giving rise to the lymphatic vasculature network and to the LN subcapsullar sinus. As revealed by *Prox1* expression, murine lymph sacs start to form at E10.5 and the entire lymphatic vasculature formation is finished by E15.5.[48] This topologically and chronologically controlled developmental process led Sabin to propose in 1909 that LN originated from the lymph sac.[46] However, using *Prox1* null and *Prox1* conditional mutant embryos, Vondenhoff *et al.* have recently revealed that the initiation of LN development does not require lymph sacs.[49] Nevertheless, hematopoietic cell clustering in prospective LN sites was less efficient in *Prox1* conditional mutant embryos, suggesting that lymphatic endothelial cells do play an important role in the generation and/or maintenance of the LN hematopoietic cell niche.[49]

Interestingly, it was recently suggested that the first signals that trigger hematopoietic cell clustering at presumptive sites of LN development may be provided by adjacent neurons.[31] Neurons in the vicinity of LN anlagen were shown to express enzymes that are essential for retinoic acid synthesis, which in turn empowers stroma cells to express CXCL13, a powerful attractant of LTi cells.[31] It remains unclear, however, whether nervous signals themselves also play a role in PP development.

B. Hematopoietic Cell Subsets in Lymph Node Development

Cells from hematopoietic origin start to colonize future LN sites early on during embryonic organogenesis.[20,32–34] Emerging embryonic fetal liver cells, believed to be the progeny of a hematopoietic progenitor $CD3^{-}CD4^{-}cKit^{+}Il7R\alpha^{+}\alpha4\beta7^{+}$, start to colonize predetermined sites between embryonic days 9.5 (E9.5) and 16.5 (E16.5) depending on the type and location of the SLO.[7,35,50–52] Hematopoietic cells include $CD4^{+}CD3^{-}Il7R\alpha^{+}c\text{-}Kit^{+}$ LTi cells, which have been identified both in humans and mice.[33,53] LTi cells are identified by expression of RORγt in conjunction with the IL7R alpha chain, the expression of surface Lymphotoxin $\alpha_1\beta_2$, and the absence of lineage markers.[33,36,54,55] LTi cells are instrumental for lymphoid organogenesis as revealed by the analysis of mutant mice that lack this cell type. Thus, mice deficient for the helix-loop-helix inhibitor Id2 or the orphan nuclear hormone RORγt lack LTi cells and as a result LNs fail to develop.[56,57]

There are additional lines of evidence supporting the concept that LTi cells are involved in the induction of lymphoid structures. Thus, adoptive transfer of LTi cells in neonatal animals with impaired SLO development was shown to rescue the organogenesis of PP and nasal-associated lymphoid tissue (NALT).[58,59] In agreement with this idea, increased LTi cell numbers generated in IL7 overexpressing mice, result in the formation of higher numbers of LN and PP.[60]

In addition to LTi cells, other hematopoietic cell types were identified in the developing LNs. Analysis of RORγt-GFP mice, revealed that not all RORγt^{+} cells express CD4.[36] This population, possibly similar to the CD3^{-} CD4^{-}Il7Rα^{+} cells previously described in E15.5–16.5 intestines, represents a significant proportion of all hematopoietic cells in primitive LN structures.[36,50,55,61] Other hematopoietic cell types, such as cells expressing CD11b, were also described in the developing LN.[36]

The role of LTi cells in LN development takes place in a well-defined window of time.[35] Nevertheless, there is increasing evidence that an LTi-like population can be found in adult mice, although the latter appear to express different genes, such as OX40 and CD30 ligands, when compared to their embryonic counterparts.[38,62–64] Adult LTi cells were shown to play an active role in immunity, and in the adult intestine they were shown to reside mainly in the cryptopatches.[36,65–67] Adult LTi cells were also implicated in the reorganization of the splenic architecture after viral infections.[64] Interestingly, recent reports have described additional diversity within adult LTi-like cells, a fraction of which expresses the NK cell receptor NKp46.[53,68–71] Mouse intestinal CD3^{-} NKp46^{+}IL7Rα^{+}NK1.1^{-} cells lack expression of cytotoxic functions, but express IL22 and the orphan nuclear hormone receptor RORγt.[69–71] Interestingly, cells with similar characteristics have also been described in human tonsils, intestinal lamina propria, and PPs.[53,68]

C. Stroma Cell Subsets in Lymph Node Development

LTo cells are of mesenchymal origin and form the scaffold of the developing LN.[1,72,73] LTo cells found at LN anlagen can be very heterogeneous, expressing different levels of cytokines, adhesion molecules, and a variety of different receptors.[72,74]

LTo cells are induced to differentiate by LTi cells through LTαβ signaling. As a consequence, LTo cells upregulate VCAM-1 and ICAM-1 on their surface.[1,2] However, VCAM-1^{+} ICAM-1^{+} LTo cells can be further subdivided into VCAM-1^{int}ICAM-1^{int} and VCAM-1^{hi}ICAM-1^{hi} according to the expression levels of these molecules.[72] Interestingly, VCAM-1^{int}ICAM-1^{int} LTo cells are specifically reduced in peripheral LNs when compared to mesenteric LNs.[72] In fact, these differences in expression profiles may provide a reasonable explanation for the differential developmental requirements between peripheral and mesenteric LNs.

During their maturation process, it is believed that LTo cells gain the capacity to attract and retain hematopoietic cells in place. This was formally proven by adoptive transfer of cell suspensions from neonatal mesenteric LN.[75] In these experiments, upon intradermal injection of these cell suspensions, stroma cells of host origin were capable of attracting hematopoietic cells from donor origin, forming ectopic lymphoid structures.[75]

An organized network of stroma cells persists throughout adulthood. Cells of mesenchymal origin represent heterogeneous populations in the adult LN, occupying different niches and mediating different functions. Follicular dendritic cells (FDCs) of mesenchymal origin are detected around the periphery and within the B cell areas, while fibroblast reticular cells (FRCs) are located in T cell areas. Marginal reticular cells (MRCs) have also been described in the outer follicular region immediately underneath the subcapsular sinus of LNs.[76]

D. Major Signaling Axes in Lymph Node Development

It is generally accepted that lymphotoxin αβ (LTαβ) expressing hematopoietic cells induce maturation of lymphotoxin receptor (LTβR) positive mesenchymal LTo cells, which in turn upregulate VCAM-1, ICAM-1, and the homeostatic cytokines and chemokines IL7, CCL19, CCL21, and CXCL13.[77,78] Importantly, in the absence of LT/LTR signaling, the early steps of LN primordia formation occur normally, but LTi cells probably fail to survive in these rudiments, hampering further LN development from E16.5 onward.[37] Accordingly, analysis of E16.5 LN anlagen from *Ltα*$^{-/-}$ mice showed a crucial role for LTα in the modulation of VCAM-1, ICAM-1, and MADCAM-1 on local mesenchymal cells, supporting the concept that lymphotoxin signaling is a determinant of LTo cell development.[79] Thus, it is believed that, upon an initial productive interaction between LTi and LTo cells, a positive feedback loop involving LTαβ/LTβR signaling promotes and sustains the formation of lymphoid organ primordia.[1,2] Reinforcing this model were the observations that combined ablation of CXCL13, CCL21, and CCL19 or CCR7 and CXCR5 resulted in failure of LN and PP development.[80,81]

Despite a crucial role for LTαβ/LTβR axis in lymphoid organogenesis, as demonstrated by disruption in the development of LN in *Lta*$^{-/-}$, *Ltb*$^{-/-}$, and *Ltbr*$^{-/-}$ mice, this signaling axis is unlikely to be sufficient to ensure SLO development.[1,2] In fact, while *in utero* agonist anti-LTβR treatment rescues LN development in *Lta*$^{-/-}$ mice, the same treatment in LTi-deficient *RORgt*$^{-/-}$ mice fails to promote SLO development.[36,82] These observations support the hypothesis that LTi cells, or other hematopoietic cell subsets, provide yet unknown signals to LTo cells in addition to those provided by the LTαβ/LTβR signaling axis. Elucidating these additional molecular and cellular axes is a key challenge for the future understanding of lymphoid organogenesis.

The TNF family member TRANCE, its receptor RANK, and a molecule involved in TRANCE signaling, TRAF 6, have also been implicated in LN development.[83–86] TRANCE is expressed by LTo cells at prospective sites of LN development and provides a survival and differentiation signal to LTi cells, which in turn upregulate LTαβ expression.[72,78] Accordingly, *Trance*-deficient embryos have reduced LTi cell numbers, which exhibit poor capacity to induce maturation of LTo cells.[84]

E. Requirements for Peyer's Patches Development

PPs are specialized SLOs located in the lamina propria of the mucosa and extending into the submucosa of the ileum. Adult PPs have a similar structure and organization as those of LNs, with B and T cells segregating in well-defined and organized regions.[87] Interestingly, this pattern of B and T lymphocyte segregation is totally independent of infection or antigen stimulation since germfree mice also display discrete B and T cell areas.[88] However, there are specialized aspects to the organization of mature PPs. Thus, the luminal side of the PP exhibits a protrusion that is covered by a layer of follicle-associated enteric epithelium containing specialized epithelial cells, designated as M cells, which play an important role in mediating immune functions.[89] Compared to LNs, PPs integrate with the lymphatic network in a slightly different manner since they lack afferent lymphatics. Nevertheless, PPs have efferent lymphatics that emerge from lymphatic sinuses on the serosal side of the PPs and drain the lymph and immune cells to the mesenteric LN and then to the thoracic duct.[90]

In mice, PP development occurs during embryonic life. PPs develop in variable numbers (from 5 to 12) at the antimesenteric side of the mid-intestine, initiating their development in the duodenum and proximal ileum; subsequently, the most distal PP primordia are formed at more or less regular intervals in the lower mid-gut.[54,91]

F. Hematopoietic Cell Subsets in Peyer's Patch Development

The cellular mechanisms implicated in PP development are well characterized and, as for LN development, PP genesis relies on interactions mostly between cells of hematopoietic and mesenchymal origin. Thus, there are clear parallels between PP and LN development; however, differential processes are found, increasingly suggesting that that they may require specialized cellular and molecular players.

Similar to LN development, the first hematopoietic cells that colonize the embryonic intestine are thought to be the progeny of a fetal liver progenitor $CD3^{-}CD4^{-}cKit^{+}IL7R\alpha^{+}\alpha4\beta7^{+}$, which is the fetal equivalent of the common lymphoid precursor that is present in adult bone marrow.[50,51] The colonization of the intestine by hematopoietic cells starts at E12.5 and is initially restricted to the pyloric/duodenum region and the caecum.[7] Although the reasons behind this discrete distribution are unknown, it is possible that the initial sites of colonization are enriched for certain chemokines that trigger the initial clustering process. Progressively, the number of hematopoietic cells in the gut increases, and by E15.5 a high number of motile hematopoietic cells are found evenly distributed throughout the intestine wall.[7,52,54,55] By E16.5, some of these cells start to aggregate at variable places in the duodenum and proximal ileum to form the first PP primordia.[7,52,54,55] Therefore, conversely to LN development, which always occurs at defined positions in the body, location of the PP primordia appears to have a stochastic element.

The initial hematopoietic population that colonizes the intestine is very heterogeneous. Similar to LN development, $CD4^+CD3^-IL7R\alpha^+c\text{-}Kit^+$ LTi cells are one of the main protagonists of PP genesis.[52,54,56] However, a distinct hematopoietic cell type named lymphoid tissue initiator (LTin) cell, which has the surface phenotype $CD4^-CD3^-c\text{-}Kit^+IL7R\alpha^-CD11c^+$, was also described in the developing gut.[7,92] From E16.5 onward, LTi and LTin cells aggregate with mesenchymal origin $VCAM\text{-}1^+ICAM\text{-}1^+$ LTo cells, signaling the onset of the PP anlagen.[7,54,55,77,93]

PPs fail to develop in the absence of LTi cells observed in helix-loop-helix inhibitor Id2 or orphan nuclear hormone RORγt-deficient mice.[56,57] Accordingly, adoptive transfer of LTi cells into neonatal mice with reduced numbers of PPs rescues the development of these enteric structures.[58] Furthermore, in transgenic models that express high levels of IL7, and therefore increased LTi cell numbers, the number of PPs is also increased.[60] Interestingly, in these transgenic E16.5 guts, LTi cells induce a continuum stretch of $VCAM\text{-}1^+$ stroma cells on the antimesenteric side of the gut, which eventually resolves into discrete patches before the neonatal period.[60]

Selective but partial ablation of LTin cells, through the administration of diphtheria toxin (DT) to CD11c-DT receptor embryos, results in impaired PP development, demonstrating that a full complement of LTin cells is required for normal PP genesis.[7] Supporting this idea is also the fact that mice deficient for the tyrosine kinase receptor RET, expressed by LTin cells, fail to form PPs despite the fact that LTi cells are present in normal numbers in the guts of these mice.[7] Taken together, these observations suggest that LTi cells are essential but not sufficient for PP development (Fig. 5).

G. Stroma Cell Subsets in Peyer's Patch Development

Prior to hematopoietic cell clustering, rare $VCAM\text{-}1^+$ patches are identified in E15.5 guts; nevertheless, VCAM-1 expression levels at this stage are clearly negligible when compared to E17.5,[54] strongly supporting the concept that complete and productive differentiation of LTo cells requires interaction with a significant number of hematopoietic cells. In fact, these interactions are key to PP development. LTin and LTi cells cluster with $VCAM\text{-}1^+ICAM\text{-}1^+$ LTo cells so that $VCAM\text{-}1^+$ patches are unequivocally identified by E16.5 as a hallmark of PP primordia formation.[7,54,55,93] During the initial phases of PP anlagen formation, all cellular components, LTo, LTin, and LTi cells, are evenly distributed within the primitive PP cluster.[7,93] Nevertheless, it is possible that

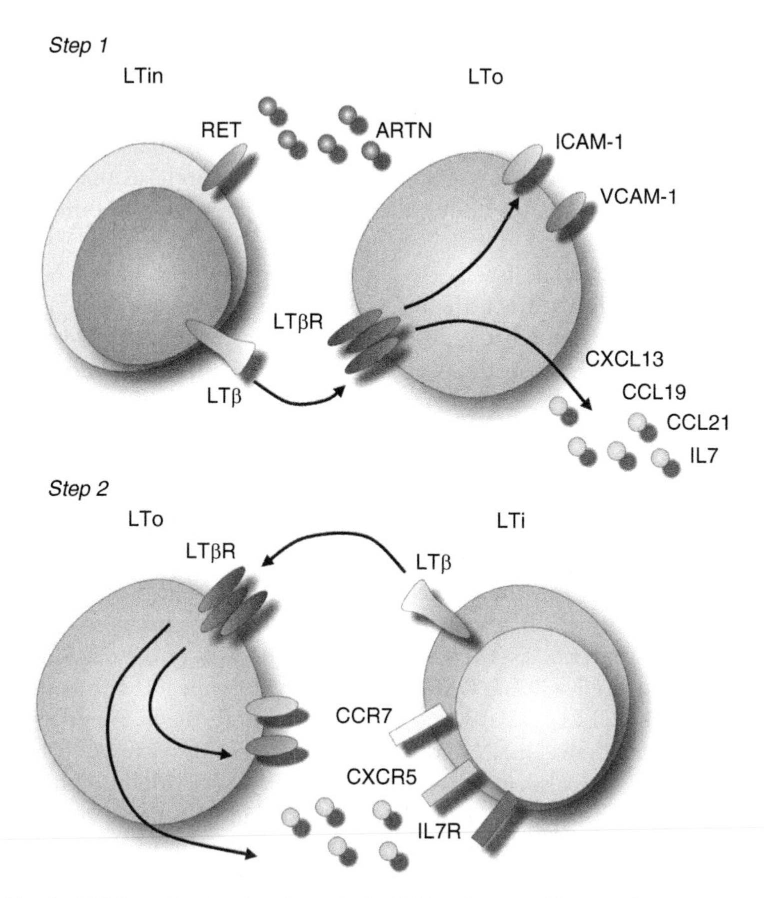

FIG. 5. Cellular and molecular players during PP development. Stroma cells at prospective PP locations produce the RET ligand ARTN. RET expressing LTin cells are then believed to provide initial maturation signals to stroma cells, possibly via LT/LTR signaling. In their turn, stroma cells start to express adhesion molecules and to produce chemokines and cytokines. The combined effect of these molecules attracts LTi cells that further activate stroma cells by the production of LT and TNF. A positive feedback loop is then created resulting in the recruitment of additional LTi cells. Adapted from Ref. 92.

LTo cells are key players in mediating the final structural organization within the PP primordia since they are among the first cell types to segregate into follicle-like structures by E18.5.[93]

Interestingly, the heterogeneity of hematopoietic cells found in PP primordia is also observed in enteric LTo cells. As a matter of fact, PP's stroma cells fall within two categories according to VCAM-1 and ICAM-1 expression and can be subdivided in VCAM-1^{hi}ICAM-1^{hi} and VCAM-1^{int}ICAM-1^{int}.[74]

The existence of these two PP LTo subsets creates an obvious parallel to LN LTo subsets.[75] Nevertheless, it is not known whether, similar to LN LTo cells, PP VCAM-1^{hi}ICAM-1^{hi} cells are the main producers of the homeostatic chemokines CXCL13, CCL21, and CCL19.[75] The observation that both hematopoietic and stroma cells are heterogeneous raise the hypothesis that different LTo cells may preferentially interact and provide different cues to different PP hematopoietic cells.

It is clear by now that, irrespective of the lymphoid organ, LTo cells have many common features. Nevertheless, despite the obvious parallels, LN and PP LTo cells exhibit remarkable differences that may well justify some of the differential requirements for the genesis of these structures. For example, LTo cells from mesenteric LNs have surface expression of TRANCE, while their PP counterparts lack surface expression of this ligand.[72,74] These distinct gene expression profiles strongly support the notion that the differential activity of LTo cells determines the key molecular signaling axes that are used in different organs, as the example of TRANCE and IL7R signaling in LN and PP, respectively, testifies.

The genetic signature of PP and LN LTo cells was further assessed by microarray analysis, which revealed additional differences between these cell types. Thus, mesenteric LN LTo cells express high levels of the cytokines and chemokines IL6, IL7, CCL7, CXCL1, and CCL11, while PP LTo cells have higher levels of CCL21, CCL19, and CXCL13.[74] Interestingly, the differential expression profile between LN and PP LTo cells is not restricted to cytokines and chemokines but also includes transcription factors. Meox2, Lhx8, and Prrx1 are highly expressed by LN LTo cells, but their functional significance is completely unclear.[74] Furthermore, it would be important to clarify whether different genetic signatures are cell autonomous or induced by interactions with different cellular players.

The PP stroma cell network persists throughout adulthood. FDCs of mesenchymal origin distribute in the periphery and within the B cell areas of PP.[94] FRCs locate in T cell areas, produce IL7, and create the conduit system, but it is currently unknown if conduits exist in PP.[95–97] MRCs have also been described in the PP subepithelial dome but their function remains elusive.[76,98,99]

H. Major Signaling Axes in Peyer's Patch Development

Similar to LN development, the signaling axis LTαβ/LTβR plays a central and essential role in PP genesis. This is clearly demonstrated by $Lta^{-/-}$, $Ltb^{-/-}$, and $Ltbr^{-/-}$ mice which fail to develop PP.[1,2] In fact, mesenchymal LTo cells, which express LTβR, upregulate VCAM-1, ICAM-1, IL7, CCL19, CCL21, and CXCL13 upon triggering by LT, thereby creating a positive feedback loop that retains and attracts additional hematopoietic cells.[77,78]

Despite the clear parallels between LN and PP development, some signaling pathways are not used similarly in the genesis of both structures. Probably, the best illustration of such differential usage is provided by IL7/IL7R and TRANCE/TRANCE-R signaling axes. Thus, while IL7R signal is determinant for PP development, as revealed by *Il7r*$^{-/-}$ mice, brachial, axillary, and mesenteric LN develop normally in these animals.[52,55,73] Furthermore, while in *Trance*$^{-/-}$ and *Traf6*$^{-/-}$ mice LN development is severely compromised, PP form normally in these mice.[83–86]

The tyrosine kinase receptor RET also plays a differential role in LN and PP genesis. This is revealed by the absence of PP in *Ret* null embryos, which have seemingly normal LN anlagen development.[7] Importantly, the RET ligand ARTN is a strong attractant of gut hematopoietic cells, as revealed by the use ARTN in explants gut cultures, which in turn induce LTin and LTi clustering and upregulation of VCAM-1 by mesenchymal cells, thus resulting in ectopic enteric lymphoid structures.[7]

It was recently suggested that the triggering event of LN genesis may be mediated by retinoic acid signals that are provided by adjacent neurons.[31] It remains unclear whether such signaling axis operates in the intestine. *Gdnf* and *Gfra1* null embryos that fail to develop a myenteric nervous plexus have normal PP development, arguing against such a hypothesis.[7,100,101] However, parasympathetic and/or sympathetic nervous axons, still present in the guts of these mutants, may provide such retinoic acid cues for PP formation.

I. Organization Within the Peyer's Patch Rudiment

In mice, the first B cell follicles are detected after birth, but the initial setup of PP architecture starts during embryonic life.[93,102,103] Cells positive for IL7Rα (possibly LTi cells) and VCAM-1$^+$ LTo cells initially distribute equally throughout the primordium at E17.5, while CD11c$^+$ (possibly LTin cells) locate at the periphery of PP primordium.[93] By E18.5, IL7Rα$^+$, CD11c$^+$, and VCAM-1$^+$ cells aggregate in irregular but discrete regions that around birth form follicle-like subregions. In these subregions, each cell type distributes differently; IL7Rα$^+$ cells occupy the center, while VCAM-1$^+$ and CD11c$^+$ cells occupy the periphery.[93] IL7Rα$^+$ and CD11c$^+$ cells have different levels CXCR5 and CCR7, and therefore it is possible that these cells may have different affinities for areas with differential expression for the respective chemokines.[93] Altogether, these observations suggest that the main cellular players in PP genesis, LTi, LTin, and LTo cells, trigger the initial architecture of PP prior to mature lymphocyte colonization. In fact, this initial architectural PP organization is totally independent of B or T cells, as demonstrated by the fact in *scid/scid* mice this process occurs in a perfectly normal manner.[93]

Homing of T and B lymphocytes into PP occurs at the level of the high endothelial venules (HEVs). The function and establishment of HEVs is partly dependent on MAdCAM-1 expression.[34] In the PP anlagen, MAdCAM-1 positive cells are initially detected at E16.5, and by E17.5 a reticulated yet nonorganized pattern of MAdCAM-1 expression is then detected.[93] The appearance of an organized network of MAdCAM-1 positive cells occurs around birth and coincides with lymphocyte colonization and the formation of follicle structures composed by IL7Rα, CD11c, and VCAM-1 positive cells.[93]

References

1. Randall TD, Carragher DM, Rangel-Moreno J. Development of secondary lymphoid organs. *Annu Rev Immunol* 2008;**26**:627–50.
2. Vondenhoff MF, Kraal G, Mebius RE. Lymphoid organogenesis in brief. *Eur J Immunol* 2007;**37**(Suppl 1):S46–52.
3. Breschet G. Le Systeme Lymphatique considere sous les rapports anatomique, physiologique et pathologique, Chez J.-B. Bailliere Libraire de l'academie Royale de Medecine. 1836.
4. Gulland G. The development of lymphatic glands. *J Pathol Bacteriol* 1894;**2**:447–85.
5. Saxer F. Uber die Entwickelung und den Bau der normalen Lymphdrusen und die Entstehung der roten und weissen Blutkorperchen. *Anat Hefte* 1896;**6**:347–532.
6. Sabin FR. *The origin and development of the lymphaticsystem. Hospital Reports Monographs New Series No V.* The Johns Hopkins Press; 1931; 65–70.
7. Veiga-Fernandes H, Coles MC, Foster KE, Patel A, Williams A, Natarajan D, et al. Tyrosine kinase receptor RET is a key regulator of Peyer's Patch organogenesis. *Nature* 2007;**446**:547–51.
8. Ranvier L. Morphologie et developpement du systeme lymphatiques. *Arch D'Anat Microsc* 1897;**1**:137–52.
9. Kling C. Studien uber die Entwicklung der Lymphdrusen beim Menshchen. *Arch F Mikr Anat* 1904;**63**:575–610.
10. Sabin F. The development of the lymphatic nodes in the pig and their relationship to the lymph hearts. *Am J Anat* 1905;**4**:355–89.
11. Lewis F. The first lymph glands in rabbit and human embryos. *Anat Rec* 1909;**3**:341–53.
12. Higgins G. The jugular lymph sac in the albino mouse. *Am J Anat* 1926;**37**:95–125.
13. Lewis FT. The development of the lymphatic system in rabbits. *Am J Anat* 1905;**5**:95–111.
14. Rothermel J. A developmental study of the medial retropharyngeal lymphatic node of the calf (Bos Taurus). *Am J Anat* 1929;**43**:461–96.
15. Block M. The blood forming tissues and blood of the newborn opossum (*Didelphys Virginiana*) I. Normal development through about the one hundredth day of life. *Ergeb Anat Entwicklungsgesch* 1964;**37**:237–366.
16. Archer OK, Sutherland DE, Good RA. The developmental biology of lymphoid tissue in the rabbit. Consideration of the role of thymus and appendix. *Lab Invest* 1964;**13**:259–71.
17. Miller JF, Block M, Rowlands Jr. DT, Kind P. Effect of thymectomy on hematopoietic organs of the opossum "embryo". *Proc Soc Exp Biol Med* 1965;**118**:916–21.
18. Hostetler JR, Ackerman GA. The relationship between the histological localization of alkaline phosphatase activity and appearance of lymphocytes in lymphocytic tissue of the embryonic and neonatal rabbit. *Anat Rec* 1966;**156**:191–201.

19. Ackerman GA. Developmental relationship between the appearance of lymphocytes and lymphopoietic activity in the thymus and lymph nodes of the fetal cat. *Anat Rec* 1967;**158**:387–99.
20. Hostetler JR, Ackerman GA. Lymphopoiesis and lymph node histogenesis in the embryonic and neonatal rabbit. *Am J Anat* 1969;**124**:57–75.
21. Maximow A. Relation of blood cells to connective tissues and endothelium. *Physiol Rev* 1924;**4**:533–63.
22. Maximow A. Untersuchungen uber Blut und Bindegewebe. I Die fuhsten Entwicklungsstadien der Blut- und Bindegewebeszellen beim Saugetierembeyo. *Arch F Mikr Anat* 1909; **lxxiii**:444.
23. Moore MA, Owen JJ. Experimental studies on the development of the thymus. *J Exp Med* 1967;**126**:715–26.
24. Owen JJ, Ritter MA. Tissue interaction in the development of thymus lymphocytes. *J Exp Med* 1969;**129**:431–42.
25. Tyan ML, Herzenberg LA. Studies on the ontogeny of the mouse immune system. II. Immunoglobulin-producing cells. *J Immunol* 1968;**101**:446–50.
26. Ackerman GA, Hostetler JR. Morphological studies of the embryonic rabbit thymus: the in situ epithelial versus the extrathymic derivation of the initial population of lymphocytes in the embryonic thymus. *Anat Rec* 1970;**166**:27–45.
27. Ashman RB, Papadimitriou JM. Development of lymphoid tissue in a marsupial, *Setonix brachyurus* (quokka). *Acta Anat (Basel)* 1975;**91**:594–611.
28. Ng KW, Hudson PJ, Power BE, Manji SS, Gummer PR, Martin TJ. Retinoic acid and tumour necrosis factor-alpha act in concert to control the level of alkaline phosphatase mRNA. *J Mol Endocrinol* 1989;**3**:57–64.
29. Reese DH, Fiorentino GJ, Claflin AJ, Malinin TI, Politano VA. Rapid induction of alkaline phosphatase activity by retinoic acid. *Biochem Biophys Res Commun* 1981;**102**:315–21.
30. Koni PA, Flavell RA. A role for tumor necrosis factor receptor type 1 in gut-associated lymphoid tissue development: genetic evidence of synergism with lymphotoxin beta. *J Exp Med* 1998;**187**:1977–83.
31. van de Pavert SA, Olivier BJ, Goverse G, Vondenhoff MF, Greuter M, Beke P, et al. Chemokine CXCL13 is essential for lymph node initiation and is induced by retinoic acid and neuronal stimulation. *Nat Immunol* 2009;**10**:1193–9.
32. Kelly KA, Scollay R. Seeding of neonatal lymph nodes by T cells and identification of a novel population of CD3-CD4+ cells. *Eur J Immunol* 1992;**22**:329–34.
33. Mebius RE, Rennert P, Weissman IL. Developing lymph nodes collect CD4+CD3- LTbeta+ cells that can differentiate to APC, NK cells, and follicular cells but not T or B cells. *Immunity* 1997;**7**:493–504.
34. Mebius RE, Streeter PR, Michie S, Butcher EC, Weissman IL. A developmental switch in lymphocyte homing receptor and endothelial vascular addressin expression regulates lymphocyte homing and permits CD4+ CD3- cells to colonize lymph nodes. *Proc Natl Acad Sci USA* 1996;**93**:11019–24.
35. Rennert PD, Browning JL, Mebius R, Mackay F, Hochman PS. Surface lymphotoxin alpha/beta complex is required for the development of peripheral lymphoid organs. *J Exp Med* 1996;**184**:1999–2006.
36. Eberl G, Marmon S, Sunshine MJ, Rennert PD, Choi Y, Littman DR. An essential function for the nuclear receptor RORgamma(t) in the generation of fetal lymphoid tissue inducer cells. *Nat Immunol* 2004;**5**:64–73.
37. Coles MC, Veiga-Fernandes H, Foster KE, Norton T, Pagakis SN, Seddon B, et al. Role of T and NK cells and IL7/IL7r interactions during neonatal maturation of lymph nodes. *Proc Natl Acad Sci USA* 2006;**103**:13457–62.

38. Eberl G, Littman DR. Thymic origin of intestinal alphabeta T cells revealed by fate mapping of RORgammat+ cells. *Science* 2004;**305**:248–51.
39. de Boer J, Williams A, Skavdis G, Harker N, Coles M, Tolaini M, et al. Transgenic mice with hematopoietic and lymphoid specific expression of Cre. *Eur J Immunol* 2003;**33**:314–25.
40. Repass JF, Laurent MN, Carter C, Reizis B, Bedford MT, Cardenas K, et al. IL7-hCD25 and IL7-Cre BAC transgenic mouse lines: new tools for analysis of IL-7 expressing cells. *Genesis* 2009;**47**:281–7.
41. Srinivas S, Watanabe T, Lin CS, William CM, Tanabe Y, Jessell TM, et al. Cre reporter strains produced by targeted insertion of EYFP and ECFP into the ROSA26 locus. *BMC Dev Biol* 2001;**1**:4.
42. Muzumdar MD, Tasic B, Miyamichi K, Li L, Luo L. A global double-fluorescent Cre reporter mouse. *Genesis* 2007;**45**:593–605.
43. Sharpe J, Ahlgren U, Perry P, Hill B, Ross A, Hecksher-Sorensen J, et al. Optical projection tomography as a tool for 3D microscopy and gene expression studies. *Science* 2002;**296**:541–5.
44. Kumar V, Scandella E, Danuser R, Onder L, Nitschke M, Fukui Y, et al. Global lymphoid tissue remodeling during a viral infection is orchestrated by a B cell-lymphotoxin-dependent pathway. *Blood* 2010;**115**(23):4725–33.
45. Sabin FR. On the origin of the lymphatic system from the veins and the development of the lymph hearts and thoracic duct in the pig. *Am J Anat* 1902;**1**:367–89.
46. Sabin FR. The lymphatic system in human embryos, with a consideration of the morphology of the system as a whole. *Am J Anat* 1909;**9**:43–91.
47. Srinivasan RS, Dillard ME, Lagutin OV, Lin FJ, Tsai S, Tsai MJ, et al. Lineage tracing demonstrates the venous origin of the mammalian lymphatic vasculature. *Genes Dev* 2007;**21**:2422–32.
48. Wigle JT, Oliver G. Prox1 function is required for the development of the murine lymphatic system. *Cell* 1999;**98**:769–78.
49. Vondenhoff MF, van de Pavert SA, Dillard ME, Greuter M, Goverse G, Oliver G, et al. Lymph sacs are not required for the initiation of lymph node formation. *Development* 2009;**136**:29–34.
50. Yoshida H, Kawamoto H, Santee SM, Hashi H, Honda K, Nishikawa S, et al. Expression of alpha(4)beta(7) integrin defines a distinct pathway of lymphoid progenitors committed to T cells, fetal intestinal lymphotoxin producer, NK, and dendritic cells. *J Immunol* 2001;**167**:2511–21.
51. Mebius RE, Miyamoto T, Christensen J, Domen J, Cupedo T, Weissman IL, et al. The fetal liver counterpart of adult common lymphoid progenitors gives rise to all lymphoid lineages, CD45+CD4+CD3- cells, as well as macrophages. *J Immunol* 2001;**166**:6593–601.
52. Adachi S, Yoshida H, Honda K, Maki K, Saijo K, Ikuta K, et al. Essential role of IL-7 receptor alpha in the formation of Peyer's patch anlage. *Int Immunol* 1998;**10**:1–6.
53. Cupedo T, Crellin NK, Papazian N, Rombouts EJ, Weijer K, Grogan JL, et al. Human fetal lymphoid tissue-inducer cells are interleukin 17-producing precursors to RORC+ CD127+ natural killer-like cells. *Nat Immunol* 2009;**10**:66–74.
54. Adachi S, Yoshida H, Kataoka H, Nishikawa S. Three distinctive steps in Peyer's patch formation of murine embryo. *Int Immunol* 1997;**9**:507–14.
55. Yoshida H, Honda K, Shinkura R, Adachi S, Nishikawa S, Maki K, et al. IL-7 receptor alpha+ CD3(-) cells in the embryonic intestine induces the organizing center of Peyer's patches. *Int Immunol* 1999;**11**:643–55.
56. Sun Z, Unutmaz D, Zou YR, Sunshine MJ, Pierani A, Brenner-Morton S, et al. Requirement for RORgamma in thymocyte survival and lymphoid organ development. *Science* 2000;**288**:2369–73.

57. Yokota Y, Mansouri A, Mori S, Sugawara S, Adachi S, Nishikawa S, et al. Development of peripheral lymphoid organs and natural killer cells depends on the helix-loop-helix inhibitor Id2. *Nature* 1999;**397**:702–6.
58. Finke D, Acha-Orbea H, Mattis A, Lipp M, Kraehenbuhl J. CD4+CD3- cells induce Peyer's patch development: role of alpha4beta1 integrin activation by CXCR5. *Immunity* 2002;**17**:363–73.
59. Fukuyama S, Hiroi T, Yokota Y, Rennert PD, Yanagita M, Kinoshita N, et al. Initiation of NALT organogenesis is independent of the IL-7R, LTbetaR, and NIK signaling pathways but requires the Id2 gene and CD3(-)CD4(+)CD45(+) cells. *Immunity* 2002;**17**:31–40.
60. Meier D, Bornmann C, Chappaz S, Schmutz S, Otten LA, Ceredig R, et al. Ectopic lymphoid-organ development occurs through interleukin 7-mediated enhanced survival of lymphoid-tissue-inducer cells. *Immunity* 2007;**26**:643–54.
61. White A, Carragher D, Parnell S, Msaki A, Perkins N, Lane P, et al. Lymphotoxin a-dependent and -independent signals regulate stromal organizer cell homeostasis during lymph node organogenesis. *Blood* 2007;**110**:1950–9.
62. Kim MY, Anderson G, White A, Jenkinson E, Arlt W, Martensson IL, et al. OX40 ligand and CD30 ligand are expressed on adult but not neonatal CD4+CD3- inducer cells: evidence that IL-7 signals regulate CD30 ligand but not OX40 ligand expression. *J Immunol* 2005;**174**:6686–91.
63. Kim MY, Toellner KM, White A, McConnell FM, Gaspal FM, Parnell SM, et al. Neonatal and adult CD4+ CD3- cells share similar gene expression profile, and neonatal cells up-regulate OX40 ligand in response to TL1A (TNFSF15). *J Immunol* 2006;**177**:3074–81.
64. Scandella E, Bolinger B, Lattmann E, Miller S, Favre S, Littman DR, et al. Restoration of lymphoid organ integrity through the interaction of lymphoid tissue-inducer cells with stroma of the T cell zone. *Nat Immunol* 2008;**9**:667–75.
65. Lane PJ, McConnell FM, Withers D, Gaspal F, Saini M, Anderson G. Lymphoid tissue inducer cells: bridges between the ancient innate and the modern adaptive immune systems. *Mucosal Immunol* 2009;**2**:472–7.
66. Pabst O, Herbrand H, Friedrichsen M, Velaga S, Dorsch M, Berhardt G, et al. Adaptation of solitary intestinal lymphoid tissue in response to microbiota and chemokine receptor CCR7 signaling. *J Immunol* 2006;**177**:6824–32.
67. Bouskra D, Brezillon C, Berard M, Werts C, Varona R, Boneca IG, et al. Lymphoid tissue genesis induced by commensals through NOD1 regulates intestinal homeostasis. *Nature* 2008;**456**:507–10.
68. Cella M, Fuchs A, Vermi W, Facchetti F, Otero K, Lennerz JK, et al. A human natural killer cell subset provides an innate source of IL-22 for mucosal immunity. *Nature* 2009;**457**:722–5.
69. Luci C, Reynders A, Ivanov II, Cognet C, Chiche L, Chasson L, et al. Influence of the transcription factor RORgammat on the development of NKp46+ cell populations in gut and skin. *Nat Immunol* 2009;**10**:75–82.
70. Satoh-Takayama N, Vosshenrich CA, Lesjean-Pottier S, Sawa S, Lochner M, Rattis F, et al. Microbial flora drives interleukin 22 production in intestinal NKp46+ cells that provide innate mucosal immune defense. *Immunity* 2008;**29**:958–70.
71. Sanos SL, Bui VL, Mortha A, Oberle K, Heners C, Johner C, et al. RORgammat and commensal microflora are required for the differentiation of mucosal interleukin 22-producing NKp46+ cells. *Nat Immunol* 2009;**10**:83–91.
72. Cupedo T, Vondenhoff MF, Heeregrave EJ, De Weerd AE, Jansen W, Jackson DG, et al. Presumptive lymph node organizers are differentially represented in developing mesenteric and peripheral nodes. *J Immunol* 2004;**173**:2968–75.
73. Mebius RE. Organogenesis of lymphoid tissues. *Nat Rev Immunol* 2003;**3**:292–303.

74. Okuda M, Togawa A, Wada H, Nishikawa S. Distinct activities of stromal cells involved in the organogenesis of lymph nodes and Peyer's patches. *J Immunol* 2007;**179**:804–11.
75. Cupedo T, Jansen W, Kraal G, Mebius RE. Induction of secondary and tertiary lymphoid structures in the skin. *Immunity* 2004;**21**:655–67.
76. Katakai T, Suto H, Sugai M, Gonda H, Togawa A, Suematsu S, et al. Organizer-like reticular stromal cell layer common to adult secondary lymphoid organs. *J Immunol* 2008;**181**:6189–200.
77. Honda K, Nakano H, Yoshida H, Nishikawa S, Rennert P, Ikuta K, et al. Molecular basis for hematopoietic/mesenchymal interaction during initiation of Peyer's patch organogenesis. *J Exp Med* 2001;**193**:621–30.
78. Yoshida H, Naito A, Inoue J, Satoh M, Santee-Cooper SM, Ware CF, et al. Different cytokines induce surface lymphotoxin-alphabeta on IL-7 receptor-alpha cells that differentially engender lymph nodes and Peyer's patches. *Immunity* 2002;**17**:823–33.
79. Vondenhoff MF, Greuter M, Goverse G, Elewaut D, Dewint P, Ware CF, et al. LTbetaR signaling induces cytokine expression and up-regulates lymphangiogenic factors in lymph node anlagen. *J Immunol* 2009;**182**:5439–45.
80. Luther SA, Ansel KM, Cyster JG. Overlapping roles of CXCL13, interleukin 7 receptor alpha, and CCR7 ligands in lymph node development. *J Exp Med* 2003;**197**:1191–8.
81. Ohl L, Henning G, Krautwald S, Lipp M, Hardtke S, Bernhardt G, et al. Cooperating mechanisms of CXCR5 and CCR7 in development and organization of secondary lymphoid organs. *J Exp Med* 2003;**197**:1199–204.
82. Rennert PD, James D, Mackay F, Browning JL, Hochman PS. Lymph node genesis is induced by signaling through the lymphotoxin beta receptor. *Immunity* 1998;**9**:71–9.
83. Dougall WC, Glaccum M, Charrier K, Rohrbach K, Brasel K, De Smedt T, et al. RANK is essential for osteoclast and lymph node development. *Genes Dev* 1999;**13**:2412–24.
84. Kim D, Mebius RE, MacMicking JD, Jung S, Cupedo T, Castellanos Y, et al. Regulation of peripheral lymph node genesis by the tumor necrosis factor family member TRANCE. *J Exp Med* 2000;**192**:1467–78.
85. Kong YY, Yoshida H, Sarosi I, Tan HL, Timms E, Capparelli C, et al. OPGL is a key regulator of osteoclastogenesis, lymphocyte development and lymph-node organogenesis. *Nature* 1999;**397**:315–23.
86. Naito A, Azuma S, Tanaka S, Miyazaki T, Takaki S, Takatsu K, et al. Severe osteopetrosis, defective interleukin-1 signalling and lymph node organogenesis in TRAF6-deficient mice. *Genes Cells* 1999;**4**:353–62.
87. Griebel PJ, Hein WR. Expanding the role of Peyer's patches in B-cell ontogeny. *Immunol Today* 1996;**17**:30–9.
88. Crabbe PA, Nash DR, Bazin H, Eyssen H, Heremans JF. Immunohistochemical observations on lymphoid tissues from conventional and germ-free mice. *Lab Invest* 1970;**22**:448–57.
89. Corr SC, Gahan CC, Hill C. M-cells: origin, morphology and role in mucosal immunity and microbial pathogenesis. *FEMS Immunol Med Microbiol* 2008;**52**:2–12.
90. Pellas TC, Weiss L. Migration pathways of recirculating murine B cells and CD4+ and CD8+ T lymphocytes. *Am J Anat* 1990;**187**:355–73.
91. Nishikawa S, Honda K, Vieira P, Yoshida H. Organogenesis of peripheral lymphoid organs. *Immunol Rev* 2003;**195**:72–80.
92. Fukuyama S, Kiyono H. Neuroregulator RET initiates Peyer's-patch tissue genesis. *Immunity* 2007;**26**:393–5.
93. Hashi H, Yoshida H, Honda K, Fraser S, Kubo H, Awane M, et al. Compartmentalization of Peyer's patch anlagen before lymphocyte entry. *J Immunol* 2001;**166**:3702–9.
94. Finke D. Induction of intestinal lymphoid tissue formation by intrinsic and extrinsic signals. *Semin Immunopathol* 2009;**31**(2):151–69.

95. Link A, Vogt TK, Favre S, Britschgi MR, Acha-Orbea H, Hinz B, et al. Fibroblastic reticular cells in lymph nodes regulate the homeostasis of naive T cells. *Nat Immunol* 2007;**8**:1255–65.
96. Gretz JE, Kaldjian EP, Anderson AO, Shaw S. Sophisticated strategies for information encounter in the lymph node: the reticular network as a conduit of soluble information and a highway for cell traffic. *J Immunol* 1996;**157**:495–9.
97. Sixt M, Kanazawa N, Selg M, Samson T, Roos G, Reinhardt DP, et al. The conduit system transports soluble antigens from the afferent lymph to resident dendritic cells in the T cell area of the lymph node. *Immunity* 2005;**22**:19–29.
98. Katakai T, Hara T, Lee JH, Gonda H, Sugai M, Shimizu A. A novel reticular stromal structure in lymph node cortex: an immuno-platform for interactions among dendritic cells, T cells and B cells. *Int Immunol* 2004;**16**:1133–42.
99. Katakai T, Hara T, Sugai M, Gonda H, Shimizu A. Lymph node fibroblastic reticular cells construct the stromal reticulum via contact with lymphocytes. *J Exp Med* 2004;**200**:783–95.
100. Cacalano G, Farinas I, Wang LC, Hagler K, Forgie A, Moore M, et al. GFRalpha1 is an essential receptor component for GDNF in the developing nervous system and kidney. *Neuron* 1998;**21**:53–62.
101. Moore MW, Klein RD, Farinas I, Sauer H, Armanini M, Phillips H, et al. Renal and neuronal abnormalities in mice lacking GDNF. *Nature* 1996;**382**:76–9.
102. Villena A, Zapata A, Rivera-Pomar JM, Barrutia MG, Fonfria J. Structure of the non-lymphoid cells during the postnatal development of the rat lymph nodes. Fibroblastic reticulum cells and interdigitating cells. *Cell Tissue Res* 1983;**229**:219–32.
103. Yoshida K, Kaji M, Takahashi T, van den Berg TK, Dijkstra CD. Host origin of follicular dendritic cells induced in the spleen of SCID mice after transfer of allogeneic lymphocytes. *Immunology* 1995;**84**:117–26.

T Follicular Helper Cells During Immunity and Tolerance

MICHELLE A. LINTERMAN* AND
CAROLA G. VINUESA†

*Cambridge Institute for Medical Research and the Department of Medicine, Addenbrooke's Hospital, Cambridge, England, United Kingdom

†Immunology Program, John Curtin School of Medical Research, Australian National University, Canberra, Australia

Helper T cells are required for the generation of a potent immune response to foreign antigens. Amongst them, T follicular helper (Tfh) cells are specialized in promoting protective, long-lived antibody responses that arise from germinal centers. Within these structures, the specificity of B cell receptors may change, due to the process of random somatic hypermutation aimed at increasing the overall affinity of the antibody response. The danger of emerging self-reactive specificities is offset by a stringent selection mechanism delegated in great part

Progress in Molecular Biology
and Translational Science, Vol. 92
DOI: 10.1016/S1877-1173(10)92009-7

to Tfh cells. Only those B cells receiving survival signals from Tfh cells can exit the germinal centers to join the long-lived pools of memory B cells and bone marrow-homing plasma cells. Thus, a crucial immune tolerance checkpoint to prevent long-term autoantibody production lies in the ability to tolerize Tfh cells and to control positive and negative selection signals delivered by this subset. This review tackles the known mechanisms that ensure Tfh tolerance, many of them shared by other T helper subsets during thymic development and priming, but others unique to Tfh cells. Amongst the latter are checkpoints at the stages of Tfh differentiation, follicular migration, growth, longevity, and quality control of selection signals. Finally, we also discuss the consequences of a breakdown in Tfh tolerance.

I. Introduction

Helper T cells are essential moderators of the immune response. Although they do not have the capability to directly target invading pathogens for destruction they orchestrate a coordinated response from multiple other cell types. To remove the threat to the host, helper T cells facilitate the production of memory antibody responses, to ensure that the host can mount a fast and effective response to subsequent infections. Helper T cells ($CD4^+$) cells provide cues in the form of soluble mediators and receptor–ligand interactions to cells of both the innate and the adaptive immune systems that trigger and modulate their effector function. Because of the extensive roles $CD4^+$ helper T cells play in the defense against pathogens it is essential that they are only capable of mounting a response against exogenous antigens and not self-tissues.

$CD4^+$ helper T cells are a heterogenous population, and to date several subsets have been characterized including: Th1, Th2, Th17, and T follicular helper (Tfh) cells. There are also regulatory $CD4^+$ T cells (Tregs) that repress the growth and function of T cell helper and cytotoxic subsets. Each type of effector T cell is controlled by a key transcriptional regulator, expresses a distinct array of cell surface molecules, and secretes "signature" cytokines, which together facilitate the specific role of that T cell subset within an arm of the immune system. When effector T cells are dysregulated, they can cause many different types of immune pathology.[1–3] Therefore, activation and differentiation of effector T cells and tight control of their function is essential for maintaining tolerance to self-antigens.

Tfh cells are distinguished from other helper subsets by their unique ability to home to B cell follicles and provide help to antigen-specific B cells that are undergoing somatic hypermutation (SHM) of their Ig V region genes and changing their affinity for antigen. Tfh-mediated signals ensure selection of B cells with higher affinity to the immunizing antigen, which can then differentiate to become long-lived plasma cells or memory B cells. Because of the possibility of the emergence of self-reactive B cell clones through the process of SHM and the longevity of selected clones, it is paramount that stringent tolerance mechanisms exist to control delivery of positive selection signals from Tfh cells to B cells. Over the last few years, it has become clear that Tfh cells are critical to prevent autoimmunity arising from germinal centers. Although the mechanisms that control Tfh development and function are still not fully understood, we outline here the principles of Tfh biology, with particular emphasis on the known checkpoints that maintain Tfh tolerance, and review the evidence that autoimmune disorders arise when these checkpoints fail.

II. T-Dependent Antibody Responses

After immunization with protein antigen or infection, antigens enter the secondary lymphoid tissues through the blood and lymphatic fluid in the form of soluble antigens, immune complexes, or coupled to dendritic cells (DCs). Once within the secondary lymphoid tissues, B cells locally scan extrafollicular DCs for antigen; if they encounter antigen, they localize outside the follicle.[4] B cells that do not encounter antigen will migrate into the follicle and slowly scan the follicular dendritic cells (FDCs) for antigens arrayed on their surface.[5]

If the naïve B cell does not encounter antigen, it will return to the circulatory system via the blood or lymph. If its cognate antigen is captured within the follicle, naïve B cells upregulate CCR7 almost threefold, without modulating CXCR5 expression, allowing them to relocate to the interface of the T cell zone and the B cell follicle (T–B border).[6–8] At the T–B border, B cells that have recognized antigen via their B cell receptors (BCR), internalize, and present peptide fragments on MHC class II, allowing them to engage in cognate interaction with primed T cells and receive T cell help.[9]

For $CD4^+$ T cells to provide B cell help, they themselves must first be primed within the secondary lymphoid tissues. DCs bearing antigen migrate from the periphery to the T zone where they present antigen, in the context of MHC class II, to the naïve T cells. T cells with a T cell receptor (TCR) specific for the presented antigen will form a stable contact with the DC, receive an additional signal through the costimulatory receptor CD28, and clonally expand within the T zone where they differentiate into effector T cells.[10] As they do this, T cells downregulate CCR7, which positions them at the T:B border. At this location, primed T cells can interact with antigen-specific B cells. Successful cognate T:B interactions induce both cell types to enter cell cycle. B cells can then differentiate along one of three pathways: (i) they can enter the follicle to seed the germinal center reaction, (ii) they can move to the medullary cords in the lymph nodes or the splenic junction zones or bridging channels to form extrafollicular foci, or (iii) they can become low affinity early memory B cells that have not participated in germinal center reactions and can be subsequently recruited in a secondary challenge with the same antigen.[11–13]

The exact cues that mediate this decision are as yet unclear; it is possible that these different outcomes are controlled by internal cellular mechanisms inherited from the stochastic potential of the individual activated B cell before the first division.[14] BCR antigen affinity appears to play a role after the first division, with high-affinity promoting extrafollicular plasmablast growth.[11]

A. The Extrafollicular Antibody Response

B cells participating in the extrafollicular response are retained in the T zone until they have undergone two rounds of cell division.[15] It is important to emphasize that the initial T:B interaction initiates the process of isotype switching, which can be as abundant in extrafollicular responses as in germinal center reactions. Indeed, γ1 and γ2a switch transcripts are first detected in B blasts soon after T cell priming, during this early proliferative phase in the outer T zone.[16] B blasts then start to upregulate CD138 and produce increased amounts of immunoglobulin for secretion, becoming "plasmablasts." Plasmablasts then migrate to extrafollicular foci where they associate with DCs expressing high levels of CD11c.[17] Migration and follicular-exclusion of plasmablasts is thought to be due to upregulation of CXCR4, and downregulation of CXCR5 and CCR7.[9] Interaction with CD11c^{high} DCs is required for the maintenance of the plasmablasts and the terminal differentiation of a proportion of these plasmablasts into plasma cells.[17] Differentiating plasmablasts upregulate activation-induced deaminase (AID), which allows immunoglobulin isotype class switching to occur. Although AID is also known to facilitate SHM in germinal center B cells, V-region mutation rates in plasma cells from extrafollicular foci are low. Plasma cells derived from the extrafollicular response survive and produce class-switched antibodies for 3 days before undergoing apoptosis *in situ*.[18]

While it is clear that the initial T:B cognate interaction is essential for both T-dependent extrafollicular and follicular antibody responses, it is not clear whether a subsequent interaction between plasmablasts and T cells is required at the sites of plasmablast growth. Extrafollicular plasmablasts express MHC class II and can be observed interacting with T cells that share antigen specificity within extrafollicular foci.[11] The role of T cells located in extrafollicular foci has not been defined; they may play a role in supporting extrafollicular antibody production. In two autoimmune mouse models, $CD4^+$ $ICOS^+$ T cells have been found to support immunoglobulin production by autoreactive B cells in extrafollicular foci.[19,20] These extrafollicular T cells share some characteristics of Tfh cells including ICOS expression and IL-21 production. However, extrafollicular T cells appear to be distinct from Tfh cells as they do not migrate toward CXCL13,[19] do not express high levels of PD-1[21,22], and do not require the expression of SAP to help B cells produce immunoglobulin.[23]

B. The Germinal Center Response

After T cell priming in the T zone, activated B cells can also migrate into the B cell follicle where only a few proliferating cells are required to initiate the germinal center response. After entry into the follicle the activated B cells migrate to its center, where they rapidly proliferate as "B blasts" within the FDC network. As these B blasts expand and migrate toward the pole of the germinal center closer to the T zone, the light and dark zones of the germinal center become more apparent. The dark zone is tightly packed with proliferating B cells (centroblasts), while the light zone contains B cells, FDC processes, and a small number of Tfh cells.[24] CXCL12 and CXCL13 provide a chemokine gradient by which germinal center B cells can migrate from between the two zones. CXCL12 is more prevalent in the dark zone, while CXCL13 is present within the light zone, accumulating on FDC processes. All germinal center B cells express CXCR5 (the receptor for CXCL13) at similar levels; thus control of CXCR4 (CXCL12 receptor) surface expression determines the localization of B cells within the germinal center.[25] Recent *in vivo* imaging studies have suggested B cell proliferation and selection are likely to occur in both dark zones and light zones.[26] While centroblasts are undergoing rapid proliferation they undergo SHM of their V-region genes. This random process can increase, decrease, or not alter the affinity of the B cell antigen receptor, or even convert an otherwise tolerant cell into a self-reactive one.[27,28]

Centroblasts decrease the level of CXCR4 after undergoing SHM, and their nondividing progeny, centrocytes, express the mutated BCR on the surface. This allows them to pick up antigen held on FDC processes, internalize it, and express the processed peptides on their surface in the context of MHC class II, for presentation to Tfh cells. Tfh cells mediate centrocyte selection, allowing the highest affinity B cells to exit the germinal center as long-lived plasma cells or memory B cells, while the rest of the centrocytes either undergo apoptosis or return to the centroblast pool for further rounds of SHM.[29] When T cell help is absent in germinal centers, these abort six days after immunization.[30] And, when the Tfh-derived survival signal for germinal center B cells—CD40L—is blocked, established germinal centers rapidly dissolve.[31] Together these studies suggest Tfh cells are essential for a productive and sustained germinal center response.

III. Germinal Center-Derived Autoimmunity

The B cell repertoire of a healthy individual contains a significant proportion of self-reactive cells.[32,33] The majority of these are unlikely to cause pathology as they secrete low-affinity IgM antibodies that bind soluble antigens.[34,35] By contrast the pathogenic antibodies from individuals with autoimmunity are normally high affinity, class-switched γ-globulins that contain mutations indicative of selection.[36,37] This suggests that these self-reactive cells are not simply by-products of B cell development; rather, like pathogen-specific B cells, it is likely they have been activated and have undergone affinity maturation in the periphery after antigen encounter. As described above, SHM is considered to be a hallmark of the germinal center response, with only low levels occurring physiologically outside the germinal center,[38] although substantial SHM has been found to occur outside the follicles in autoimmune MRL^{lpr} mice.[39] There is convincing evidence to show that SHM can convert B cells previously specific to foreign antigen into cells with self-reactive specificities.[27] The presence of somatically mutated autoantibodies in autoimmune disease suggests that at least a proportion might indeed be derived from germinal center.[40]

The presence of spontaneous germinal centers occurs concurrently with the development of systemic autoimmunity in numerous mouse strains,[41] and in various human systemic and organ-specific autoimmune diseases the frequency of ectopic germinal centers is high. These tertiary lymphoid structures arise in anatomical locations close to sources of autoantigens and have been shown to be a source of autoantibodies.[42–44] Further evidence of defective germinal center tolerance in autoimmunity comes from studies of naturally occurring B cells with a self-reactive BCR, which are not deleted during B cell development and circulate in the peripheral blood of healthy controls and patients with the autoimmune disorder systemic lupus erythematosus (SLE). Tracking these cells in SLE patients demonstrated that, unlike healthy controls, these autoreactive cells participate in the germinal center response, and could enter the memory B cell compartment.[45] Together, these studies strongly suggest that autoimmunity can arise within the B cell follicle, although specifically how this occurs has yet to be fully described.

IV. Germinal Center Tolerance

The random nature of SHM within the germinal center means that B cells with self-reactive specificity can, and likely do, arise during a response to foreign antigen and germinal center-derived autoantibodies cause a range of autoimmune diseases.[46] Therefore, it is essential that there are peripheral tolerance mechanisms in place to censor self-reactive germinal center B cells to ensure these cells do not exit the germinal center as long-lived high-affinity plasma cells or memory B cells, which could contribute to life-long autoimmunity.[47] The majority of germinal center B cells (~95%) are programmed to undergo apoptosis *in situ*[46]; those that live and subsequently egress from the germinal center do so because they have received three sets of survival signals: (i) via their BCR, (ii) from FDCs, and (iii) from Tfh cells. Consequently, these independent signals are in themselves key tolerance checkpoints: their thresholds and mode of delivery need to be controlled for autoimmunity not to occur. Examples of BCR-mediated germinal center B cell survival are provided by mice that are genetically deficient for key signaling molecules downstream of the BCR—that is SPIB, CD45, PI3K, and TC21—which are highly susceptible to apoptosis, suggesting that centrocytes with an altered BCR that does not recognize antigen will undergo apoptosis.[48] Interaction of germinal center B cells with FDCs also prolongs germinal center B cell survival, possibly due to extending the stimulus given via the BCR. This FDC:B cell interaction is mediated by various pairs of adhesion molecules including lymphocyte function-associated antigen 1 (LFA-1) interacting with intercellular adhesion molecule 1 (ICAM-1), and very late antigen 4 (VLA-4) interacting with vascular cell-adhesion molecule 1 (VCAM-1).[49] Recent work has highlighted the importance of these signals *in vivo*: mice deficient in Dock8 (dedicator of cytokinesis 8)—a member of a family of Rho–Rac GTP-exchange factors—fail to accumulate ICAM-1 in the B cell immunological synapse required for interaction with FDCs, and in these mice germinal centers cannot be sustained.[50] Interaction of centrocytes with FDCs not only provides them with survival signals but it also allows them to collect antigen to present to T cells. Tfh cells engage in cognate interactions with centrocytes and provide pro-survival signals and possibly pro-death signals to germinal center B cells[51]; the evidence for this will be discussed below. The role of Tfh cells in censoring germinal center B cells makes them an essential peripheral tolerance checkpoint for preventing germinal center-derived autoimmunity, and therefore, defective Tfh cell formation, tolerization, expansion, migration, function, and death are plausible mechanisms by which germinal center-derived autoimmunity can occur. Below we will review the evidence suggesting a stringently tolerized Tfh cell compartment is essential for preventing self-reactivity, and we will describe the emerging Tfh tolerance checkpoints.

V. T Follicular Helper Cells

The first illustration of the requirement of T cell help in the germinal center was the observation that tonsillar Tfh contain large amounts of CD40L that they can rapidly express on their surface and provide centrocytes with pro-survival signals in conjunction with those derived from BCR ligation.[52–56] Since these initial studies, Tfh cells have been shown to be a subset of helper T cells that have been implicated in the provision of help to germinal center B cells within the B cell follicle[51] and can persist after primary antigen encounter as memory T cells, where they help facilitate a quick response to subsequent antigen exposure.[57] Moreover, the recent identification of Bcl-6 as the Tfh transcriptional regulator by three separate groups, including ours, has clearly revealed Tfh cells are indeed a separate lineage from other helper subsets, specialized in provision of help to B cells.[58–60]

Tfh cells have a transcriptional profile distinct from Th1, Th2, Th17, and Treg cells[61,62] and can form in the absence of key Th1, Th2, and Th17 molecules (*Il4*$^{-/-}$, *Ifnγ*$^{-/-}$, *Stat6*$^{-/-}$, *Stat4*$^{-/-}$, *Il17*$^{-/-}$, *Il17f*$^{-/-}$ and doubly deficient *Rora*$^{-/-}$ *Rorc*$^{-/-}$ mice) suggesting that their formation and maintenance is independent of other helper T cell lineages.[63] The Tfh cell subset is distinguished by expressing the highest levels of CXCR5.[64–66] Some of the molecules highly expressed by Tfh cells facilitate their helper function and their anatomical localization. These molecules include: CXCR5, CD200, programmed cell death 1 (PD-1), CD40L, ICOS, SH2 domain 1A (SAP), IL-10, and IL-21.[51]

VI. Tfh Cells Mediate Selection of Germinal Center B Cells

Fully differentiated Tfh cells in the germinal center provide pro-survival signals to high-affinity centrocytes required for their differentiation into long-lived plasma cells or memory B cells. Tfh are also likely to direct lower affinity centrocytes for recycling to the centroblast pool for further rounds of SHM within the germinal center.[30,51,67] How Tfh cells regulate which centrocytes egress from the germinal center as plasma or memory cells has not been fully elucidated. There is evidence that Tfh cells can provide pro-survival differentiation signals by provision of CD40L signals (positive selection) and cytokines—predominantly IL-21 and IL-4—and can also trigger centrocyte death (negative selection) by ligation of their death receptor CD95 (Fig. 1).

A. Positive Selection

Tfh cells form cognate interactions with germinal center B cells and provide pro-survival signals. Arguably, CD40L is the most essential Tfh-derived signal for germinal center B cell survival. *In vitro*, transient CD40L signals promote germinal center B cell differentiation into memory B cells.[55,64] CD40L signals also support immunoglobulin production by germinal center B cells.[64] Evidence of an important role for CD40L in germinal center B cell selection *in vivo* comes from studies in which CD40:CD40L interactions are blocked after the onset of germinal center reactions in mice. This blockade rapidly disrupts established germinal center, indicating CD40L are essential for the survival and renewal of germinal center B cells.[31]

Other important Tfh-derived signals important for positive selection of germinal center B cells are IL-21 and IL-4. Mice deficient in IL-4R signaling developed smaller sized germinal centers in the presence of virtually intact extrafollicular responses indicating IL-4 participates in T cell-mediated selection of germinal center B cells essential to maintain germinal centers.[68] Recent evidence of a role for IL-21 in germinal center maintenance comes from studies in our laboratory showing IL-21 acts directly in B cells to maintain germinal center B cell numbers.[69] Perhaps not surprisingly, IL-21 was also found to be required for optimal affinity maturation, consistent with the idea that T cells are required to promote reentry into cell cycle of selected centrocytes so they can acquire an increased number of mutations in their Ig V region genes. The mechanism by which IL-21 exerts this effect appears to be at least in

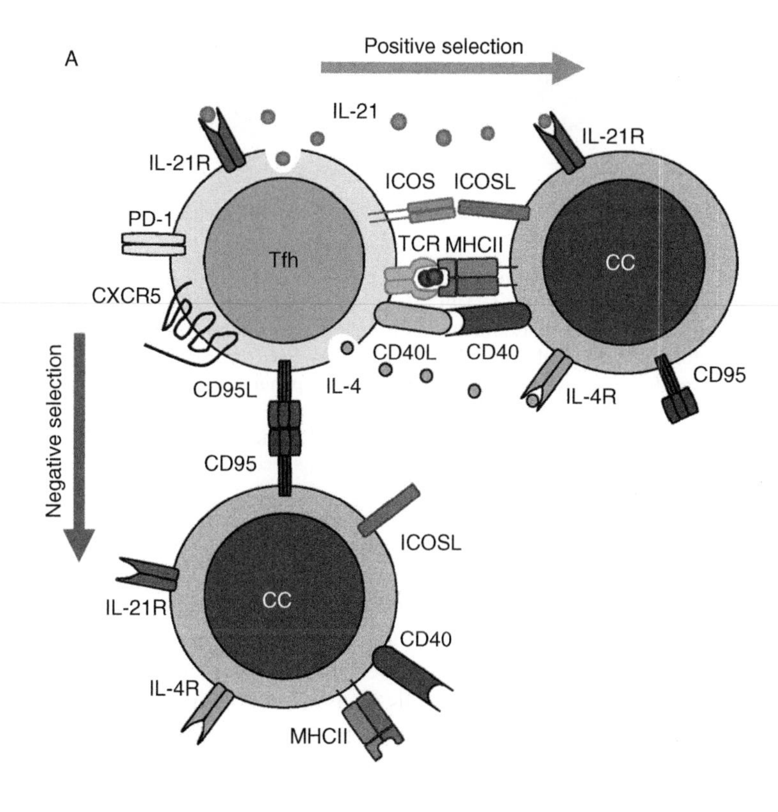

FIG. 1. (Continued)

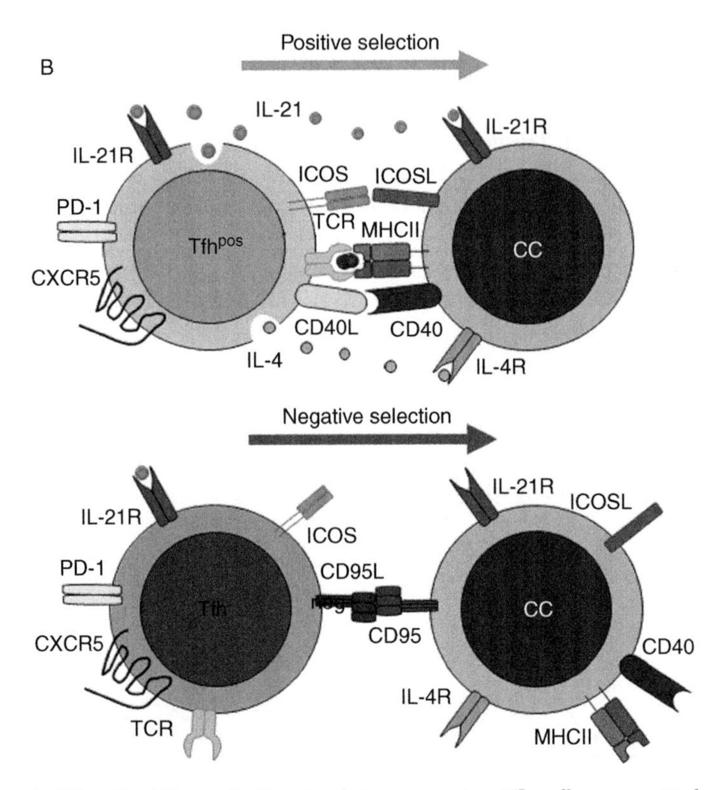

FIG. 1. Tfh cells deliver selection signals to centrocytes. Tfh cells can positively provide costimulation to centrocytes through CD40L, IL-21, and IL-4 during stable cognate interactions. During this interaction Tfh cells also receive costimulatory signals through engaging ICOSL from B cells and autocrine IL-21 production. Tfh cells might also have the capacity to negatively select centrocytes through the provision of CD95L. (A) The same population of Tfh cells might provide both positive and negative selection signals to centrocytes; postive signals delivered during cognate interaction, and negative signals delivered in the absence of cognate interaction. (B) Alternatively, the pro-survival and pro-death signals might be delivered by specialized subsets of Tfh cells within the germinal center.

part though the maintenance of high Bcl-6 expression in germinal center B cells. This is consistent with our previous finding that mice in which Bcl-6 gene-dose has been halved also have fewer germinal center B cells.[23]

How Tfh cells "select" the B cells that require help has not yet been determined experimentally. It has been suggested that competition for T cell help is a mechanism by which centrocyte selection is achieved.[70] Germinal center B cells are 5–20 times more abundant in the follicle than T cells and are constantly moving through the network of FDCs, where they appear to scavenge antigen presented on the surface of these cells to present to Tfh cells.[70–72] Germinal center B cells are so numerous that those with highest affinity are predicted to out compete low-affinity centrocytes for T cell help.[73]

Experimental evidence, at least *in vitro*, has shown B cells with higher affinity antigen receptors can recruit more T cell help than their lower affinity counterparts,[74] and although T cells can contact multiple B cells simultaneously, they can only form a productive synapse with one cell at a time.[75] Taken together, tight control of the number of Tfh cells creates an environment where centrocytes must compete for T cell help. This regulation may be a critical mediator of germinal center tolerance. B cells with higher antigen affinity might be able to collect more antigen from FDCs or even have the ability to acquire antigen from other centrocytes with lower affinity. The ability to capture more antigen may provide high-affinity centrocytes with a selective advantage by means of being able to form a greater number of cognate interactions with Tfh cells, outcompeting lower affinity centrocytes for T cell help and survival signals. The contacts between follicular B and T cells are usually short, with only one third of T cells moving at a speed consistent with a cognate T:B interaction.[76] This suggests that very few B cells engage T cells in a productive conjugation. However, the mechanism by which Tfh cells discriminate which centrocytes to provide help to and thus maintain tolerance in the germinal center has yet to be elucidated.

B. Negative Selection

Within the germinal center a substantial proportion of Tfh cells have been observed dragging apoptotic B cell blebs,[76] suggesting that they may have engaged in a cognate interaction with a live centrocyte and then delivered a death signal to that cell. Tfh cells may induce germinal center B cell apoptosis through ligation of the death receptor, CD95/FAS, expressed at high levels on the surface of germinal center B cells.[77] The ligand for CD95 (CD95L) is also very highly expressed on tonsillar Tfh cells compared with other T cell subsets,[78] suggesting this is the most likely source of CD95L that induces B cell death during T:B interactions in the germinal center. The CD95 death receptor pathway is essential for maintaining immune tolerance: In the absence of either CD95 or CD95L mice develop a severe lymphoproliferative lupus-like autoimmune disease,[79–84] and, a large part of the disease burden can be attributed to the B cell compartment.[85–89] Furthermore, loss of CD95 signaling on cells that have undergone isotype class switching—that is germinal center B cells and extrafollicular plasma cells—recapitulates the lymphoproliferative phenotype of pan-CD95-deficent animals and results in the accumulation of spontaneous germinal center B cells within the lymph node.[89] Finally, it has also been shown that CD95-induced apoptosis is important for limiting the egress of memory B cells from the germinal center.[90] Together, these data suggest that

CD95-induced cell death is required to regulate the emergence of self-reactive B cells from the germinal center in the absence of foreign antigen to prevent autoimmunity, and that Tfh may be the source of CD95L that mediate apoptosis.

C. Balancing Tfh-Derived Signals that Promote Germinal Center B Cell Death Versus Survival

Tfh cells have the capacity to engage B cells in either life or death interactions through provision of CD40L and CD95L, respectively. The evidence described above suggests there is a role for both positive and negative Tfh-derived signals in the process of centrocyte selection. But it seems paradoxical that the same population of Tfh cells regulates both the life and death of centrocytes. Applying Ockham's razor, the simpler explanation would be that positive selection requires the establishment of a cognate interaction, but negative selection does not. However, it is likely to be more complex than this: *in vivo* imaging has demonstrated that most (96%) T:B interactions in the germinal center are short and do not result in stable interactions, and that B cells subsequently go on to contact numerous Tfh cells rather than undergoing apoptosis after a single interaction[76] (Fig. 1A).

An alternative hypothesis is that the well-described heterogeneity of Tfh cells serves the purpose of diversifying their functions so that one subpopulation can support centrocyte survival and another one can mediate apoptosis. Within $CD4^{+}$ Tfh cells, several different subsets have been described including NKT cells and Tregs. Furthermore, among conventional $CD4^{+}$ $CXCR5^{hi}$ $PD1^{hi}$ Tfh cells, two discrete subsets exist that differ in expression of CD57 and preformed CD40L.[51] $CD8^{+}$ T cells can also be found in germinal centers, and these might exert cytotoxic properties. It is clear that further research needs to be undertaken to understand Tfh-mediated selection in germinal centers. Despite the existing uncertainties regarding the mechanism of selection, recent experimental evidence has demonstrated that a breakdown in Tfh cell tolerance can result in aberrant germinal center responses and autoimmunity (Fig. 1B).

VII. Breaks in Tfh Tolerance Result in Autoimmunity

As discussed above, a break in germinal center tolerance would result in the generation of self-reactive B cells that are able to secrete pathogenic autoantibodies. Multiple autoimmune-prone mouse strains exhibit spontaneous germinal center formation and autoantibody production.[41] Furthermore, in a number of these strains it has been shown that there is aberrant spontaneous expansion of Tfh cells that correlates with a lupus-like phenotype and antinuclear antibody development.[91–94] Work from our group has suggested that this is a causal association. *Sanroque* mice that are homozygous for the "*san*" allele of *Roquin* exhibit spontaneous Tfh and germinal center formation and develop a lupus-like disease characterized by antinuclear antibody production and glomeruonephritis.[93] To determine whether the spontaneous Tfh formation is responsible for a break in germinal center tolerance and the autoimmune phenotype of *sanroque* mice, we specifically neutralized Tfh formation/function by making *sanroque* mice deficient in the SLAM-associated protein (SAP), which is required for Tfh formation and function (discussed in detail below).[95] In the absence of SAP, spontaneous Tfh and germinal center formation, autoantibody production and end-stage renal pathology were completely abrogated in *sanroque* mice. Furthermore, halving the gene dose of *Bcl6* specifically reduces the spontaneous Tfh and germinal center formation and ameliorates the lupus-like phenotype of *sanroque* mice. Conversely, adoptive transfer of spontaneous Tfh, but not other effector T cells, from *sanroque* mice into wild type recipients induced a germinal center without immunization.[23] Together these data suggest that aberrant accumulation of Tfh cells leads to positive selection of self-reactive germinal center B cells and autoimmunity in *sanroque* mice.

In addition to regulating the numbers of Tfh cells within the germinal center, the quality and quantity of Tfh-derived stimulatory signals also need to be tightly controlled. As mentioned above, IL-21 is produced at the highest levels by Tfh cells,[61] and is required to maintain normal germinal center B cell numbers and optimal affinity maturation.[69] Excessive production of this cytokine appears to have a role in the lupus-like pathology of both BXSB/Yaa and *MRL*^{lpr} mice. However, in autoimmune settings, inappropriate IL-21 production is not limited to Tfh cells in germinal centers, but can be derived from extrafollicular T cells as well,[19,20] which are likely to be related to Tfh cells. In another murine model of autoimmunity, the BXD2 mouse strain in which spontaneous germinal center formation has also been described, somatically mutated and class-switched pathogenic autoantibodies are produced, and the mice develop erosive arthritis and glomerulonephritis.[96,97] This pathology is associated with accumulation of Tfh cells in germinal centers, which produce abnormally high levels of IL-17. The excess of IL-17 results in prolonged retention of germinal center B cells in follicles and contributes to the immune pathology in this strain.[98] In the lupus-prone NZB/W F1 mouse strain $CD4^+CXCR5^+ICOS^+$ Tfh cells also produce excessive IL-17 that contributes to the formation of self-reactive antibodies.[99] ICOS blockade in these mice also reduces spontaneous Tfh and germinal center formation and ameliorates the lupus pathology.[100] Together, these studies demonstrate that the quality of the "help" delivered by Tfh cells to centrocytes is also important to ensure that germinal center tolerance is not broken.

In addition to the accumulation of Tfh cells in the germinal centers of secondary lymphoid tissues, lupus-prone *sanroque* mice described above also exhibit an increase in $CXCR5^+$ $ICOS^{hi}$ $PD\text{-}1^{hi}$ $CD4^+$ cells in peripheral blood that correlates closely with the numbers of their spleen and lymph node counterparts.[101] Simpson et al. examined this "circulating Tfh-like" (cTfh) population in humans with systemic autoimmunity and discovered that 20–30% of patients with SLE and Sjogren's syndrome have an expanded cTfh population. This population is stable over time and positively correlates with disease severity—elevated autoantibody titers and end organ damage.[102] This suggests that dysregulated Tfh formation/maintenance may contribute to autoimmune pathology, not only in mice but also in humans.

VIII. Tolerizing Tfh Cells

A number of tolerance checkpoints are in place from the very early stages of T cell development in the thymus, through to their activation in the periphery, and terminal differentiation into short-lived effectors or longer lived memory subsets. These checkpoints consist of a combination of cell-extrinsic and cell-intrinsic mechanisms that ensure that T cells are tolerant of self-antigens and autoimmunity does not occur. Many of these mechanisms, particularly those occurring during thymic development through to the stage of T cell priming are shared by all $CD4^+$ T helper subsets. Nevertheless, specialized tolerance checkpoints that control Tfh cell differentiation, migration, growth, lifespan, and function also exist and appear crucial to prevent autoimmunity of germinal center origin. While this field is still in its infancy, we will outline some of the shared and unique mechanisms that are in place to minimize selection of autoreactive germinal center B cells by Tfh cells in germinal centers.

A. Central Tolerance

T cell precursors migrate from the bone marrow to the thymus where they begin to rearrange their antigen receptor genes to diversify the specificity of the TCR. Genes encoding the TCR are assembled by the random process of V(D)J gene rearrangement, facilitated by the recombination-activating gene (RAG) 1 and 2 enzymes, generating T cells whose antigen receptors can recognize a wide variety of exogenous and endogenous peptides. The stochastic nature of RAG-mediated rearrangements means it is essential that these cells are rigorously selected to ensure that self-reactive thymocytes are censored: this process is known as central tolerance. T cell precursors enter the thymus with the potential to develop into $\alpha\beta$ or $\gamma\delta$ T cells; Tfh are derived from cells of the $\alpha\beta$ lineage. Double negative ($CD4^-CD8^-$) pro-$\alpha\beta$ T cells rearrange their β minigene elements by first bringing together the $D\beta$ and $J\beta$ elements, then joining the Vβ segment to allow expression of the rearranged TCRβ, in conjunction with the surrogate α-chain, as a pre-TCR receptor on the cell surface. Where cells that have a productive TCRβ receive survival signals from thymic epithelial cells (TECs), those that do not, undergo apoptosis. These nonhematopoietic cells express both MHC class I and class II, and as thymocytes move through this network of stromal cells they interact with MHC:peptide complexes expressed on the TECs. Thymocytes with very high affinity to this complex are deleted in a process known as negative selection and cells that receive a low-affinity signal are stimulated to survive, known as positive selection.[103] T cells then express both the CD4 and CD8 coreceptors—the "double positive stage" ($CD4^+CD8^+$)—and then they recombine their Vα and Jα elements to allow surface expression of TCRαβ. Rearrangement of the TCRα and does not cease until the developing thymocyte receives positive selection signals from cortical TECs. Therefore, cells that are not positively selected, such as self-reactive cells, can edit their TCR by maintaining RAG expression and rearranging the TCRα chain in an attempt to generate a productive TCR, and have a second chance at being positively selected.[104] Double positive cells that receive a survival signal from MHC class I are directed into the CD8 lineage and lose expression of CD4; likewise those thymocytes that are MHC class II restricted become $CD4^+CD8^-$ cells. However, most thymocytes (~80%) express a TCR that will not recognize self-MHC and without receiving a crucial survival signal will die by neglect.

Once the T cells have completed maturation in the thymus, most will emerge as naïve cells that go on to be activated in the periphery where, upon antigen stimulation, they proliferate and differentiate into effector T cells. Central tolerance in the thymus is very effective; however, a small proportion of self-reactive T cells will exit the thymus, and require further censoring in the periphery.[105–107] In the absence of intact peripheral tolerance mechanisms these self-reactive T cells can become activated and trigger autoimmunity.

B. Peripheral Tolerance

Most of the T cells that exit the thymus will bear TCRs that recognize exogenous antigens; however, those self-reactive T cells that have escaped central tolerance need to be moderated in the periphery. Peripheral tolerance can be induced through a cell-extrinsic mechanism where autoreactive cells are suppressed by regulatory T cells (Tregs) or via cell-intrinsic mechanisms that modify or remove the helper T cells to prevent immunopathology. After $CD4^+$ T cells exit the thymus they are naïve and require antigen to be presented on MHC class II by antigen-presenting cells to become activated. If the cells are not activated under optimal conditions, they will be rendered functionally unresponsive or anergic; if naïve cells are fully activated in the presence of costimulation they go on to proliferate and acquire effector function where they will provide help to other cells to clear antigen. After the effector phase of the immune response most of the cells will die by apoptosis and the immune response will resolve. Like other effector T cells, Tfh cells are subject to these peripheral tolerance mechanisms, and a breakdown of any of these checkpoints could result in a self-reactive Tfh population.

1. T Cell Priming and Anergy Induction

T cell activation is tightly regulated by positive and negative signals. In concert with recognition of the peptide MHC complex the T cell requires an additional signal for T cell activation.[108] CD28 is constitutively expressed on both naïve and effector T cells. Its ligands CD80/86 are upregulated on APCs in response to pathogenic stimuli. CD28 has a consensus sequence in its extracellular domain that specifically binds the ligands CD80/86.[109] Interaction of CD28 with CD80/86 is required for T cell proliferation, differentiation, and cytokine production.[110] CD80/86 are upregulated on APCs in response to danger signals, ensuring that signaling through CD28 only occurs in response to a pathogenic insult. When a T cell recognizes MHC:peptide complexes in the absence of CD28 ligation, as in the case of self-antigen, the cell undergoes apoptosis or is rendered anergic. Once a T cell enters the state of anergy it cannot respond to subsequent TCR-ligation.[111] Thus, T cell priming requires two activation signals; "signal 1" in the form of the MHC–peptide complex delivered in conjunction with "signal 2," CD28 costimulation. The dependence on pathogen-induced signals for CD28 ligation ensures that T cells are only primed in response to exogenous antigens. In pathological situations in which the requirement for CD28 costimulation is bypassed, self-reactive T cells can be primed in response to foreign antigen and trigger autoimmunity.

What are the biochemical differences downstream of TCR signaling in normally activated versus anergic T cells? When priming occurs, signal one in conjunction with CD28 costimulation (signal two) activates two important signaling pathways: (i) the calcium signaling pathway that allows translocation of the nuclear factor of activated T-cells (NFAT), a transcription factor, into the nucleus; and (ii) the Ras/MAP kinase pathway that causes upregulation of the AP-1 transcriptional regulator complex. In activated T cells NFAT and AP-1 form complexes that bind to composite sites in the promoters of genes and turn on a transcriptional program that enables full T cell activation. By contrast, in anergic cells in which the TCR has been ligated in the absence of CD28 costimulation, the calcium signaling pathway is activated but the RAS/MAP kinase pathway is not. NFAT is thus activated in the absence of the AP-1 complex and translocates to the nucleus where it binds to and induces expression of anergy-associated genes.[112]

Anergy was first described *in vitro* when a T cell has been presented with antigen under suboptimal conditions; either by ligation of the TCR in the absence of costimulation or by a low-affinity TCR interaction with costimulation that renders the cell functionally unresponsive to further stimuli from the anergizing antigen.[113–118] Anergy can be induced *in vivo* using super antigens or by injection of soluble antigen intravenously, oral administration of antigen, or chronic antigen exposure.[119–124] The common endpoint is functional unresponsiveness although the cellular and molecular mechanisms that induce this state differ depending on the anergizing stimulus.[125,126] There are at least three reported instances where naïve T cell activation has been shown to occur in the absence of CD28 costimulation and therefore to bypass anergy induction in self-reactive T cells. These include mice in which mutated Roquin causes ICOS overexpression on naïve T cells, mice deficient in Cbl-b, and Traf6-deficient mice. In all these cases described below mice develop lupus-like autoimmunity.

Icos arose by gene duplication of *Cd28* and both costimulators share common signaling pathways: they are both capable of activating PI3K and stimulation through either receptor in conjunction with TCR results in the upregulation of an almost identical set of genes, the two main exceptions being IL-2 and ICOS, which are upregulated by CD28, but not by ICOS signaling.[127] A fundamental difference however is their controlled compartmentalization: While CD28 is expressed on all T cells and mainly plays a role in the priming of naïve T cells, ICOS is not expressed on naive T cells and mainly plays a role in the regulation of effector cell maturation, function, and lifespan. Unlike CD28 ligands, ICOSL is constitutively expressed and is therefore not regulated by pathogen-derived signals.[128] We have recently provided evidence that excessive and uncontrolled ICOS expression contributes to autoimmunity. The *sanroque* mouse strain overexpresses ICOS on both naïve and effector / memory T cells, suggesting that unrestrained ICOS expression might be required for aberrant immune activation in *sanroque* mice. Reducing the magnitude of ICOS expression in *sanroque* mice, simply by halving the gene dose, ameliorated some autoimmune manifestations of this strain.[129] It is therefore likely that ICOS overexpression results in inappropriate activation of self-reactive naïve $CD4^+$ T cells, as T cells are able to receive signal 1 (MHC–self-peptide) and signal 2 (ICOSL) in the absence of pathogen-derived stimuli and CD28 costimulation, providing a mechanism by which rare self-reactive T cells that have escaped thymic deletion are primed. To investigate this possibility, we generated $Roquin^{san/san}$ $Cd28^{-/-}$ and $Roquin^{san/san}$ $Cd28^{-/-}$ $Icosl^{-/-}$ mice. After immunization with T-dependent antigens or bacterial infection, which are CD28 mediated, the response to this immunological challenge became independent of CD28 and dependent on ICOS when *Roquin* is mutated.[21] Taken together, this work illustrates the requirement of wild type Roquin to repress ICOS to ensure that activation of T cells is solely dependent on CD28, and thus on danger signals that induce expression of CD28 ligands.

The E3 ubiquitin ligases Cbl-b and TRAF6 are upregulated in anergic T cells.[130,131] T cells deficient in either Cbl-b or TRAF6 are resistant to anergy induction *in vitro* and *in vivo*, and do not require CD28 costimulation for T cell activation *in vitro*.[132–136] Targeted genetic ablation of murine Cbl-b results in the development of autoimmunity: Cbl-b-deficient mice develop spontaneous systemic autoimmunity with age characterized by lymphocytic infiltrates in multiple organs and development of antinuclear antibodies.[132] In addition these mice are highly susceptible to the induction of experimental autoimmune encephomyelitis (EAE), collagen-induced arthritis (CIA), and autoimmune diabetes.[131,133,137] In rats, Cbl-b represses the development of spontaneous diabetes.[138]

T cells deficient for Cbl-b proliferate and produce IL-2 in the absence of CD28 costimulation suggesting that loss of this E3 ligase uncouples T cell priming from CD28 ligation. Furthermore, doubly deficient $Cd28^{-/-}$ $Cblb^{-/-}$ mice develop spontaneous autoimmunity with the same frequency as $Cd28^{+/+}$ $Cblb^{-/-}$ animals demonstrating that the break in tolerance occurs in the absence of CD28 ligation.[139] This suggests that loss of Cbl-b uncouples T cell priming from CD28 costimulation resulting in a break in anergy. However, although genetic ablation of *Cblb* in CD28-deficient mice can restore T cell proliferation, IL-2 production, and T-dependent antibody production, it does not rescue the production of the cytokines IFNγ and IL-4, the upregulation of ICOS, or the Tfh/germinal center response.[139] This demonstrates that not all pathways downstream of CD28 are under the control of Cbl-b and that the break in anergy/tolerance is not simply uncoupling T cell activation from CD28 signaling. The autoimmune phenotype and T cell hyperproliferation of *Cbl-b* deficient mice cannot be attributed to impaired generation or function of regulatory T cells. Tregs that lack Cbl-b can suppress effector T cells: however, Cbl-b deficient effector T cells are partially resistant to Treg-mediated suppression, suggesting that loss of *Cbl-b* intrinsically dysregulates effector T cells in more than one way.[140,141]

Genetic ablation of *Traf6* in lymphocytes results in a severe fatal autoimmune disease with T cell infiltration in multiple organs.[134] TRAF6 has a crucial role in maintaining central tolerance. However, this is due to the requirement of TRAF6 within the medullary epithelial cells, where it is essential for their differentiation, organization, and regulation of AIRE expression. TRAF6 is not required in thymocytes themselves for effective negative selection.[142] Furthermore, loss of TRAF6 in T cells alone results in increased T and B cell numbers, splenomegaly, lymphadenopathy, lymphocytic infiltrates in multiple organs, development of anti-dsDNA antibodies and hyper IgM, IgG, and IgE. TRAF6-deficiency in T cells does not alter the number or function of regulatory T cells. However, these TRAF6-deficent T cells, like $Cblb^{-/-}$ T cells, are also more resistant to Treg-mediated suppression.[135] Together, the role of Roquin, Cbl-b, and TRAF-6 demonstrates that uncoupling T cell priming from pathogen-derived signals can result in autoimmunity due to T cell intrinsic events, and reveals that CD28 is a critical checkpoint in peripheral T cell tolerance.

2. Activation-Induced Cell Death

Activation-induced cell death (AICD) is a phenomenon that mediates the contraction of antigen-activated T cells and reflects an increased sensitivity to apoptosis of previously activated T cells upon restimulation through the TCR. Several forms of AICD have been described *in vivo* and *in vitro*, in which T cells die at varying intervals after initial stimulation depending on the nature of the infection or culture conditions.[143] In the periphery, but not in the thymus, AICD is mediated through interaction of CD95-trimers with homotrimers of its ligand (CD95L, also known as CD178, TNFL6, or FasL). CD95 is a member of the tumor necrosis factor receptor superfamily that contains a death domain within its cytoplasmic tail. Upon ligation, this death domain recruits an adaptor and initiates an active caspase cascade that targets cell substrates for cleavage and induces cell death. The repetitive engagement of the TCR *in vitro* stimulates T cells to undergo AICD. This coincides with upregulation of the ligand for CD95, which only appears on T cells after activation. AICD can be effectively blocked by inhibiting the CD95/CD95L pathway.[144–147] Repeated administration of soluble antigen to TCR-transgenic CD95-deficent mice demonstrated that splenic, but not thymic, antigen-specific T cells required CD95 to undergo cell death.[148] As repetitive stimulation of the TCR mimics the repeated presentation of self-antigens *in vivo*, CD95-mediated cell death is a plausible mechanism by which autoreactive T cells are permanently removed from the periphery.[149]

The critical role of CD95-induced cell death in maintaining immunological tolerance came from the study of mice with mutations in the genes encoding CD95 and its ligand. The CD95L-mutant MRL^{gld} and CD95-mutant MRL^{lpr} strains have an almost identical phenotype: Pathology includes enlarged spleen and lymph nodes, lymphocytic infiltrates in multiple organs, glomerulonephritis, and antinuclear antibodies. Both strains have multiple cellular abnormalities including, but not limited to T cells, B cells, and macrophages. The most striking cellular abnormality is the presence of a $CD4^-CD8^-Thy1.1^+$ subset of T cells, not observed in wild type mice. Although the phenotype is caused by a monogenic defect, the genetic background of the strain modifies the severity of disease, with the MRL background resulting in the most severe pathology.[79–84]

CD95 is widely expressed on multiple cell types, including DCs, B cells, and activated T cells; so dissecting the contribution of each cell type to disease progression is important to understand where tolerance to self is broken. Lack of B cells in *lpr* mice altered the expansion of activated and memory T cells indicating that T and B cells act in concert to contribute to pathogenesis.[85–87]

Autoantibodies derived from *lpr* mice are high affinity and have undergone SHM and class switching. These mutated autoantibodies have been shown to arise from extrafollicular foci.[39,150] Conditional deletion of CD95 in either T cells or B cells alone results in defective lymphocyte homeostasis and tissue destruction, demonstrating that loss of death receptor-mediated cell death on either cell type is sufficient to break tolerance.[88,89]

The role of T cells for initiating production autoantibodies was demonstrated by making MRL^{lpr} rheumatoid factor (RF) transgenic mice T cell deficient ($Tcrb^{-/-}$). In the absence of T cell help, the extrafollicular autoantibody response could not be initiated, though it could be rescued by transfer of antichromatin IgG2a to *MRL-lpr* $Tcrb^{-/-}$ RF-transgenic mice, demonstrating that while T cells were required to initiate pathology, they were not required for its progression. In addition, loss of T cells causes decreased number of mutations accumulating in the extrafollicular foci. It is as yet not clear whether T cells promote SHM or the survival of mutated self-reactive B cells.[151] A recent study has described a subset of extrafollicular T cells in MRL^{lpr} (nontransgenic) mice that promotes IgG responses through ICOS and IL-21 signaling. Whether these factors stimulate the B cells to undergo SHM at increased frequency within extrafollicular foci, promoting expansion of mutated B cells or allowing their survival, is yet to be determined.[19] Together, these studies demonstrate a role for CD95-induced apoptosis in the maintenance of peripheral T tolerance and the prevention of autoimmune disease.

C. Mechanisms that Specifically Limit Differentiation into Tfh Cells

The tolerance mechanisms described above are shared by all peripheral $CD4^+$ T cells. As T cells differentiate into one of the several possible helper subsets, they become subject to a range of more specialized tolerance checkpoints destined to control their specific functions. These tolerance checkpoints are particularly critical in the case of Tfh cells, due to their potential to cause long-lasting autoimmunity if they aberrantly select mutated self-reactive germinal center B cells that can live for decades in an individual as bone marrow plasma cells or memory B cells. These checkpoints operate to limit their differentiation, expansion, lifespan, and provision of helper signals (Fig. 2).

Besides strong TCR ligation, signaling downstream of several stimulatory and cytokine receptors has been shown to be essential for primed T cells to differentiate into bona fide Tfh cells and expand in adequate numbers.[152] This includes, ICOS- and OX40-mediated costimulation, signals through IL-21R and/or IL-6R, and other putative B cell-derived signals that are only delivered after formation of stable T:B cognate interactions. Not surprisingly, excess of any of the signals above has been associated with autoimmunity and these will be reviewed below.

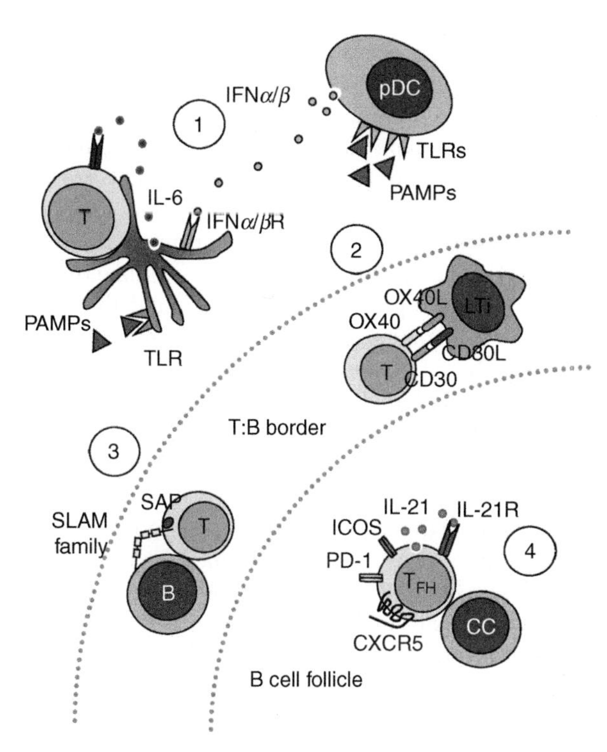

FIG. 2. Checkpoints in Tfh cell tolerance. (1) In order to differentiate into Tfh cells, T cells need to receive a strong signal via their TCR from DCs that have been stimulated in response to foreign antigen to express the ligands for CD28, CD80/86. These DCs must also be responsive to pathogen-induced type 1 interferons to secrete IL-6 (mice) or IL-12 (humans). (2) After T cell priming, T cells also require additional costimulatory signals through OX40L and CD30L from $CD3^{-}CD4^{+}$ accessory cells to ensure upregulation of CXCR5. (3) Stable cognate interactions between primed T and B cells, mediated by the SLAM-family of receptors, are essential for follicular entry. During these interactions T cells might receive additional costimulatory signals that promote their entry into the germinal center. (4) Tfh cells in the germinal center are tightly controlled by enhanced susceptibility to apoptosis, Treg-mediated suppression, and costimulatory signals through ICOS, CTLA-4, and PD-1. Together, these checkpoints ensure that the Tfh population is tolerant to self-antigen and, therefore can only engage in cognate interactions with somatically mutated centrocytes that share the same foreign antigen specificity.

The cues above induce Bcl-6 upregulation, which drives terminal Tfh differentiation and maintenance of this phenotype. Through the repression of various microRNAs and protein-coding mRNA transcripts, Bcl6 orchestrates a transcriptional profile that allows Tfh development, expansion, follicular homing, and B cell helper function.[58–60] Within B cells, mutual repression of Bcl-6 and Blimp-1 regulates the generation and maintenance of germinal center B cell (high Bcl-6 and low Blimp-1) and plasma cell (high Blimp1 and low Bcl6) programs.[153–155] This antagonism between Bcl-6 and Blimp-1 also regulates the fate of $CD4^+$ effectors: Blimp-1 upregulation prevents Tfh development *in vivo* through its ability to repress Bcl-6.[59]

Bcl6 confers a Tfh phenotype partly through repressing the transcriptional regulators of the Th1, Th2, and Th17 helper cell subsets, *Tbx21*, *Gata3*, and *Rorc*, respectively.[58,60,156,157] In human tonsil Tfh cells, Bcl-6 can bind the promoters of Tbet (*Tbx21*) and Rorγt (*Rorc*)[58]; in mice, there is evidence of Bcl-6-mediated repression of T-bet and Gata-3.[60,156] Unlike repression of Tbet and Rorγt, Bcl-6 does not repress *Gata3* mRNA, and thus Bcl-6-mediated repression must occur posttrancriptionally.[156] The signature cytokines of Th1, Th2, and Th17 cells, IFNγ, IL-4, and IL-17, are all produced by Tfh cells but at much lower levels than non-Tfh effector cells.[58–60,158,159]

Overexpression of Bcl6 results in repression of genes involved in Tfh homing, such as CCR7, and causes upregulation of key Tfh molecules including ICOS, PD-1, CXCR4, and CXCR5.[58–60]. As Bcl6 is a transcriptional repressor it is perhaps counter-intuitive that Tfh-associated genes go up during forced Bcl6 expression. This may be at least in part explained by the finding that Bcl-6 overexpression represses over 50% (34) of all microRNAs expressed at significant levels in activated T cells, including the microRNA 17–92 cluster. These miRNAs have been predicted to target the Tfh molecules upregulated in Bcl-6 expressing cells for translational inhibition. Overexpression of the miRNA cluster 17–92 in mouse lymphocytes led to a reduction in surface CXCR5 expression. This suggests that expression of Bcl6 in cells that have received Tfh differentiation cues results in downregulation of the microRNA 17–92 cluster, and facilitates surface CXCR5 expression and follicular homing.[58]

1. Setting a Threshold for TCR Signaling Strength

The strength of the signal received through the TCR signal plays an essential role in directing T cells into the Tfh lineage. T cells that bind peptide/MHCII with high affinity are over represented in the Tfh cell subset compared to other antigen specific helper T cells. These cells also have a more restricted TCR repertoire than other effector T cells. Furthermore, in a competitive environment where the affinity of specific binding of T cells is known, T cells with greater affinity for antigen preferentially enter the Tfh lineage over other T helper cell subsets. The mechanism by which increased antigen binding preferentially directs cells into the Tfh lineage has yet to be elucidated. The increased signaling cascade downstream of the TCR may influence the production of key Tfh cytokines such as IL-21, favoring Tfh differentiation in cells with the highest affinity.[158] It is plausible this TCR signaling threshold is an important tolerance checkpoint: a T cell that has successfully undergone negative selection in the thymus will be unlikely to bind self-antigen with high affinity. Thus, the requirement for strong TCR binding for Tfh differentiation is likely to prevent weakly cross-reactive T cells from entering germinal centers (Fig. 2, checkpoint 1).

2. Limiting the Production of Cytokines Required for Tfh Differentiation

The timing and exact cues that induce Bcl-6 expression in cells that will become Tfh during the response to protein antigen have not yet been defined. *In vitro*, in conjunction with TCR and CD28 both IL-6 and IL-21 can induce Bcl6 expression in naïve T cells, indicating that both these cytokines could play a role in commitment to the Tfh lineage.[60] Indeed, Tfh cells are reduced in the absence of IL-21 signaling,[63,160] and our studies indicate that they do form but are not maintained, and the population declines faster. This indicates that IL-21 signaling is an important autocrine maintenance factor for Tfh cells.[69]

Functional redundancy between IL-21 and IL-6 has been described in Th17 cell differentiation.[161–165] It is therefore possible that these two cytokines may also play redundant roles in Tfh cell formation. Furthermore, when $CD11c^+$ DCs cannot produce IL-6 due to impaired responsiveness to type 1 interferons, the formation of Tfh cells is selectively impaired after immunization with protein antigen in adjuvant, with other T helper subsets forming in normal numbers.[166] This implicates type 1 interferons and IL-6 as key inducers of Tfh cells, likely through causing Bcl6 upregulation. IL-6 also induces the production of IL-21 in $CD4^+$ T cells. Thus, stimulation of T cells first by IL-6, which induces both Bcl6 and IL-21 and then by IL-21, which both supports its own production[167] and that of Bcl6, is likely to act as a positive feedback loop that regulates the sequential induction and continued expression of Bcl6 that is essential for Tfh formation and maintenance. In humans, it appears that production of IL-12 by DCs, rather than IL-6, is required for Tfh differentiation.[157,168]

Not surprisingly, excessive production of IL-21, IL-6, IL-12, and type I interferons have all been associated with lupus-like phenotypes and autoantibody production. In humans, SLE patients have increased IFNα in the serum that positively correlates with anti-dsDNA antibody titers and disease severity.[169] In mice, type I interferons have been shown to contribute to the pathogenesis of lupus-like disease.[170] This appears to be in part due to IFNα/β inappropriately acting on T cells: loss of type I interferon signaling completely abrogates the anti-dsDNA and ssDNA antibodies from class-switched T-dependent isotypes, but not T-independent isotypes in NZB mice.[171] Furthermore, administration of type I interferons to nonautoimmune humans or preautoimmune lupus-prone mice can result in the production of autoantibodies and hasten the onset of autoimmunity.[172,173]

IL-6 has also been implicated in autoimmune disease; IL-6 has a role in the generation of pathogenic Th17 cells and autoimmunity derived from this population.[174] IL-6 may facilitate Th-17-mediated autoimmunity by stopping the conversion of Th17 cells to Treg cells.[175,176] Excessive production of IL-6 by B cells in *Lyn*$^{-/-}$ mice is required for autoantibody formation and lupus-like pathology of this strain.[177] Because of the multifaceted role for IL-6 in the immune system further work is required to dissect out the contribution of IL-6 in Tfh-related pathology.

Unlike lymphocytes, the cells of innate immune system are capable of recognizing foreign antigens through expression of pattern recognition receptors (PRR) that recognize pathogen-associated molecular patterns (PAMPs) that are common to microorganisms. These PAMPs are not expressed by self-antigens and expression of PRR allows innate cells to discriminate self-antigen from microbial infections. After recognition of PAMPs, APCs upregulate MHC class II, and secrete cytokines and costimulatory molecules, allowing APCs that have been exposed to foreign antigen to prime T cells to respond to that antigen.[178] For differentiation of T cells into tolerant Tfh cells, it is essential that production of IL-6, IL-12, and type I interferons by APCs occurs in response to foreign but not self-antigen. Type I interferons can be induced by ligation of TLRs 7 and 9 on plasmacytoid DCs, or in conventional DCs and macrophages by stimulation of TLR 3, 4, and 9. IFNα/β can also be induced independent of the TLRs through the RNA helicases, RIG-I (retinoic acid–inducible gene I), and MDA5 (melanoma differentiation-associated gene 5) recognizing cytosolic viral or bacterial nucleic acids. The TLR-independent pathway does not occur in plasmacytoid DCs, only in conventional DCs and macrophages.[179] In addition to stimulating the pro-inflammatory type I interferons the expression of cytokines that signal from the innate to the adaptive systems, including IL-6 and IL-12, are also dependent on that cell receiving signals from TLRs.[178] This interdependence of Tfh stimulating cytokines on pathogen inducible signals ensures that Tfh cells only develop in response to foreign antigens. Breakdown of the control on cytokines has immediate implications for autoimmunity, and while there is certainly more than one mechanism by which these cytokines contribute to disease pathogenesis, their ability to promote Tfh formation and control their expansion undoubtedly contributes to pathogenesis (Fig. 2, checkpoint 1).

3. Limiting OX40 and CD30 Signals that Facilitate Follicular Entry

Ligation of OX40 and CD30 are important for generation of Tfh cells. OX40 signaling has been shown to promote CXCR5 expression[180] and localization of Tfh cells to the follicle.[91,181] Tfh cells have fourfold higher expression of OX40 compared with other naïve T cells and other T effector cells.[158] OX40-

deficent mice and CD30-deficient mice can still form Tfh cells, but doubly deficient *Ox40*$^{-/-}$ *Cd30*$^{-/-}$ T cells have an impaired ability to enter the B cell follicles, suggesting a functional redundancy of these two molecules.[182–184] $CD4^{+}CD3^{-}$ accessory cells also known as lymph node-inducing cells (Lti) are an important source of OX40L and CD30L in the secondary lymphoid tissues, and they localize at the T:B border and within the follicles. These accessory cells are antigen-presenting cells required to support the adaptive immune system's ability to generate memory responses.[185] Unlike conventional APCs, the $CD3^{-}CD4^{+}$ cells express low levels of CD80/86[181] and interact with primed T cells at the T:B border and with Tfh cells within the germinal center. Here they provide OX40L and CD30L to both Tfh cell precursors outside the follicle and Tfh cells in the germinal center.[185]

Controlling the precise amount of OX40 and OX40L signals on T cells and APCs respectively is an important checkpoint for regulating the numbers of T cells that enter or are retained in the follicle: overexpression of OX40L results in excessive numbers of $CD4^{+}$ T cells entering the germinal center,[91] and antinuclear antibody production[94] (Fig. 2, checkpoint 2).

4. Controlling the Stability of T:B Cognate Interactions

Cognate T:B cell interaction at the T:B junction is essential for Tfh cell differentiation. In both B cell-deficient (μMT) mice and transgenic mice with a B cell repertoire with a different antigen specificity to the T cell compartment there is an absolute failure to form Tfh cells.[59] Formation of Tfh cells in the absence of antigen-specific interactions with B cells can be corrected by overexpression of Bcl6, suggesting that the interaction of Tfh precursors and B cells is important for Bcl6 upregulation and lineage commitment.[59] The B cell-derived signals that determine Tfh formation are not known but there is evidence that ICOSL signals play a role: Tfh cells were reduced by ~50% in mice lacking ICOSL expression on B cells.[63] The stability of T:B interactions appears to be a key determinant of Tfh fate, and this requirement has been demonstrated in a series of studies using SAP-deficient mice.

SAP, encoded by the *Sh2d1a* gene, is a small adapter protein that consists of a single SH2 (SRC homology 2) domain that couples the SLAM-family receptors to intracellular signaling cascades. SAP is expressed by multiple cell types in the immune system including T cells, B cells, natural killer cells, NKT cells, eosinophils, and platelets.[95,186] The SLAM-family of receptors, SLAM (CD150), LY9 (CD229, SLAMF3), 2B4, CD84 (SLAMF5), Ly108 (SLAMF6, NTBA in humans), and CRACC (CS1, CD319, SLAMF7), are self-ligands, with the exception of 2B4 that binds CD48. There is also evidence that SAP may also signal downstream of the TCR, independent of SLAM receptors.[187]

Characterization of SAP-deficient mice has revealed the important role of this small adaptor to the production of T-dependent antibody responses. After immunization with T-dependent antigens *Sap*$^{-/-}$ mice fail to form germinal centers, and have reduced numbers of antigen-specific memory B cells and long-lived plasma cells.[188–192] The defects in germinal center formation and humoral immunity appear to be caused by absence of SAP expression in T cells, rather than B cells[188–191,193]; although there have been some suggestions SAP might also mediate signaling in B cells.[192]

The precise stage of the humoral immune response in which SAP acts has recently been defined: SAP is an absolute requirement for functional Tfh formation but is dispensible for the B cell helper function of T cells in the extrafollicular foci. T cell help is required for the follicular and extrafollicular responses to T-dependent antigens. *Sap*$^{-/-}$ T cells are capable of helping B cells become IgG1^{+} antibody forming cells in a contact-dependent *in vitro* system,[190] and wild type B cells adoptively transferred into *Sap*$^{-/-}$ mice are capable of forming antigen-specific plasma cells 5 days after immunization.[23] These studies demonstrate that *Sap*$^{-/-}$ T cells are capable of supporting antibody production and that T-dependent extrafollicular antibody production is independent of SAP expression in T cells. Elegant *in vivo* imaging studies have demonstrated that SAP-deficient CD4^{+} T cells are able to form stable interactions with DCs but *Sap*$^{-/-}$ T cells, unlike wild type CD4^{+} cells, were unable to form long-lived cognate interactions with antigen-presenting B cells. The consequence of this is that fewer *Sap*$^{-/-}$ CD4^{+} T cells enter the germinal center and those that do are not retained there.[194] The implications of this work are twofold: First, they specifically implicate SAP and possibly the SLAM-family of receptors in mediating the interaction between T and B cells outside the follicle, thereby determining which T cells develop into Tfh cells. Second, this study suggests that interaction time between a T and a B cell determines whether a T cell will help B cells outside the follicle or become a Tfh cell. As mentioned above, T cells that enter the Tfh lineage have the highest peptide binding affinity compared to other effector T cells suggesting they have been selected on the basis of signal strength.[158] Thus, the length of time for which T–B conjugates are maintained may be important for T cells to reach a particular signaling threshold before entering the follicle.

The requirement of SAP-mediated stable T:B interactions for Tfh formation emerges as yet another tolerance mechanism by which weakly cross-reactive B cells may fail to enter germinal centers due to the inability to sustain prolonged contacts with B cells. Indeed, loss of SAP protein ameliorates the spontaneous lupus-like pathology of the NZB/W F1 Sle1b, *MRL*^{lpr}, and *Sanroque* (*Roquin*$^{san/san}$) mouse strains[23,195,196] and reduces the pathology of pristine-induced lupus, with undetectable kidney disease.[197] However, *Sap*$^{-/-}$ mice develop EAE, a Th1/Th17-mediated disease, at the same frequency as wild type mice but with more severe pathology.[197] This highlights a specific role of

Tfh cells in SLE, and suggests that these cells are not required for EAE development and may even be protective. Furthermore, it demonstrates that the interactions between T and B cells outside the follicle are important for regulating the generation of Tfh cells (Fig. 2, checkpoint 3).

D. Control of Tfh Migration into Follicles

Migration of T cells into the germinal center requires expression of the right balance of homing receptors. After T cell priming almost all $CD4^+$ T cells transiently express CXCR5,[198] but Tfh cells are the only subset to retain expression of this chemokine receptor during the effector phase. Despite increased CXCR5 expression, coexpression of the chemokine receptor CCR7 prevents recently primed cells from homing to the CXCL13 produced by the FDCs and entering the B cell follicle,[22] as the ligands for CCR7, CCL19, and CCL21, have the highest concentration in the adjacent T cell zone.[199] The balance of the chemokine receptors CCR7 and CXCR5 regulates the entry of Tfh cell into the follicle.[199] After immunization with protein–antigen in adjuvant antigen-specific T cells readily enter the follicle. However, when mice are exposed to the same antigen during tolerizing conditions, in the absence of adjuvant, T cells can only be observed in the interfollicular areas, and not in the follicle.[200] This suggests that if a T cell encounters antigen in the absence of optimal APC activation, as is the case of self-antigen, T cells cannot enter the follicles and act as Tfh cells. Thus, exclusion of self-reactive T cells from the germinal center is an important checkpoint for Tfh cell tolerance.

E. Mechanisms that Limit Growth and/or Survival of Tfh Cells in Follicles

1. Tfh Cells Have Heightened Sensitivity to Apoptosis

Controlled cell suicide, known as apoptosis, is an essential process in almost every biological system. As described previously, Apoptosis is indispensable for the development and maintenance of a tolerant helper T cell repertoire. Interference with cellular apoptosis results in immune pathology.[201] Apoptosis can be triggered via two distinct pathways; the mitochondrial death pathway and the death receptor pathway.

Apoptosis via the mitochondrial death pathway results from the permeabilization of the outer mitochondrial membrane and is triggered by various endogenous stimuli including; growth factor withdrawal, extracellular stress, various drugs, and irradiation. This process is regulated by the BCL-2 protein family, whose members fall into three categories as directly pro-apoptotic (BAX and BAK), directly antiapoptotic (BCL-2, BCL-X_L, BCL-W, MCL1, A1, and BOO), or those that integrate the signals from the extracellular environment and release the inhibitory signals of the antiapoptotic Bcl-2 family members on the pro-apoptotic proteins BAX and BAK, allowing apoptosis to occur (BIM, BAD, BIK, BID, Harakiri, NOXA, Puma, and BMF).

Tonsillar Tfh cells have decreased expression of antiapoptotic BCL-2 and Bcl-XL and increased expression of the pro-apoptotic BCL-2 family members BAK, BID, and BAD. Because of this, Tfh cells are more prone to undergo apoptosis after TCR stimulation than naïve or memory helper T cells.[78] Histological staining has revealed that, unlike interfollicular T cells, human tonsillar Tfh cells do not express BCL-2,[202] which is consistent with the role of Bcl6, the lineage transcription factor of Tfh cells, in repressing Bcl-2 expression.[203] Overexpression of Bcl-2 in the entire hematopoietic compartment (VavP-Bcl2 mice), but not in B cells alone (Eμ-Bcl2 mice), results in spontaneous germinal center formation development accompanied by an increase in B cells that produce class-switched somatically mutated immunoglobulin. Genetic ablation of CD4 cells in VavP-Bcl2 mice abrogates spontaneous germinal center formation.[204] Together, these studies suggest that Tfh cells normally express a combination of Bcl-2 family members that renders them sensitive to apoptosis. Interfering with this powerful tolerance mechanism may promote germinal center formation and/or persistence in the absence of exogenous antigen.

It is also likely that Tfh cells formed after immunization die through the process of Fas/CD95-mediated AICD, described above: Human tonsillar Tfh cells express higher levels of CD95 and its ligand than naïve or memory helper T cells, and are more sensitive to CD95 triggered cell death than these other helper populations.[78] Thus, AICD of Tfh cells in the follicle could be an important mechanism to regulate the size of the Tfh population and the germinal center response.

2. Treg-Mediated Suppression

Within the germinal center there is also evidence that Tfh cells are also subjected to dominant tolerance mechanisms. In both mouse and humans, a subset of $CD4^{+}CD25^{+}$ regulatory T cells has been identified within the germinal center. These Treg cells are capable of suppressing the effects of Tfh cells and may play a critical role in moderating Tfh function, including IL-17 expression by Tfh cells, thus moderating the provision of help that Tfh cells can provide to B cells within the germinal center.[99,205]

3. Limiting Costimulation in the Follicles

Three members of the CD28 family of costimulatory receptors, cytotoxic T lymphocyte antigen 4 (CTLA-4), ICOS, and PD-1 have been suggested to provide signals to T cells within the germinal center. All these receptors have evolved to perform separate roles during the immune response. The shared ligands for CD28 and CTLA-4—CD80/86, ICOSL, and the ligands for PD-1 (PD-1L1/L2) are expressed at high levels within the germinal center, where they can ligate Tfh cells and regulate their function.

Germinal center B cells express elevated levels of CD86.[206] While this could simply be a reflection of B cell activation in this microenvironment, it is likely that CD86 expression exerts a regulatory role in the germinal center ligating CD28, CTLA-4, or both. Elegant work from Walker et al. used CTLA-4 Ig transgenic mice to demonstrate that delivery of CD28 signals on the day after immunization is sufficient to initiate the germinal center response but further CD28 signals are not required for its maintenance, suggesting that a major role of CD28 in the germinal center response is allowing T cell priming and generation of Tfh cells. In this system continuous blockade of CTLA-4 ligation induced abnormal persistence of germinal centers, suggesting that this receptor is essential for curtailing germinal center reactions.[207] It is also possible that CTLA-4 blockade was affecting the function of Tregs in the germinal centers.[208] In wild type mice, administration of anti-CD86 antibody during an established germinal center response results in a decrease in SHM and memory B cell development,[31] indicating that the signals received by Tfh cells through CD86 receptor itself (expressed on a fraction of activated T cells), or through ligating CTLA-4 or CD28 can alter the quality of help they can deliver to centrocytes.

ICOS is expressed on all helper T cell subsets, and at highest levels on Tfh cells.[209] *Icos*$^{-/-}$ mice form small germinal centers after immunization, suggesting that Tfh cells require signaling through ICOS:ICOSL interactions for their formation or maintenance.[210–212] Expression of ICOSL on B cells is required for formation of the Tfh cell population,[63] and ICOS signals have also been shown to regulate Tfh cell longevity.[213] ICOSL is expressed on naïve B cells; after activation, surface expression of ICOSL decreases to nearly undetectable amounts. In response to CD40 signals B cells reexpress ICOSL on their surface.[214] Thus, germinal center B cells that have established cognate interactions with CD40L-expressing T cells are likely to be a rich source of ICOSL in the follicle and may be responsible for maintaining Tfh cells. In *sanroque* mice, ICOS is overexpressed on naïve and effector/memory T cells and Tfh cells accumulate in the absence of exogenous antigen,[93] resulting in a lupus-like pathology.[23] Halving the gene dose of *Icos* in *sanroque* mice also reduced the spontaneous formation of Tfh cells by ~50%.[129] Together, these findings demonstrate that control of ICOS signaling is required to control Tfh cell numbers and in doing so, prevent autoimmunity.

Another CD28-family member, programmed cell death-1 (PD-1), is expressed at very high levels on Tfh cells that are localized to the light zone of the germinal center.[22,215,216] PD-1 is an inhibitory costimulatory receptor encoded by the gene *Pcdc-1*. Many different cell types including T cells, B cells, monocytes, natural killer T cells, and DCs express PD-1. It is not expressed on naïve T cells—rather it is upregulated after activation, and the ligation effects can be observed within a few hours after T cell activation.[217,218] PD-1 can be ligated by PD-L1, PD-L2, and CD80.[219–221] As with the other costimulatory receptors, PD-1 signaling acts to modify the signals given to the T cell through antigen stimulation and other accessory molecules. After PD-1 ligation, two residues of its cytoplasmic tail are phosphorylated, and the SH2-domain containing tyrosine phosphatases 1 and 2 (SHP 1 and 2) are recruited to bind its ITIM and ITSM motifs, allowing them access to the TCR complex.[218,222,223]

PD-1 can only exert inhibitory effects when it is concentrated in the immunological synapse, where it modifies the levels of phosphorylation of the TCR complex molecules CD3ζ, ZAP10, and PKCθ due to SHP recruitment, thus attenuating signaling. PD-1 activation inhibits PI3K activity, resulting in the downstream suppression of Akt and the downregulation of Bcl-xL, IL-2, and IFNγ.[224,225] Intriguingly, PD-1 ligation also has specific effects on attenuating the effects of the key Tfh molecules, ICOS and IL-21, while has little or no effects on CD28 costimulation events.[225] This suggests that one of the major functions of PD-1 is regulation of the "quality" of the Tfh response: negative regulation of Tfh cells to ensure that they do not provide inappropriate help to self-reactive B cells (Fig. 2, checkpoint 4).

The expression of high levels of PD-1 by Tfh cells may also alter the cellular interactions within the germinal center. Engagement of PD-1 by PD-L1 abrogates the TCR-induced stop signal, inhibiting the interaction of PD-1$^+$ T cells and DCs. This has been shown to be important for preventing autoimmune diabetes in mice.[226] The ligands PD-L1 and PD-L2 are abundantly expressed in the germinal center on T cells, DCs, FDCs, but not on B cells.[227] Given the cellular localization of these ligands, and the high expression of PD-1 by Tfh cells, an additional role for PD-1 ligation on Tfh cells may be to discourage T cells interacting with any non-B cells in the germinal center, ensuring that the Tfh population, that is outnumbered by B cells 20:1 in the germinal center, can provide help and selection signals to centrocytes.

IX. Concluding Remarks

The germinal center response to foreign antigen provides the host with long lasting high-affinity immunity to pathogens. The stochastic nature of affinity maturation in the germinal center means that during a response to foreign antigen, self-reactive B lymphocytes can arise *de novo* from this microenviroment and have the potential to generate persistent life-long autoimmune disease. Thus, it is vital that the germinal center response is tightly regulated to ensure only centrocytes with high affinity for exogenous antigen are selected to egress as long-lived plasma cells or memory cells. Tfh cells are central to the selection of germinal center B cells; they can provide positive selection signals through CD40L in combination with IL-21 and IL-4, and can also provide negative selection signals via CD95L. For Tfh cells to deliver pro-survival signals to centrocytes, they need to form a cognate interaction, which means that the Tfh cell population must itself be rigorously subjected to peripheral tolerance mechanisms to ensure that an autoreactive Tfh cell and self-reactive centrocyte do not engage in a productive conjugation. Understanding how Tfh cells develop and are tolerized, as well as the mechanisms that control Tfh function, has the potential to provide new insights into the pathogenesis of autoantibody-driven diseases and ultimately allow the development of more specific therapies.

References

1. Dong C. TH17 cells in development: an updated view of their molecular identity and genetic programming. *Nat Rev Immunol* 2008;**8**:337–48.
2. King C. New insights into the differentiation and function of T follicular helper cells. *Nat Rev Immunol* 2009;**9**:757–66.
3. Liew FY. T(H)1 and T(H)2 cells: a historical perspective. *Nat Rev Immunol* 2002;**2**:55–60.
4. Qi H, Egen JG, Huang AY, Germain RN. Extrafollicular activation of lymph node B cells by antigen-bearing dendritic cells. *Science* 2006;**312**:1672–6.
5. Miller MJ, Wei SH, Parker I, Cahalan MD. Two-photon imaging of lymphocyte motility and antigen response in intact lymph node. *Science* 2002;**296**:1869–73.
6. Okada T, Ngo VN, Ekland EH, Forster R, Lipp M, Littman DR, et al. Chemokine requirements for B cell entry to lymph nodes and Peyer's patches. *J Exp Med* 2002;**196**:65–75.
7. Ekland EH, Forster R, Lipp M, Cyster JG. Requirements for follicular exclusion and competitive elimination of autoantigen-binding B cells. *J Immunol* 2004;**172**:4700–8.
8. Okada T, Miller MJ, Parker I, Krummel MF, Neighbors M, Hartley SB, et al. Antigen-engaged B cells undergo chemotaxis toward the T zone and form motile conjugates with helper T cells. *PLoS Biol* 2005;**3**:e150.
9. Reif K, Ekland EH, Ohl L, Nakano H, Lipp M, Forster R, et al. Balanced responsiveness to chemoattractants from adjacent zones determines B-cell position. *Nature* 2002;**416**:94–9.
10. McHeyzer-Williams LJ, Malherbe LP, McHeyzer-Williams MG. Checkpoints in memory B-cell evolution. *Immunol Rev* 2006;**211**:255–68.
11. Chan TD, Gatto D, Wood K, Camidge T, Basten A, Brink R. Antigen affinity controls rapid T-dependent antibody production by driving the expansion rather than the differentiation or extrafollicular migration of early plasmablasts. *J Immunol* 2009;**183**:3139–49.
12. Toyama H, Okada S, Hatano M, Takahashi Y, Takeda N, Ichii H, et al. Memory B cells without somatic hypermutation are generated from Bcl6-deficient B cells. *Immunity* 2002;**17**:329–39.
13. Inamine A, Takahashi Y, Baba N, Miyake K, Tokuhisa T, Takemori T, et al. Two waves of memory B-cell generation in the primary immune response. *Int immunol* 2005;**17**:581–9.
14. Hawkins ED, Markham JF, McGuinness LP, Hodgkin PD. A single-cell pedigree analysis of alternative stochastic lymphocyte fates. *Proc Natl Acad Sci USA* 2009;**106**:13457–62.
15. MacLennan IC, Toellner KM, Cunningham AF, Serre K, Sze DM, Zuniga E, et al. Extrafollicular antibody responses. *Immunol Rev* 2003;**194**:8–18.

16. Toellner KM, Luther SA, Sze DM, Choy RK, Taylor DR, MacLennan IC, et al. T helper 1 (Th1) and Th2 characteristics start to develop during T cell priming and are associated with an immediate ability to induce immunoglobulin class switching. *J Exp Med* 1998;**187**:1193–204.
17. Garcia De Vinuesa C, Gulbranson-Judge A, Khan M, O'Leary P, Cascalho M, Wabl M, et al. Dendritic cells associated with plasmablast survival. *Eur J Immunol* 1999;**29**:3712–21.
18. Smith KGC, Hewitson TD, Nossal GJ, Tarlinton DM. The phenotype and fate of the antibody-forming cells of the splenic foci. *Eur J Immunol* 1996;**26**:444–8.
19. Odegard JM, Marks BR, Diplacido LD, Poholek AC, Kono DH, Dong C, et al. ICOS-dependent extrafollicular helper T cells elicit IgG production via IL-21 in systemic autoimmunity. *J Exp Med* 2008;**205**:2873–86.
20. Bubier JA, Sproule TJ, Foreman O, Spolski R, Shaffer DJ, Morse 3rd HC, et al. A critical role for IL-21 receptor signaling in the pathogenesis of systemic lupus erythematosus in BXSB-Yaa mice. *Proc Natl Acad Sci USA* 2009;**106**:1518–23.
21. Linterman MA, Rigby RJ, Wong R, Silva D, Withers D, Anderson G, et al. Roquin differentiates the specialized functions of duplicated T cell costimulatory receptor genes CD28 and ICOS. *Immunity* 2009;**30**:228–41.
22. Haynes NM, Allen CD, Lesley R, Ansel KM, Killeen N, Cyster JG. Role of CXCR5 and CCR7 in follicular Th cell positioning and appearance of a programmed cell death gene-1high germinal center-associated subpopulation. *J Immunol* 2007;**179**:5099–108.
23. Linterman MA, Rigby RJ, Wong RK, Yu D, Brink R, Cannons JL, et al. Follicular helper T cells are required for systemic autoimmunity. *J Exp Med* 2009;**206**:561–76.
24. MacLennan IC. Germinal centers. *Ann Rev Immunol* 1994;**12**:117–39.
25. Allen CD, Ansel KM, Low C, Lesley R, Tamamura H, Fujii N, et al. Germinal center dark and light zone organization is mediated by CXCR4 and CXCR5. *Nat Immunol* 2004;**5**:943–52.
26. Hauser AE, Shlomchik MJ, Haberman AM. *In vivo* imaging studies shed light on germinal-centre development. *Nat Rev* 2007;**7**:499–504.
27. Diamond B, Scharff MD. Somatic mutation of the T15 heavy chain gives rise to an antibody with autoantibody specificity. *Proc Natl Acad Sci USA* 1984;**81**:5841–4.
28. Atassi MZ, Casali P. Molecular mechanisms of autoimmunity. *Autoimmunity* 2008;**41**:123–32.
29. Kelsoe G. Life and death in germinal centers (redux). *Immunity* 1996;**4**:107–11.
30. de Vinuesa CG, Cook MC, Ball J, Drew M, Sunners Y, Cascalho M, et al. Germinal centers without T cells. *J Exp Med* 2000;**191**:485–94.
31. Han S, Hathcock K, Zheng B, Kepler TB, Hodes R, Kelsoe G. Cellular interaction in germinal centers. Roles of CD40 ligand and B7-2 in established germinal centers. *J Immunol* 1995;**155**:556–67.
32. Wardemann H, Yurasov S, Schaefer A, Young JW, Meffre E, Nussenzweig MC. Predominant autoantibody production by early human B cell precursors. *Science* 2003;**301**:1374–7.
33. Primi D, Hammarstrom L, Smith CI, Moller G. Characterization of self-reactive B cells by polyclonal B-cell activators. *J Exp Med* 1977;**145**:21–30.
34. Dighiero G, Lymberi P, Holmberg D, Lundquist I, Coutinho A, Avrameas S. High frequency of natural autoantibodies in normal newborn mice. *J Immunol* 1985;**134**:765–71.
35. Cote RJ, Morrissey DM, Houghton AN, Thomson TM, Daly ME, Oettgen HF, et al. Specificity analysis of human monoclonal antibodies reactive with cell surface and intracellular antigens. *Proc Natl Acad Sci USA* 1986;**83**:2959–63.
36. Radic MZ, Weigert M. Genetic and structural evidence for antigen selection of anti-DNA antibodies. *Ann Rev Immunol* 1994;**12**:487–520.
37. Winkler TH, Fehr H, Kalden JR. Analysis of immunoglobulin variable region genes from human IgG anti-DNA hybridomas. *Eur J Immunol* 1992;**22**:1719–28.
38. McHeyzer-Williams MG, McLean MJ, Lalor PA, Nossal GJ. Antigen-driven B cell differentiation *in vivo*. *J Exp Med* 1993;**178**:295–307.

39. William J, Euler C, Christensen S, Shlomchik MJ. Evolution of autoantibody responses via somatic hypermutation outside of germinal centers. *Science* 2002;**297**:2066–70.
40. Ray SK, Putterman C, Diamond B. Pathogenic autoantibodies are routinely generated during the response to foreign antigen: a paradigm for autoimmune disease. *Proc Natl Acad Sci USA* 1996;**93**:2019–24.
41. Luzina IG, Atamas SP, Storrer CE, daSilva LC, Kelsoe G, Papadimitriou JC, et al. Spontaneous formation of germinal centers in autoimmune mice. *J Leukoc Biol* 2001;**70**:578–84.
42. Armengol MP, Juan M, Lucas Martin A, Fernandez-Figueras MT, Jaraquemada D, Gallart T, et al. Thyroid autoimmune disease: demonstration of thyroid antigen-specific B cells and recombination-activating gene expression in chemokine-containing active intrathyroidal germinal centers. *Am J Pathol* 2001;**159**:861–73.
43. Salomonsson S, Jonsson MV, Skarstein K, Brokstad KA, Hjelmstrom P, Wahren-Herlenius M, et al. Cellular basis of ectopic germinal center formation and autoantibody production in the target organ of patients with Sjogren's syndrome. *Arthritis Rheum* 2003;**48**:3187–201.
44. Sims GP, Shiono H, Willcox N, Stott DI. Somatic hypermutation and selection of B cells in thymic germinal centers responding to acetylcholine receptor in myasthenia gravis. *J Immunol* 2001;**167**:1935–44.
45. Cappione 3rd A, Anolik JH, Pugh-Bernard A, Barnard J, Dutcher P, Silverman G, et al. Germinal center exclusion of autoreactive B cells is defective in human systemic lupus erythematosus. *J Clin Invest* 2005;**115**:3205–16.
46. Vinuesa CG, Sanz I, Cook MC. Dysregulation of germinal centres in autoimmune disease. *Nat Rev* 2009;**9**:845–57.
47. Tarlinton D, Radbruch A, Hiepe F, Dorner T. Plasma cell differentiation and survival. *Curr Opin Immunol* 2008;**20**:162–9.
48. Tarlinton D. B-cell memory: are subsets necessary? *Nat Rev* 2006;**6**:785–90.
49. Koopman G, Keehnen RM, Lindhout E, Newman W, Shimizu Y, van Seventer GA, et al. Adhesion through the LFA-1 (CD11a/CD18)-ICAM-1 (CD54) and the VLA-4 (CD49d)-VCAM-1 (CD106) pathways prevents apoptosis of germinal center B cells. *J Immunol* 1994;**152**:3760–7.
50. Randall KL, Lambe T, Johnson A, Treanor B, Kucharska E, Domaschenz H, et al. Dock8 mutations cripple B cell immunological synapses, germinal centers and long-lived antibody production. *Nat Immunol* 2009;**10**:1283–91.
51. Vinuesa CG, Tangye SG, Moser B, Mackay CR. Follicular B helper T cells in antibody responses and autoimmunity. *Nat Rev* 2005;**5**:853–65.
52. Liu YJ, Joshua DE, Williams GT, Smith CA, Gordon J, MacLennan IC. Mechanism of antigen-driven selection in germinal centres. *Nature* 1989;**342**:929–31.
53. Velardi A, Mingari MC, Moretta L, Grossi CE. Functional analysis of cloned germinal center CD4+ cells with natural killer cell-related features. Divergence from typical T helper cells. *J Immunol* 1986;**137**:2808–13.
54. Hsu SM, Cossman J, Jaffe ES. Lymphocyte subsets in normal human lymphoid tissues. *Am J Clin Pathol* 1983;**80**:21–30.
55. Casamayor-Palleja M, Khan M, MacLennan IC. A subset of CD4+ memory T cells contains preformed CD40 ligand that is rapidly but transiently expressed on their surface after activation through the T cell receptor complex. *J Exp Med* 1995;**181**:1293–301.
56. Casamayor-Palleja M, Feuillard J, Ball J, Drew M, MacLennan IC. Centrocytes rapidly adopt a memory B cell phenotype on co-culture with autologous germinal centre T cell-enriched preparations. *Int Immunol* 1996;**8**:737–44.
57. Fazilleau N, Eisenbraun MD, Malherbe L, Ebright JN, Pogue-Caley RR, McHeyzer-Williams L, et al. Lymphoid reservoirs of antigen-specific memory T helper cells. *Nat Immunol* 2007;**8**:753–61.

58. Yu D, Rao S, Tsai LM, Lee SK, He Y, Sutcliffe EL, et al. The transcriptional repressor Bcl-6 directs T follicular helper cell lineage commitment. *Immunity* 2009;**31**(3):457–68.
59. Johnston RJ, Poholek AC, Ditoro D, Yusuf I, Eto D, Barnett B, et al. Bcl6 and Blimp-1 are reciprocal and antagonistic regulators of T follicular helper cell differentiation. *Science* 2009;**325**(5943):1006–10.
60. Nurieva RI, Chung Y, Martinez GJ, Yang XO, Tanaka S, Matskevitch TD, et al. Bcl6 mediates the development of T follicular helper cells. *Science* 2009;**325**(5943):1001–5.
61. Chtanova T, Tangye SG, Newton R, Frank N, Hodge MR, Rolph MS, et al. T follicular helper cells express a distinctive transcriptional profile, reflecting their role as non-Th1/Th2 effector cells that provide help for B cells. *J Immunol* 2004;**173**:68–78.
62. Kim CH, Lim HW, Kim JR, Rott L, Hillsamer P, Butcher EC. Unique gene expression program of human germinal center T helper cells. *Blood* 2004;**104**:1952–60.
63. Nurieva RI, Chung Y, Hwang D, Yang XO, Kang HS, Ma L, et al. Generation of T follicular helper cells is mediated by interleukin-21 but independent of T helper 1, 2, or 17 cell lineages. *Immunity* 2008;**29**:138–49.
64. Schaerli P, Willimann K, Lang AB, Lipp M, Loetscher P, Moser B. CXC chemokine receptor 5 expression defines follicular homing T cells with B cell helper function. *J Exp Med* 2000;**192**:1553–62.
65. Breitfeld D, Ohl L, Kremmer E, Ellwart J, Sallusto F, Lipp M, et al. Follicular B helper T cells express CXC chemokine receptor 5, localize to B cell follicles, and support immunoglobulin production. *J Exp Med* 2000;**192**:1545–52.
66. Kim JR, Lim HW, Kang SG, Hillsamer P, Kim CH. Human CD57+ germinal center-T cells are the major helpers for GC-B cells and induce class switch recombination. *BMC Immunol* 2005;**6**:3.
67. Toellner KM, Jenkinson WE, Taylor DR, Khan M, Sze DM, Sansom DM, et al. Low-level hypermutation in T cell-independent germinal centers compared with high mutation rates associated with T cell-dependent germinal centers. *J Exp Med* 2002;**195**:383–9.
68. Cunningham AF, Serre K, Toellner KM, Khan M, Alexander J, Brombacher F, et al. Pinpointing IL-4-independent acquisition and IL-4-influenced maintenance of Th2 activity by CD4 T cells. *Eur J Immunol* 2004;**34**:686–94.
69. Linterman MA, Beaton L, Yu D, Ramiscal RR, Srivastava M, Hogan JJ, et al. IL-21 acts directly on B cells to regulate Bcl-6 expression and germinal center responses. *J Exp Med* 2010;**207**(2):353–63.
70. Allen CD, Okada T, Cyster JG. Germinal-center organization and cellular dynamics. *Immunity* 2007;**27**:190–202.
71. Hauser AE, Junt T, Mempel TR, Sneddon MW, Kleinstein SH, Henrickson SE, et al. Definition of germinal-center B cell migration *in vivo* reveals predominant intrazonal circulation patterns. *Immunity* 2007;**26**:655–67.
72. Schwickert TA, Lindquist RL, Shakhar G, Livshits G, Skokos D, Kosco-Vilbois MH, et al. *In vivo* imaging of germinal centres reveals a dynamic open structure. *Nature* 2007;**446**:83–7.
73. Meyer-Hermann ME, Maini PK, Iber D. An analysis of B cell selection mechanisms in germinal centers. *Math Med Biol* 2006;**23**:255–77.
74. Batista FD, Neuberger MS. Affinity dependence of the B cell response to antigen: a threshold, a ceiling, and the importance of off-rate. *Immunity* 1998;**8**:751–9.
75. Kupfer H, Monks CR, Kupfer A. Small splenic B cells that bind to antigen-specific T helper (Th) cells and face the site of cytokine production in the Th cells selectively proliferate: immunofluorescence microscopic studies of Th-B antigen-presenting cell interactions. *J Exp Med* 1994;**179**:1507–15.
76. Allen CD, Okada T, Tang HL, Cyster JG. Imaging of germinal center selection events during affinity maturation. *Science* 2007;**315**:528–31.

77. Smith KGC, Nossal GJ, Tarlinton DM. FAS is highly expressed in the germinal center but is not required for regulation of the B-cell response to antigen. *Proc Natl Acad Sci USA* 1995;**92**:11628–32.
78. Marinova E, Han S, Zheng B. Human germinal center T cells are unique Th cells with high propensity for apoptosis induction. *Int Immunol* 2006;**18**:1337–45.
79. Watanabe-Fukunaga R, Brannan CI, Copeland NG, Jenkins NA, Nagata S. Lymphoproliferation disorder in mice explained by defects in Fas antigen that mediates apoptosis. *Nature* 1992;**356**:314–7.
80. Takahashi T, Tanaka M, Brannan CI, Jenkins NA, Copeland NG, Suda T, et al. Generalized lymphoproliferative disease in mice, caused by a point mutation in the Fas ligand. *Cell* 1994;**76**:969–76.
81. Cohen PL, Eisenberg RA. Lpr and gld: single gene models of systemic autoimmunity and lymphoproliferative disease. *Ann Rev Immunol* 1991;**9**:243–69.
82. Adachi M, Suematsu S, Suda T, Watanabe D, Fukuyama H, Ogasawara J, et al. Enhanced and accelerated lymphoproliferation in Fas-null mice. *Proc Natl Acad Sci USA* 1996;**93**:2131–6.
83. Mixter PF, Russell JQ, Budd RC. Delayed kinetics of T lymphocyte anergy and deletion in lpr mice. *J Autoimmun* 1994;**7**:697–710.
84. Mountz JD, Bluethmann H, Zhou T, Wu J. Defective clonal deletion and anergy induction in TCR transgenic lpr/lpr mice. *Semin Immunol* 1994;**6**:27–37.
85. Shlomchik MJ, Madaio MP, Ni D, Trounstein M, Huszar D. The role of B cells in lpr/lpr-induced autoimmunity. *J Exp Med* 1994;**180**:1295–306.
86. Fukuyama H, Adachi M, Suematsu S, Miwa K, Suda T, Yoshida N, et al. Requirement of Fas expression in B cells for tolerance induction. *Eur J Immunol* 2002;**32**:223–30.
87. Chan O, Shlomchik MJ. A new role for B cells in systemic autoimmunity: B cells promote spontaneous T cell activation in MRL-lpr/lpr mice. *J Immunol* 1998;**160**:51–9.
88. Hao Z, Hampel B, Yagita H, Rajewsky K. T cell-specific ablation of Fas leads to Fas ligand-mediated lymphocyte depletion and inflammatory pulmonary fibrosis. *J Exp Med* 2004;**199**:1355–65.
89. Hao Z, Duncan GS, Seagal J, Su YW, Hong C, Haight J, et al. Fas receptor expression in germinal-center B cells is essential for T and B lymphocyte homeostasis. *Immunity* 2008;**29**:615–27.
90. Takahashi Y, Ohta H, Takemori T. Fas is required for clonal selection in germinal centers and the subsequent establishment of the memory B cell repertoire. *Immunity* 2001;**14**:181–92.
91. Brocker T, Gulbranson-Judge A, Flynn S, Riedinger M, Raykundalia C, Lane P. CD4 T cell traffic control: *in vivo* evidence that ligation of OX40 on CD4 T cells by OX40-ligand expressed on dendritic cells leads to the accumulation of CD4 T cells in B follicles. *Eur J Immunol* 1999;**29**:1610–6.
92. Subramanian S, Tus K, Li QZ, Wang A, Tian XH, Zhou J, et al. A Tlr7 translocation accelerates systemic autoimmunity in murine lupus. *Proc Natl Acad Sci USA* 2006;**103**:9970–5.
93. Vinuesa CG, Cook MC, Angelucci C, Athanasopoulos V, Rui L, Hill KM, et al. A RING-type ubiquitin ligase family member required to repress follicular helper T cells and autoimmunity. *Nature* 2005;**435**:452–8.
94. Murata K, Nose M, Ndhlovu LC, Sato T, Sugamura K, Ishii N. Constitutive OX40/OX40 ligand interaction induces autoimmune-like diseases. *J Immunol* 2002;**169**:4628–36.
95. Schwartzberg PL, Mueller KL, Qi H, Cannons JL. SLAM receptors and SAP influence lymphocyte interactions, development and function. *Nat Rev* 2009;**9**:39–46.
96. Hsu HC, Wu Y, Yang P, Wu Q, Job G, Chen J, et al. Overexpression of activation-induced cytidine deaminase in B cells is associated with production of highly pathogenic autoantibodies. *J Immunol* 2007;**178**:5357–65.

97. Hsu HC, Zhou T, Kim H, Barnes S, Yang P, Wu Q, et al. Production of a novel class of polyreactive pathogenic autoantibodies in BXD2 mice causes glomerulonephritis and arthritis. *Arthritis Rheum* 2006;**54**:343–55.
98. Hsu HC, Yang P, Wang J, Wu Q, Myers R, Chen J, et al. Interleukin 17-producing T helper cells and interleukin 17 orchestrate autoreactive germinal center development in autoimmune BXD2 mice. *Nat Immunol* 2008;**9**:166–75.
99. Wu HY, Quintana FJ, Weiner HL. Nasal anti-CD3 antibody ameliorates lupus by inducing an IL-10-secreting CD4+ CD25− LAP+ regulatory T cell and is associated with down-regulation of IL-17+ CD4+ ICOS+ CXCR5+ follicular helper T cells. *J Immunol* 2008;**181**:6038–50.
100. Hu YL, Metz DP, Chung J, Siu G, Zhang M. B7RP-1 blockade ameliorates autoimmunity through regulation of follicular helper T cells. *J Immunol* 2009;**182**:1421–8.
101. Simpson N, Gatenby PA, Wilson A, Malik S, Fulcher DA, Tangye SG, et al. Expansion of circulating T cells resembling follicular helper T cells is a fixed phenotype that identifies a subset of severe systemic lupus erythematosus. *Arthritis Rheum* 2010;**62**:234–44.
102. Simpson N, Gatenby PA, Wilson A, Malik S, Fulcher DA, Tangye SG, et al. Expansion of circulating T cells resembling TFH cells is a fixed phenotype that identifies a subset of severe systemic lupus erythematosus. *Arthritis Rheum* 2009; [in press].
103. Bevan MJ. In thymic selection, peptide diversity gives and takes away. *Immunity* 1997;**7**:175–8.
104. McGargill MA, Derbinski JM, Hogquist KA. Receptor editing in developing T cells. *Nat Immunol* 2000;**1**:336–41.
105. Zehn D, Bevan MJ. T cells with low avidity for a tissue-restricted antigen routinely evade central and peripheral tolerance and cause autoimmunity. *Immunity* 2006;**25**:261–70.
106. Bouneaud C, Kourilsky P, Bousso P. Impact of negative selection on the T cell repertoire reactive to a self-peptide: a large fraction of T cell clones escapes clonal deletion. *Immunity* 2000;**13**:829–40.
107. Liu GY, Fairchild PJ, Smith RM, Prowle JR, Kioussis D, Wraith DC. Low avidity recognition of self-antigen by T cells permits escape from central tolerance. *Immunity* 1995;**3**:407–15.
108. Lafferty KJ, Cunningham AJ. A new analysis of allogeneic interactions. *Aust J Exp Biol Med Sci* 1975;**53**:27–42.
109. Peach RJ, Bajorath J, Brady W, Leytze G, Greene J, Naemura J, et al. Complementarity determining region 1 (CDR1)- and CDR3-analogous regions in CTLA-4 and CD28 determine the binding to B7-1. *J Exp Med* 1994;**180**:2049–58.
110. Bluestone JA. New perspectives of CD28-B7-mediated T cell costimulation. *Immunity* 1995;**2**:555–9.
111. Lenschow DJ, Walunas TL, Bluestone JA. CD28/B7 system of T cell costimulation. *Ann Rev Immunol* 1996;**14**:233–58.
112. Macian F, Garcia-Cozar F, Im SH, Horton HF, Byrne MC, Rao A. Transcriptional mechanisms underlying lymphocyte tolerance. *Cell* 2002;**109**:719–31.
113. Lamb JR, Skidmore BJ, Green N, Chiller JM, Feldmann M. Induction of tolerance in influenza virus-immune T lymphocyte clones with synthetic peptides of influenza hemagglutinin. *J Exp Med* 1983;**157**:1434–47.
114. Jenkins MK, Schwartz RH. Antigen presentation by chemically modified splenocytes induces antigen-specific T cell unresponsiveness *in vitro* and *in vivo*. *J Exp Med* 1987;**165**:302–19.
115. Sloan-Lancaster J, Evavold BD, Allen PM. Induction of T-cell anergy by altered T-cell-receptor ligand on live antigen-presenting cells. *Nature* 1993;**363**:156–9.
116. Quill H, Schwartz RH. Stimulation of normal inducer T cell clones with antigen presented by purified Ia molecules in planar lipid membranes: specific induction of a long-lived state of proliferative nonresponsiveness. *J Immunol* 1987;**138**:3704–12.

117. Korb LC, Mirshahidi S, Ramyar K, Sadighi Akha AA, Sadegh-Nasseri S. Induction of T cell anergy by low numbers of agonist ligands. *J Immunol* 1999;**162**:6401–9.
118. Schwartz RH. A cell culture model for T lymphocyte clonal anergy. *Sciences (New York)* 1990;**248**:1349–56.
119. Rocha B, von Boehmer H. Peripheral selection of the T cell repertoire. *Sciences (New York)* 1991;**251**:1225–8.
120. Rammensee HG, Kroschewski R, Frangoulis B. Clonal anergy induced in mature V beta 6+ T lymphocytes on immunizing Mls 1b mice with Mls-1a expressing cells. *Nature* 1989;**339**:541–4.
121. Asai K, Hachimura S, Kimura M, Toraya T, Yamashita M, Nakayama T, et al. T cell hyporesponsiveness induced by oral administration of ovalbumin is associated with impaired NFAT nuclear translocation and p27kip1 degradation. *J Immunol* 2002;**169**:4723–31.
122. Frauwirth KA, Alegre ML, Thompson CB. CTLA-4 is not required for induction of CD8(+) T cell anergy *in vivo*. *J Immunol* 2001;**167**:4936–41.
123. Adler AJ, Huang CT, Yochum GS, Marsh DW, Pardoll DM. *In vivo* CD4+ T cell tolerance induction versus priming is independent of the rate and number of cell divisions. *J Immunol* 2000;**164**:649–55.
124. Tanchot C, Barber DL, Chiodetti L, Schwartz RH. Adaptive tolerance of CD4+ T cells *in vivo*: multiple thresholds in response to a constant level of antigen presentation. *J Immunol* 2001;**167**:2030–9.
125. Chiodetti L, Choi S, Barber DL, Schwartz RH. Adaptive tolerance and clonal anergy are distinct biochemical states. *J Immunol* 2006;**176**:2279–91.
126. Choi S, Schwartz RH. Molecular mechanisms for adaptive tolerance and other T cell anergy models. *Semin Immunol* 2007;**19**:140–52.
127. Riley JL, Mao M, Kobayashi S, Biery M, Burchard J, Cavet G, et al. Modulation of TCR-induced transcriptional profiles by ligation of CD28, ICOS, and CTLA-4 receptors. *Proc Natl Acad Sci USA* 2002;**99**:11790–5.
128. Greenwald RJ, Freeman GJ, Sharpe AH. The B7 family revisited. *Ann Rev Immunol* 2005;**23**:515–48.
129. Yu D, Tan AH, Hu X, Athanasopoulos V, Simpson N, Silva DG, et al. Roquin represses autoimmunity by limiting inducible T-cell co-stimulator messenger RNA. *Nature* 2007;**450**:299–303.
130. Heissmeyer V, Macian F, Im SH, Varma R, Feske S, Venuprasad K, et al. Calcineurin imposes T cell unresponsiveness through targeted proteolysis of signaling proteins. *Nat Immunol* 2004;**5**:255–65.
131. Jeon MS, Atfield A, Venuprasad K, Krawczyk C, Sarao R, Elly C, et al. Essential role of the E3 ubiquitin ligase Cbl-b in T cell anergy induction. *Immunity* 2004;**21**:167–77.
132. Bachmaier K, Krawczyk C, Kozieradzki I, Kong YY, Sasaki T, Oliveira-dos-Santos A, et al. Negative regulation of lymphocyte activation and autoimmunity by the molecular adaptor Cbl-b. *Nature* 2000;**403**:211–6.
133. Chiang YJ, Kole HK, Brown K, Naramura M, Fukuhara S, Hu RJ, et al. Cbl-b regulates the CD28 dependence of T-cell activation. *Nature* 2000;**403**:216–20.
134. Chiffoleau E, Kobayashi T, Walsh MC, King CG, Walsh PT, Hancock WW, et al. TNF receptor-associated factor 6 deficiency during hemopoiesis induces Th2-polarized inflammatory disease. *J Immunol* 2003;**171**:5751–9.
135. King CG, Kobayashi T, Cejas PJ, Kim T, Yoon K, Kim GK, et al. TRAF6 is a T cell-intrinsic negative regulator required for the maintenance of immune homeostasis. *Nat Med* 2006;**12**:1088–92.
136. King CG, Buckler JL, Kobayashi T, Hannah JR, Bassett G, Kim T, et al. Cutting edge: requirement for TRAF6 in the induction of T cell anergy. *J Immunol* 2008;**180**:34–8.

137. Gronski MA, Boulter JM, Moskophidis D, Nguyen LT, Holmberg K, Elford AR, et al. TCR affinity and negative regulation limit autoimmunity. *Nat Med* 2004;**10**:1234–9.
138. Yokoi N, Komeda K, Wang HY, Yano H, Kitada K, Saitoh Y, et al. Cblb is a major susceptibility gene for rat type 1 diabetes mellitus. *Nat Genet* 2002;**31**:391–4.
139. Krawczyk CM, Jones RG, Atfield A, Bachmaier K, Arya S, Odermatt B, et al. Differential control of CD28-regulated *in vivo* immunity by the E3 ligase Cbl-b. *J Immunol* 2005;**174**:1472–8.
140. Wohlfert EA, Callahan MK, Clark RB. Resistance to CD4+CD25+ regulatory T cells and TGF-beta in Cbl-b$^{-/-}$ mice. *J Immunol* 2004;**173**:1059–65.
141. Loeser S, Loser K, Bijker MS, Rangachari M, van der Burg SH, Wada T, et al. Spontaneous tumor rejection by cbl-b-deficient CD8+ T cells. *J Exp Med* 2007;**204**:879–91.
142. Akiyama T, Maeda S, Yamane S, Ogino K, Kasai M, Kajiura F, et al. Dependence of self-tolerance on TRAF6-directed development of thymic stroma. *Sciences (New York)* 2005;**308**:248–51.
143. Lenardo MJ. The molecular regulation of lymphocyte apoptosis. *Semin Immunol* 1997;**9**:1–5.
144. Fas SC, Fritzsching B, Suri-Payer E, Krammer PH. Death receptor signaling and its function in the immune system. *Curr Dir Autoimmun* 2006;**9**:1–17.
145. Brunner T, Mogil RJ, LaFace D, Yoo NJ, Mahboubi A, Echeverri F, et al. Cell-autonomous Fas (CD95)/Fas-ligand interaction mediates activation-induced apoptosis in T-cell hybridomas. *Nature* 1995;**373**:441–4.
146. Ju ST, Panka DJ, Cui H, Ettinger R, el-Khatib M, Sherr DH, Stanger BZ, Marshak-Rothstein A. Fas(CD95)/FasL interactions required for programmed cell death after T-cell activation. *Nature* 1995;**373**:444–8.
147. Yang Y, Mercep M, Ware CF, Ashwell JD. Fas and activation-induced Fas ligand mediate apoptosis of T cell hybridomas: inhibition of Fas ligand expression by retinoic acid and glucocorticoids. *J Exp Med* 1995;**181**:1673–82.
148. Singer GG, Abbas AK. The fas antigen is involved in peripheral but not thymic deletion of T lymphocytes in T cell receptor transgenic mice. *Immunity* 1994;**1**:365–71.
149. Walker LS, Abbas AK. The enemy within: keeping self-reactive T cells at bay in the periphery. *Nat Rev Immunol* 2002;**2**:11–9.
150. Shlomchik MJ, Marshak-Rothstein A, Wolfowicz CB, Rothstein TL, Weigert MG. The role of clonal selection and somatic mutation in autoimmunity. *Nature* 1987;**328**:805–11.
151. Herlands RA, Christensen SR, Sweet RA, Hershberg U, Shlomchik MJ. T cell-independent and toll-like receptor-dependent antigen-driven activation of autoreactive B cells. *Immunity* 2008;**29**:249–60.
152. Linterman MA, Vinuesa CG. Signals that influence T follicular helper cell differentiation and function. *Semin Immunopathol* 2010;**32**(2):183–96.
153. Shaffer AL, Yu X, He Y, Boldrick J, Chan EP, Staudt LM. BCL-6 represses genes that function in lymphocyte differentiation, inflammation, and cell cycle control. *Immunity* 2000;**13**:199–212.
154. Shaffer AL, Lin KI, Kuo TC, Yu X, Hurt EM, Rosenwald A, et al. Blimp-1 orchestrates plasma cell differentiation by extinguishing the mature B cell gene expression program. *Immunity* 2002;**17**:51–62.
155. Reljic R, Wagner SD, Peakman LJ, Fearon DT. Suppression of signal transducer and activator of transcription 3-dependent B lymphocyte terminal differentiation by BCL-6. *J Exp Med* 2000;**192**:1841–8.
156. Kusam S, Toney LM, Sato H, Dent AL. Inhibition of Th2 differentiation and GATA-3 expression by BCL-6. *J Immunol* 2003;**170**:2435–41.

157. Ma CS, Suryani S, Avery DT, Chan A, Nanan R, Santner-Nanan B, et al. Early commitment of naïve human CD4(+) T cells to the T follicular helper (T(FH)) cell lineage is induced by IL-12. *Immunol Cell Biol* 2009;**87**(8):590–600.
158. Fazilleau N, McHeyzer-Williams LJ, Rosen H, McHeyzer-Williams MG. The function of follicular helper T cells is regulated by the strength of T cell antigen receptor binding. *Nat Immunol* 2009;**10**:375–84.
159. Reinhardt RL, Liang HE, Locksley RM. Cytokine-secreting follicular T cells shape the antibody repertoire. *Nat Immunol* 2009;**10**:385–93.
160. Vogelzang A, McGuire HM, Yu D, Sprent J, Mackay CR, King C. A fundamental role for interleukin-21 in the generation of T follicular helper cells. *Immunity* 2008;**29**:127–37.
161. Zhou L, Ivanov II, Spolski R, Min R, Shenderov K, Egawa T, et al. IL-6 programs T(H)-17 cell differentiation by promoting sequential engagement of the IL-21 and IL-23 pathways. *Nat Immunol* 2007;**8**:967–74.
162. Nurieva R, Yang XO, Martinez G, Zhang Y, Panopoulos AD, Ma L, et al. Essential autocrine regulation by IL-21 in the generation of inflammatory T cells. *Nature* 2007;**448**:480–3.
163. Korn T, Bettelli E, Gao W, Awasthi A, Jager A, Strom TB, et al. IL-21 initiates an alternative pathway to induce proinflammatory T(H)17 cells. *Nature* 2007;**448**:484–7.
164. Coquet JM, Chakravarti S, Smyth MJ, Godfrey DI. Cutting edge: IL-21 is not essential for Th17 differentiation or experimental autoimmune encephalomyelitis. *J Immunol* 2008;**180**:7097–101.
165. Sonderegger I, Kisielow J, Meier R, King C, Kopf M. IL-21 and IL-21R are not required for development of Th17 cells and autoimmunity *in vivo*. *Eur J Immunol* 2008;**38**:1833–8.
166. Cucak H, Yrlid U, Reizis B, Kalinke U, Johansson-Lindbom B. Type I interferon signaling in dendritic cells stimulates the development of lymph-node-resident T follicular helper cells. *Immunity* 2009;**31**:491–501.
167. Suto A, Kashiwakuma D, Kagami S, Hirose K, Watanabe N, Yokote K, et al. Development and characterization of IL-21-producing CD4+ T cells. *J Exp Med* 2008;**205**:1369–79.
168. Schmitt N, Morita R, Bourdery L, Bentebibel SE, Zurawski SM, Banchereau J, et al. Human dendritic cells induce the differentiation of interleukin-21-producing T follicular helper-like cells through interleukin-12. *Immunity* 2009;**31**:158–69.
169. Bengtsson AA, Sturfelt G, Truedsson L, Blomberg J, Alm G, Vallin H, et al. Activation of type I interferon system in systemic lupus erythematosus correlates with disease activity but not with antiretroviral antibodies. *Lupus* 2000;**9**:664–71.
170. Braun D, Geraldes P, Demengeot J. Type I Interferon controls the onset and severity of autoimmune manifestations in lpr mice. *J Autoimmun* 2003;**20**:15–25.
171. Santiago-Raber ML, Baccala R, Haraldsson KM, Choubey D, Stewart TA, Kono DH, et al. Type-I interferon receptor deficiency reduces lupus-like disease in NZB mice. *J Exp Med* 2003;**197**:777–88.
172. Belardelli F, Gresser I. The neglected role of type I interferon in the T-cell response: implications for its clinical use. *Immunol Today* 1996;**17**:369–72.
173. Mathian A, Weinberg A, Gallegos M, Banchereau J, Koutouzov S. IFN-α induces early lethal lupus in preautoimmune (New Zealand Black × New Zealand White) F1 but not in BALB/c mice. *J Immunol* 2005;**174**:2499–506.
174. Sonderegger I, Iezzi G, Maier R, Schmitz N, Kurrer M, Kopf M. GM-CSF mediates autoimmunity by enhancing IL-6-dependent Th17 cell development and survival. *J Exp Med* 2008;**205**:2281–94.
175. Korn T, Mitsdoerffer M, Croxford AL, Awasthi A, Dardalhon VA, Galileos G, et al. IL-6 controls Th17 immunity *in vivo* by inhibiting the conversion of conventional T cells into Foxp3 + regulatory T cells. *Proc Natl Acad Sci USA* 2008;**105**:18460–5.
176. Diveu C, McGeachy MJ, Cua DJ. Cytokines that regulate autoimmunity. *Curr Opin Immunol* 2008;**20**:663–8.

177. Tsantikos E, Oracki SA, Quilici C, Anderson GP, Tarlinton DM, Hibbs ML. Autoimmune disease in Lyn-deficient mice is dependent on an inflammatory environment established by IL-6. *J Immunol* 2010;**184**:1348–60.
178. Iwasaki A, Medzhitov R. Regulation of adaptive immunity by the innate immune system. *Science* 2010;**327**:291–5.
179. Baccala R, Hoebe K, Kono DH, Beutler B, Theofilopoulos AN. TLR-dependent and TLR-independent pathways of type I interferon induction in systemic autoimmunity. *Nat Med* 2007;**13**:543–51.
180. Flynn S, Toellner KM, Raykundalia C, Goodall M, Lane P. CD4 T cell cytokine differentiation: the B cell activation molecule, OX40 ligand, instructs CD4 T cells to express interleukin 4 and upregulates expression of the chemokine receptor, Blr-1. *J Exp Med* 1998;**188**:297–304.
181. Kim MY, Gaspal FM, Wiggett HE, McConnell FM, Gulbranson-Judge A, Raykundalia C, et al. CD4(+)CD3(-) accessory cells costimulate primed CD4 T cells through OX40 and CD30 at sites where T cells collaborate with B cells. *Immunity* 2003;**18**:643–54.
182. Gaspal FM, Kim MY, McConnell FM, Raykundalia C, Bekiaris V, Lane PJ. Mice deficient in OX40 and CD30 signals lack memory antibody responses because of deficient CD4 T cell memory. *J Immunol* 2005;**174**:3891–6.
183. Chen AI, McAdam AJ, Buhlmann JE, Scott S, Lupher Jr. ML, Greenfield EA, et al. Ox40-ligand has a critical costimulatory role in dendritic cell: T cell interactions. *Immunity* 1999;**11**:689–98.
184. Kopf M, Ruedl C, Schmitz N, Gallimore A, Lefrang K, Ecabert B, et al. OX40-deficient mice are defective in Th cell proliferation but are competent in generating B cell and CTL Responses after virus infection. *Immunity* 1999;**11**:699–708.
185. Lane PJ, Gaspal FM, Kim MY. Two sides of a cellular coin: CD4(+)CD3- cells regulate memory responses and lymph-node organization. *Nat Rev* 2005;**5**:655–60.
186. Calpe S, Wang N, Romero X, Berger SB, Lanyi A, Engel P, et al. The SLAM and SAP gene families control innate and adaptive immune responses. *Adv Immunol* 2008;**97**:177–250.
187. Li C, Schibli D, Li SS. The XLP syndrome protein SAP interacts with SH3 proteins to regulate T cell signalling and proliferation. *Cell Signal* 2009;**21**:111–9.
188. Cannons JL, Yu LJ, Jankovic D, Crotty S, Horai R, Kirby M, et al. SAP regulates T cell-mediated help for humoral immunity by a mechanism distinct from cytokine regulation. *J Exp Med* 2006;**203**:1551–65.
189. Crotty S, Kersh EN, Cannons J, Schwartzberg PL, Ahmed R. SAP is required for generating long-term humoral immunity. *Nature* 2003;**421**:282–7.
190. Kamperschroer C, Roberts DM, Zhang Y, Weng NP, Swain SL. SAP enables T cells to help B cells by a mechanism distinct from Th cell programming or CD40 ligand regulation. *J Immunol* 2008;**181**:3994–4003.
191. McCausland MM, Yusuf I, Tran H, Ono N, Yanagi Y, Crotty S. SAP regulation of follicular helper CD4 T cell development and humoral immunity is independent of SLAM and Fyn kinase. *J Immunol* 2007;**178**:817–28.
192. Morra M, Barrington RA, Abadia-Molina AC, Okamoto S, Julien A, Gullo C, et al. Defective B cell responses in the absence of SH2D1A. *Proc Natl Acad Sci USA* 2005;**102**:4819–23.
193. Veillette A. Immune regulation by SLAM family receptors and SAP-related adaptors. *Nat Rev Immunol* 2006;**6**:56–66.
194. Qi H, Cannons JL, Klauschen F, Schwartzberg PL, Germain RN. SAP-controlled T-B cell interactions underlie germinal centre formation. *Nature* 2008;**455**:764–9.
195. Jennings P, Chan A, Schwartzberg P, Wakeland EK, Yuan D. Antigen-specific responses and ANA production in B6.Sle1b mice: a role for SAP. *J Autoimmun* 2008;**31**:345–53.

196. Komori H, Furukawa H, Mori S, Ito MR, Terada M, Zhang MC, et al. A signal adaptor SLAM-associated protein regulates spontaneous autoimmunity and Fas-dependent lymphoproliferation in MRL-Faslpr lupus mice. *J Immunol* 2006;**176**:395–400.
197. Hron JD, Caplan L, Gerth AJ, Schwartzberg PL, Peng SL. SH2D1A regulates T-dependent humoral autoimmunity. *J Exp Med* 2004;**200**:261–6.
198. Ansel KM, McHeyzer-Williams LJ, Ngo VN, McHeyzer-Williams MG, Cyster JG. *In vivo*-activated CD4 T cells upregulate CXC chemokine receptor 5 and reprogram their response to lymphoid chemokines. *J Exp Med* 1999;**190**:1123–34.
199. Hardtke S, Ohl L, Forster R. Balanced expression of CXCR5 and CCR7 on follicular T helper cells determines their transient positioning to lymph node follicles and is essential for efficient B-cell help. *Blood* 2005;**106**:1924–31.
200. Kearney ER, Pape KA, Loh DY, Jenkins MK. Visualization of peptide-specific T cell immunity and peripheral tolerance induction *in vivo*. *Immunity* 1994;**1**:327–39.
201. Bouillet P, O'Reilly LA. CD95, BIM and T cell homeostasis. *Nat Rev* 2009;**9**:514–9.
202. Schenka AA, Muller S, Fournie JJ, Capila F, Vassallo J, Delsol G, et al. CD4+ T cells downregulate Bcl-2 in germinal centers. *J Clin Immunol* 2005;**25**:224–9.
203. Saito M, Novak U, Piovan E, Basso K, Sumazin P, Schneider C, et al. BCL6 suppression of BCL2 via Miz1 and its disruption in diffuse large B cell lymphoma. *Proc Natl Acad Sci USA* 2009;**106**:11294–9.
204. Egle A, Harris AW, Bath ML, O'Reilly L, Cory S. VavP-Bcl2 transgenic mice develop follicular lymphoma preceded by germinal center hyperplasia. *Blood* 2004;**103**:2276–83.
205. Lim HW, Hillsamer P, Kim CH. Regulatory T cells can migrate to follicles upon T cell activation and suppress GC-Th cells and GC-Th cell-driven B cell responses. *J Clin Invest* 2004;**114**:1640–9.
206. Damoiseaux JG, Yagita H, Okumura K, van Breda Vriesman PJ. Costimulatory molecules CD80 and CD86 in the rat; tissue distribution and expression by antigen-presenting cells. *J Leukoc Biol* 1998;**64**:803–9.
207. Walker LS, Wiggett HE, Gaspal FM, Raykundalia CR, Goodall MD, Toellner KM, et al. Established T cell-driven germinal center B cell proliferation is independent of CD28 signaling but is tightly regulated through CTLA-4. *J Immunol* 2003;**170**:91–8.
208. Wing K, Onishi Y, Prieto-Martin P, Yamaguchi T, Miyara M, Fehervari Z, et al. CTLA-4 control over Foxp3+ regulatory T cell function. *Science* 2008;**322**:271–5.
209. Hutloff A, Dittrich AM, Beier KC, Eljaschewitsch B, Kraft R, Anagnostopoulos I, et al. ICOS is an inducible T-cell co-stimulator structurally and functionally related to CD28. *Nature* 1999;**397**:263–6.
210. Dong C, Juedes AE, Temann UA, Shresta S, Allison JP, Ruddle NH, et al. ICOS co-stimulatory receptor is essential for T-cell activation and function. *Nature* 2001;**409**:97–101.
211. McAdam AJ, Greenwald RJ, Levin MA, Chernova T, Malenkovich N, Ling V, et al. ICOS is critical for CD40-mediated antibody class switching. *Nature* 2001;**409**:102–5.
212. Tafuri A, Shahinian A, Bladt F, Yoshinaga SK, Jordana M, Wakeham A, et al. ICOS is essential for effective T-helper-cell responses. *Nature* 2001;**409**:105–9.
213. Akiba H, Takeda K, Kojima Y, Usui Y, Harada N, Yamazaki T, et al. The role of ICOS in the CXCR5+ follicular B helper T cell maintenance *in vivo*. *J Immunol* 2005;**175**:2340–8.
214. Liang L, Porter EM, Sha WC. Constitutive expression of the B7h ligand for inducible costimulator on naive B cells is extinguished after activation by distinct B cell receptor and interleukin 4 receptor-mediated pathways and can be rescued by CD40 signaling. *J Exp Med* 2002;**196**:97–108.
215. Iwai Y, Okazaki T, Nishimura H, Kawasaki A, Yagita H, Honjo T. Microanatomical localization of PD-1 in human tonsils. *Immunol Lett* 2002;**83**:215–20.

216. Dorfman DM, Brown JA, Shahsafaei A, Freeman GJ. Programmed death-1 (PD-1) is a marker of germinal center-associated T cells and angioimmunoblastic T-cell lymphoma. *Am J Surg Pathol* 2006;**30**:802–10.
217. Agata Y, Kawasaki A, Nishimura H, Ishida Y, Tsubata T, Yagita H, et al. Expression of the PD-1 antigen on the surface of stimulated mouse T and B lymphocytes. *Int Immunol* 1996;**8**:765–72.
218. Chemnitz JM, Parry RV, Nichols KE, June CH, Riley JL. SHP-1 and SHP-2 associate with immunoreceptor tyrosine-based switch motif of programmed death 1 upon primary human T cell stimulation, but only receptor ligation prevents T cell activation. *J Immunol* 2004;**173**:945–54.
219. Butte MJ, Keir ME, Phamduy TB, Sharpe AH, Freeman GJ. Programmed death-1 ligand 1 interacts specifically with the B7-1 costimulatory molecule to inhibit T cell responses. *Immunity* 2007;**27**:111–22.
220. Latchman Y, Wood CR, Chernova T, Chaudhary D, Borde M, Chernova I, et al. PD-L2 is a second ligand for PD-1 and inhibits T cell activation. *Nat Immunol* 2001;**2**:261–8.
221. Freeman GJ, Long AJ, Iwai Y, Bourque K, Chernova T, Nishimura H, et al. Engagement of the PD-1 immunoinhibitory receptor by a novel B7 family member leads to negative regulation of lymphocyte activation. *J Exp Med* 2000;**192**:1027–34.
222. Okazaki T, Maeda A, Nishimura H, Kurosaki T, Honjo T. PD-1 immunoreceptor inhibits B cell receptor-mediated signaling by recruiting src homology 2-domain-containing tyrosine phosphatase 2 to phosphotyrosine. *Proc Natl Acad Sci USA* 2001;**98**:13866–71.
223. Sheppard KA, Fitz LJ, Lee JM, Benander C, George JA, Wooters J, et al. PD-1 inhibits T-cell receptor induced phosphorylation of the ZAP70/CD3zeta signalosome and downstream signaling to PKCtheta. *FEBS Lett* 2004;**574**:37–41.
224. Saunders PA, Hendrycks VR, Lidinsky WA, Woods ML. PD-L2:PD-1 involvement in T cell proliferation, cytokine production, and integrin-mediated adhesion. *Eur J Immunol* 2005;**35**:3561–9.
225. Bennett F, Luxenberg D, Ling V, Wang IM, Marquette K, Lowe D, et al. Program death-1 engagement upon TCR activation has distinct effects on costimulation and cytokine-driven proliferation: attenuation of ICOS, IL-4, and IL-21, but not CD28, IL-7, and IL-15 responses. *J Immunol* 2003;**170**:711–8.
226. Fife BT, Pauken KE, Eagar TN, Obu T, Wu J, Tang Q, et al. Interactions between PD-1 and PD-L1 promote tolerance by blocking the TCR-induced stop signal. *Nat Immunol* 2009;**10**:1185–92.
227. Brown JA, Dorfman DM, Ma FR, Sullivan EL, Munoz O, Wood CR, et al. Blockade of programmed death-1 ligands on dendritic cells enhances T cell activation and cytokine production. *J Immunol* 2003;**170**:1257–66.

Section IV

Preventing T Cell-Dependent Autoimmunity

Thymic Selection and Lineage Commitment of $CD4^+Foxp3^+$ Regulatory T Lymphocytes

PAOLA ROMAGNOLI[*,†,‡] AND
JOOST P.M. VAN MEERWIJK[*,†,‡,§]

*Tolerance and Autoimmunity Section, Centre de Physiopathologie de Toulouse Purpan, Institut National de la santé et de la Recherche Medicale (Inserm) U563, Toulouse, France

†University Paul Sabatier, Toulouse, France

‡IFR150, Institut Fédératif de Recherche Bio-Médicale de Toulouse, Toulouse, France

§Institut Universitaire de France and Faculty of Life-Sciences (UFR-SVT), University Paul Sabatier, Toulouse, France

Regulatory T lymphocytes play a central role in the control of a variety of immune-responses. Their absence in humans and in experimental animal models leads to severe autoimmune and inflammatory disorders. Consistent with their major role in prevention of autoimmune pathology, their repertoire is enriched in autospecific cells. Probably the majority of regulatory T cells develop in the thymus. How T cell-precursors choose between the conventional versus regulatory T cell lineages remains an unanswered question. More is known about selection of regulatory T cell precursors. Positive selection of these cells is favored by high affinity interactions with MHC class II/peptide ligands expressed by thymic epithelial or dendritic cells. They are also known

Progress in Molecular Biology
and Translational Science, Vol. 92
DOI: 10.1016/S1877-1173(10)92010-3

1877-1173/10 $35.00

to be relatively resistant to negative selection. These two parameters allow for the generation of the autoreactive regulatory T cell repertoire, and clearly distinguish selection-criteria of conventional versus regulatory T cell-precursors. It will now be important to elucidate the molecular mechanisms involved in the intrathymic choice of the regulatory T cell-lineage.

I. Introduction

T lymphocytes play a major role in the protection of the organism from attack by various infectious agents as well as from malignant transformations. Most T cells differentiate in the thymus from precursors of hematopoietic origin. Thymic T cell differentiation involves a complex combination of commitment to the various lineages developing in this organ, somatic mutations, and stringent selection of precursors. Somatic rearrangements of the genes encoding the α and β chains of the clonotypic heterodimeric T cell receptor (TCR) for antigen determine the T cell's specificity. The random nature of these rearrangements allows for the generation of a virtually limitless repertoire of receptors capable of interacting with rapidly evolving infectious agents. A sizeable but still minor proportion of developing T cells will be able to recognize peptides presented by the highly polymorphic class I or II major histocompatibility (MHC) molecules of the organism.[1] These useful T cell precursors are positively selected, that is, they are saved from cell death programmed approximately 3.5 days after their generation.[2] The random nature of *Tcr*-gene rearrangements also inevitably implies the generation of T cells potentially reactive to self-peptides presented by MHC molecules. Such autospecific conventional T cell (Tconv) precursors must be censored to avoid development of potentially lethal autoimmune pathology. During thymic development of the T cell repertoire, it is therefore rendered tolerant to "self" by negative selection of autospecific precursors.[3] These cells either die by apoptosis or become functionally anergic. These two mechanisms assure so-called "recessive tolerance," that is a state of tolerance that cannot be imposed upon a nontolerant T-cell population. Despite the quantitatively impressive nature of the process of negative selection,[4] some autospecific T cells fully mature and migrate to the periphery.[5] The activity of these potentially very dangerous T cells is controlled by various peripheral tolerance mechanims.[6] Probably the most important of these mechanisms is assured by active suppression by regulatory T lymphocytes (Treg). This process assures "dominant tolerance," that is, a state of tolerance that can be imposed upon a nontolerant T-cell population.[7] Treg not only control autospecific T cells but also mediate maternal tolerance to the semi-allogeneic fetus, fine-regulate immune responses to infectious agents, and avoid chronic inflammation, for example, in the gut.[8–10] Whereas a variety of Treg populations have been described,[11,12] the most extensively studied regulatory T cells are characterized by expression of the αβ heterodimeric TCR, the CD4 coreceptor, and the Foxp3 transcription factor. We address here the mechanisms underlying thymic generation of the latter Treg population and highlight the differences in development of Treg versus Tconv.

The exclusively thymic origin of T lymphocytes remains a matter of controversy. For example, intestinal T cells expressing the CD8αα homodimeric coreceptor are thought to develop locally in specialized structures[13,14] but other investigators challenge this conclusion.[15] However, athymic mice do not have T lymphocytes in secondary lymphoid organs as lymph nodes, spleen, and tonsils. Treg therefore must have a thymic origin, but this does not mean that their regulatory phenotype is imprinted in this organ. For example, several effector T lymphocyte populations (i.e., Th1, Th2, and Th17) can be distinguished based on secretion of distinct sets of cytokines (i.e., IFN-γ, IL-4, and IL-17, respectively). Differentiation of these effector cells takes place in peripheral lymphoid organs under the influence of cytokines (i.e., IL-12, IL-4, and TGF-β/IL-6, respectively). Similarly, $CD4^+Foxp3^+$ Treg can differentiate from Tconv under influence of TGF-β outside the thymus.[16] However, these induced Treg (iTreg) clearly constitute a separate lineage from thymus-derived, so-called "natural" Treg (nTreg).[17] Evidence for a thymic origin of Treg was provided by seminal observations in two distinct experimental settings.

Thymectomy of mice at days 2–4 after birth (but not later) was shown to cause autoimmune-pathology.[18,19] Similarly, Sakaguchi and colleagues found that autoimmune pathology developed in (athymic) nude mice reconstituted with T cells depleted of $CD25^+$ cells, and therefore hypothesized that $CD25^+$ cells may have regulatory capacity.[20] Injection of $CD4^+CD25^+$ T cells indeed prevented the autoimmune manifestations developing in neonatally thymectomized mice.[21] They also showed that $CD4^+CD25^+$ cells appear in the spleen at 4 days after birth, probably explaining the autoimmunity developing in mice thymectomized before.[21] Evidence that Treg develop in the thymus with delayed kinetics was later provided by the laboratory of Rudensky using mice in which green fluorescent protein is expressed under control of the *Foxp3*-promoter.[22] Together, these data suggest that neonatal thymectomy causes autoimmune pathology by a quantitative defect in Treg/Tconv ratios.

Also early experiments on induction of transplantation tolerance provided good evidence that the thymus produces regulatory T lymphocytes. In chimeric chicks, quail thymic epithelium induced tolerance to quail donor tissue grafted later on. Since in these chimeras T cells developed in chick as well as in quail thymi, it was hypothesized that xenoreactive T cells developing in the chick thymus must be controlled by Treg developing in the quail thymus. Similar data were later obtained in mice.[23]

These and other data very strongly suggested that Treg can develop in the thymus. Importantly, functional Treg are indeed found in this organ.[24,25] However, since peripheral T cells can recirculate to the thymus,[26] the presence of Treg in the thymus does not necessarily mean that they develop locally. These cells may represent Tconv activated in the periphery (e.g., in presence of TGF-β), differentiated into Treg, and then recirculated back to the thymus. In lymphopenic mice increased proportions of Treg were indeed found in the thymus suggesting that they recirculated there from the periphery.[27] However, decisive evidence that these cells can develop in the thymus was obtained by the demonstration that $CD4^{+}Foxp3^{+}$ Treg develop in *in vitro* cultures of thymi from E15.5 embryos.[28] Two distinct differentiation pathways therefore contribute to the Treg pool, one occurring inside the thymus and one in secondary lymphoid organs upon appropriate stimulation.

The search for cell surface markers that allows definition and purification of Treg has proven particularly long and arduous.[19,29,30] The finding that Treg constitutively express high levels of CD25 (the IL-2 receptor α chain), led to the formal demonstration of the existence and importance of Treg and opened the possibility of studying their development and function.[20] However, in contrast to laboratory rodents maintained in pathogen free environments, in organisms that are subjected to infectious agents CD25 cannot be used as a unique marker of Treg because it is upregulated on all activated T cells. A significant step forward in this field was made by the identification of the forkhead/winged helix transcription factor Foxp3 as the gene mutated in the scurfy mice and in humans affected with the immunodysfunction polyendocrinopathy enteropathy X-linked (IPEX) syndrome, both characterized by lethal autoimmune pathology.[31,32] Foxp3 was later shown to be strictly required for thymic development of functional Treg.[33–35] Whereas Foxp3 was initially thought to be the master switch in Treg lineage commitment, more recently it became clear that in its absence Treg-like cells develop in the thymus. It therefore is presently viewed as a Treg-lineage consolidation factor.[36] These results promoted an active search in the molecular mechanisms responsible for the induction and maintenance of Foxp3 expression.[37]

II. The Treg-Repertoire

Tconv are involved in the effector phase of immune responses to infectious agents and therefore need to be able to recognize foreign peptides presented by self-MHC molecules. Precursors recognizing self-peptides presented by self-MHC, which are potentially very dangerous for the organism, are negatively selected in the thymus. In contrast, the Treg's main function is to defend the organism from immune-attacks by auto-specific lymphocytes. One may therefore expect that these cells be activated by the very same peptide/MHC ligands as autospecific Tconv, that is, self-peptides presented by self-MHC molecules. This hypothesis was first tested by determining the frequency of peripheral $CD4^+CD25^+$ regulatory T cells recognizing syngeneic or allogeneic APC, using limited dilution analysis.[38] The results of this work indeed indicated that, as compared to the Tconv repertoire, the repertoire of Treg is enriched in self-reactive cells. Importantly, Fisson *et al.* identified a tissue-specific subset of Treg rapidly expanding upon recognition of self-antigens in draining lymph nodes.[39] However, the initial limiting dilution experiments also showed that a fraction of Treg recognized nonself antigens. Indeed, in several experimental settings of infection, the presence and immunomodulatory action of Treg specific to parasite or virus antigens were documented.[40,41] It was also shown that Treg prevent maternal immune-responses to the semi-allogeneic fetus and that they can be used to prevent rejection of allogeneic bone-marrow, skin, and heart-allografts.[42–44] Taken together these functional data demonstrate that the natural Treg repertoire contains cells that can recognize self as well as non self-antigens, in line with their physiological function of keeping immune responses to self, pathogens, and the fetus under control.

The issue of self-reactivity of the Treg population was subsequently addressed by sequencing the TCR used by Treg versus Tconv. TCRβ transgenic animals were used to limit the diversity to the TCRα chain.[45] In other studies, in addition to the fixed TCRβ gene a TCRα minilocus was used to restrict diversity even more.[46,47] The analysis of the TCRα rearrangements of Treg and Tconv in these three different transgenic systems led to the general conclusion that the Treg repertoire is diverse and broad and that it is largely shaped in the thymus. Whereas a limited overlap between the Treg and Tconv TCR-repertoires was found in the TCRβ-transgenic system, a more important overlap was reported in studies using TCRβ/TCRα-minilocus transgenic models, leading to a conflictual interpretation of results. Hsieh *et al.* proposed that TCR expressed by Treg recognize self-antigens with higher affinity than TCR on Tconv.[45] This hypothesis was tested by transducing Tconv with vectors containing sequences encoding Treg (or Tconv)-derived TCR. T cells expressing Treg-derived TCR expanded more efficiently upon adoptive transfer into syngeneic mice than T cells expressing Tconv-TCR. Moreover, the former but not the latter T cells induced wasting disease. These data suggested that Treg express autospecific TCR, but failed to exclude the possibility that they recognize antigens derived from commensals. Indeed, the wasting disease induced by injection of naïve T lymphocytes in the absence of Treg does not occur in germfree mice.[48] In contrast to the conclusion that the TCR-repertoires of Treg and Tconv are very distinct, Pacholczyk *et al.* and Wong *et al.* found a significant degree of overlap.[46,47,49] Moreover, Pacholczyk generated hundreds of T cell hybrids by fusing Treg with hybridoma cells, and assessed their response to activation with autologous antigen-presenting cells (APCs). Virtually none of these hybrids responded in this experimental setting, leading the authors to conclude that the Treg repertoire contains only very limited numbers of autospecific cells.[49] However, they did not provide an explanation for the observation that Treg proliferate in limiting dilution conditions when stimulated with autologous APCs.[38] Therefore, it remains unclear if the TCR-repertoire of Treg is substantially different from that of Tconv, though the reactivity of Treg to self-antigens has been firmly established. If confirmed, this paradox may be due to differential sensitivity of Treg and Tconv to TCR-mediated signals. Modulation of TCR-sensitivity during T cell development is controlled by miRNA181a,[50,51] and similar mechanisms may explain the potential differences between Treg and Tconv.

A very interesting recent report shows that the TCR diversity of the Treg (and to a lesser extent Tconv) repertoire is highly restricted in autoimmune type I diabetes-prone NOD mice as compared to the reference C57Bl/6 strain.[52] Analysis of the Treg repertoire in C57Bl/6.H2^{g7} congenic mice expressing the NOD's MHC class II allele I-A^{g7}, critically involved in diabetes-susceptibility, showed reduced diversity. The authors concluded that part of the striking difference in TCR-repertoire is due to structural aspects of the interaction of the TCR with I-A^{g7} peptide complexes. These observations raise the intriguing possibility that defects ("holes") in the Treg TCR-repertoire may be involved in susceptibility to autoimmune pathologies.

In conclusion, although it is difficult to reconcile all cited reports given the different experimental systems and methods utilized, a unifying line can be drawn from them. Different sets of TCR are clearly found in the repertoires of Treg and Tconv. These at least partly different repertoires probably explain the higher self-reactivity of the Treg repertoire. Importantly, this conclusion does not imply that Treg cannot recognize nonself antigens (e.g., derived from infectious agents or allografts). The observation that the Treg repertoire contains a substantial fraction of autoreactive cells raises several questions on thymic selection of their precursors and on the mechanisms underlying their lineage commitment.

III. Thymic Selection of the Treg-Repertoire

Radioresistant stromal elements in the thymus, that is, the thymic epithelium, express the MHC molecules recognized by immature T cell precursors during the process of positive selection. As far as the Tconv repertoire is concerned, this process enriches the repertoire in cells capable of recognizing antigens presented by self-MHC molecules.[53,54] Also development of the Treg repertoire requires interaction of precursors with MHC molecules expressed by thymic epithelial cells (TEC). The first evidence for this was the observation that transplantation of quail thymic epithelium into chick embryos led to immunological tolerance to quail tissue transplanted later on.[55] This observation was later extended to the mouse.[56] This seminal work led to the hypothesis that T cells with immunoregulatory capacity are selected upon interaction with MHC molecules expressed by thymic epithelium. This interaction was thought to be of sufficient strength to activate the suppressor effector function of Treg.[57] The analysis of TCR/ligand doubly transgenic mice provided the first proof of this concept. Jordan *et al.* generated doubly transgenic mice expressing a high affinity TCR specific to a hemagglutinin (HA) peptide presented by the MHC class II molecule I–E^d, as well as the HA protein. In these mice, increased percentages and numbers of Treg developed.[58] In bone marrow chimeras in which the HA-transgene was exclusively expressed on radioresistant elements, increased proportions of Treg developed. By contrast, expression of HA exclusively by radiosensitive stromal elements did not allow for increased Treg development. These data therefore clearly showed that recognition of agonist ligand expressed by TEC favors Treg development in the thymus. Similar results were obtained in other TCR/ligand doubly transgenic systems.[59–65] Interpretation of these results is complicated by the observation that in the rare reports where absolute numbers of thymic Treg were provided, their increase in the presence of agonist ligand was relatively limited.[66] Since recognition of agonist ligand by Tconv precursors leads to negative selection, in TCR/ligand doubly transgenic mice, massive deletion took place. Thymic homeostasis was therefore very much perturbed. Moreover, in one report it was shown that the increase in specific Treg was paralleled by a similar increase in Treg not expressing the transgenic TCR.[61] It was therefore important to extend these observations to an experimental system in which Treg developed from a precursor population with a naturally diverse TCR-repertoire. This was done by Ribot and colleagues who analyzed hematopoietic chimeras in which endogenous superantigens were exclusively presented by TEC. In these chimeras, two to four-fold increased numbers of superantigen-specific Treg (but not Tconv) developed (as compared to control chimeras). This observation indicated that expression of agonist ligand by TEC considerably enhances positive selection of Treg from precursors with a normally diverse TCR repertoire.[67]

Combined, the data currently available in the literature clearly shows that interaction with agonist ligand expressed by TEC favors development of Treg. In the thymus, two regions can be readily distinguished: the cortex and the medulla. The epithelial cells in these regions have distinct phenotypes and, as far as Tconv development is concerned, functions. Whereas cortical thymic epithelial cells (cTEC) positively select Tconv precursors, medullary thymic epithelial cells (mTEC) contribute to negative selection. Moreover, APCs of hematopoietic origin, mainly dendritic cells (DCs) which are mostly but not exclusively found in the medullary region, also contribute to negative selection of autospecific T cell precursors. The data indicating a role for agonist ligand in selection of the Treg repertoire therefore inspired subsequent studies aimed at better identifying the APC dedicated to Treg selection.

A. Role of the Thymic Cortex in Treg Development

Hematopoietic precursors enter the thymus at the cortico-medullary junction and then migrate to the outer cortex. During this migration, they intensively proliferate and start their maturation process by rearranging their TCR genes and expressing the co-receptors CD4 and CD8. $CD4^+CD8^+$ (double positive) thymocytes express fully assembled $\alpha\beta$ T-cell receptor (TCR) complexes on their cell surface and are ready to undergo positive selection. Whereas $CD4^+CD8^+$ thymocytes bearing TCR that do not recognize host MHC molecules die by neglect, $CD4^+CD8^+$ cells that bear TCR able to engage self-MHC molecules receive a survival signal and can continue their differentiation process. Thus positive selection selects precursors that are potentially useful to the host's immune system. In transgenic mice expressing transgenic MHC class I or II molecules exclusively by cTEC, mature $CD4^-CD8^+$ and $CD4^+CD8^-$ T cells developed.[68–70] Whereas these observations showed that cortical MHC expression is sufficient to allow differentiation of mature T lymphocytes, they failed to show that cTEC mediate positive selection, that is, selection of self-MHC restricted T cells. Taking advantage of the observation that positive selection of BV6 containing TCR-expressing thymocytes is enhanced by expression of the MHC class II molecule I–E[71] and of transgenic mice in which I–E expression was compartmentalized, Benoist and Mathis formally showed that positive selection is mediated by cTEC.[72] cTEC are specialized APC that express a different set of proteases than other thymic APC, for example, cathepsin L,[73] Thymus specific serine protease (TSSP)[74,75] and the β5t subunit of the proteasome.[76] The expression of distinct proteases appears to contribute to the generation of a unique set of MHC-bound peptides specifically implicated in positive selection.[73,76,77] cTEC also display constitutive macro-autophagy[78] allowing them to present, by MHC class II molecules, a wide variety of endogenous peptides.

The role of cTEC in the generation of Treg was initially analyzed in transgenic mice expressing MHC class II molecules only on cortical epithelial cells. In these mice, Treg develop normally in the thymus, even if a reduction in their numbers was observed. Importantly, Treg isolated from these transgenic animals *in vitro* inhibited the proliferation of Tconv with efficiency comparably to Treg isolated from wt mice. These results show that interaction of precursors with MHC class II molecules expressed by cTEC supports the generation of functional Treg.[79] To address the contribution of thymic cortical epithelium in shaping the self-reactive Treg repertoire, Ribot *et al.* analyzed the specificity of Treg and Tconv developing in transgenic mice expressing a single MHC class II/peptide ligand only on cTEC.[80] Interestingly, in these mice substantially increased numbers of Treg (but not Tconv) reactive to the single MHC/peptide ligand were selected from a precursor population with a naturally diverse TCR repertoire. These data indicate that the cortex contributes to positive selection of the autoreactive Treg repertoire. To assess if additional signals from other thymic compartments are required for full Treg differentiation, Liston *et al.* analyzed Treg development in mice in which cortico-medullary migration of thymocytes was blocked either genetically or pharmacologically.[81] Normal numbers of Treg developed and they were found accumulating in the cortex. These data indicate that cTEC are fully competent in mediating Treg lineage commitment and differentiation. In conclusion, the interaction between Treg precursors and cTEC is sufficient to support the generation of functional Treg and is at least in part responsible for shaping of the autoreactive Treg repertoire.

B. Role of the Thymic Medulla in Treg Development

After receiving a survival signal via a successful TCR–MHC interaction, $CD4^+CD8^+$ thymocytes down-regulate one of their coreceptors and express the chemokine receptor CCR7 which allows them to rapidly migrate to the medulla to continue their maturation process. It has been elegantly shown that only $CD4^+CD8^-$ and $CD4^-CD8^+$ thymocytes can enter the thymic medullary area.[82] The medulla is composed of different stromal cells, of which medullary epithelial cells (mTEC) and DCs are the most represented and studied. Interaction of autospecific Tconv-precursors with MHC/peptide complexes expressed by mTEC leads to induction of T cell anergy and, to a lesser extent to apoptosis.[83] Recognition of high affinity ligand on DC, results in death of the T cell precursor.[4] These processes of functional inactivation and deletion of auto-reactive Tconv are referred to as negative selection. Negative selection neutralizes T cells that, once in the periphery, could cause autoimmunity. It has indeed been shown that blocking the entry of thymocytes into the medulla and/or affecting its cellular organization, leads to severe systemic autoimmunity.[84–86] Moreover, T cells developing in mutant mice expressing MHC class I or II molecules exclusively in the cortex are highly "autoreactive".[69,70,87]

mTEC have the unique property of ectopically expressing tissue restricted antigens.[88] This seminal observation provided the conceptual framework for how tolerance to various peripheral tissue antigens can be achieved in the thymus. Interestingly, the transcription of part of these tissue restricted antigens is controlled by the autoimmune regulator gene AIRE.[89] In humans, mutations in this gene cause the monogenically transmitted autoimmune polyendocrinopathy candidiasis ectodermal dystrophy (APECED) syndrome.[90,91] Studies in mice with targeted disruption of the AIRE gene elegantly showed that a reduction in promiscuous gene expression of mTEC results in the generation of an autoreactive T cell repertoire and consequently in autoimmune-pathology.[92–96] Evidence from TCR-transgenic mouse models shows that expression of neo self-antigens led to deletion of specific thymocytes, and that in AIRE-deficient mice deletion of TCR-transgenic thymocytes was severely impaired.[97–99] Combined, these data demonstrate the prominent role of mTEC and AIRE in purging the developing Tconv repertoire of autoreactive cells, and therefore in induction of recessive tolerance.

The role of mTEC and AIRE in Treg differentiation remains much less clear. Given that the Treg repertoire is enriched in autoreactive cells and that Treg precursors can be selected upon interaction with agonist ligand, it is tempting to speculate that mTEC (and AIRE) play an important role in Treg development. Two potential mechanisms have been evoked. First, it was concluded that precursors of Treg (as compared to Tconv) are more resistant to negative selection mediated by mTEC. Indeed, preferential survival of Treg-precursors was observed in TCR/ligand doubly transgenic mice.[58–65] Importantly, lack of negative selection of Treg-precursors by mTEC was also observed in wt mice, using MHC class II transfer from epithelium to developing T cells as a tool to visualize the interactions of thymocytes with stromal elements.[100] Second, Treg-precursors may also be positively selected upon interaction with agonist ligand expressed by mTEC, for example, under control of AIRE. Expression of a neo self-antigen under control of the AIRE promoter resulted in selection of TCR-transgenic Foxp3$^+$ Treg precursors, demonstrating that AIRE$^+$ mTEC can induce Treg differentiation.[62] If the T cell repertoire leaving the AIRE-deficient thymus causes autoimmune-disease because of defective Treg development, complementation with a wt thymus should prevent pathology. To test this hypothesis, athymic nude mouse were grafted with wt and AIRE-deficient thymi. These mice still developed autoimmune manifestations. Moreover, AIRE° Treg prevented colitis as efficiently as wt Treg.[99] Combined, these observations showed that mTEC do not induce deletion but rather positive selection of Treg-precursors. However, they failed to prove that AIRE deficiency perturbs thymic selection of the Treg repertoire. Potential effects of AIRE-deficiency on Treg selection may have been masked in these studies by the strongly defective negative selection of Tconv precursors. More detailed analysis will therefore be required to assess if AIRE deficiency leads to holes in the Treg repertoire, thus contributing to autoimmune manifestations.

DCs are a heterogeneous population composed of conventional DC (cDC) and plasmacytoid DC (pDC). In the thymus, cDC can be subdivided to autochthonous and migratory DC. Whereas autochthonous cDC reside in the thymus, migratory cDC come from the periphery. Autochthonous and migratory cDC mainly localize in the medulla where they are intermingled. Whether autochthonous cDC are dedicated to process and present thymic self-antigens, for example, derived from apoptotic thymocytes,[101] and migratory cDC are more specialized in presenting "peripheral" antigens (e.g., high molecular weight blood proteins), is still a matter of investigation.[102,103] Importantly, it has been shown that antigen can be transferred unidirectionally from mTEC to DC.[104,105] The mechanism involved in the transfer is not yet clear, but this process clearly contributes to the efficacy of tolerance induction by promoting deletion of autoreactive thymocytes specific to tissue-restricted antigens expressed by mTEC.[106] In contrast to the growing body of information on the function of thymic cDC, the function of pDC in the thymus remains largely unexplored.

Cells of hematopoietic origin (in the thymus DC)[107] have since long been known to induce tolerance to self and nonself antigens.[108] The quantitative impact of thymic negative selection by DC was addressed in bone-marrow chimeras in which these cells did not express MHC molecules. In such chimeras, a two to three fold increase in the generation of T cells was observed, indicating that DC delete half to two-thirds of developing thymocytes.[4] When in similar chimeras Treg development was analyzed, a substantial increase in the number of Treg was also found, indicating that autospecific Treg-precursors are sensitive to DC mediated deletion.[38] Deletion of ligand-specific Treg precursors was also observed in other systems.[109] In contrast, in humans it was found that TSLP-treated DC promoted the induction of Treg *in vitro*.[110] These data raised the question as to whether the sensitivity to DC mediated deletion of Tconv and Treg is different and if small amounts of antigen presented by DC could promote Treg differentiation. To address these issues, careful titrations of agonist ligand presented by DC *in vitro* and *in vivo* were performed, and the effect on Treg differentiation studied.[111–113] Wirnsberger *et al.* used an *in vitro* system to compare the capacity of different populations of thymic APC to convert $CD4^+CD8^-$ thymocytes into Treg. The authors found that mTEC induced Treg differentiation and, surprisingly, showed that cDC can also do so, albeit at 100-fold lower ligand density.[111] In a recent report of Atibalentja *et al.* similar titrations were performed *in vivo* in TCR transgenic mice. The authors injected different doses of antigen i.v. and analyzed negative selection and development of Foxp3-expressing $CD4^+CD8^-$ thymocytes. Thus, they showed that the amount of ligand required to generate Treg is fivefold lower than that required for DC-mediated deletion. Consistent with earlier published data,[38] the authors found that Treg-precursors are sensitive to DC-mediated deletion. Importantly, Treg precursors were found to be substantially (at least 10-fold) less sensitive to deletion than Tconv precursors. Thus, they provided an explanation for the paradoxical finding that DC can not only delete Treg-precursors but can also induce Treg-differentiation.[113] Taken together, the data currently available in the literature indicate an original dual role for thymic DC in Treg-development: these cells induce Treg differentiation at low, and delete Treg precursors at higher agonist ligand density.

C. A Dedicated APC for Treg Differentiation?

In conclusion, thymic selection of Treg appears to be a highly flexible and accommodating process. Treg precursors are positively selected upon interaction with high affinity ligand expressed by distinct thymic stromal components, that is, cTEC, mTEC, cDC, and pDC. It appears therefore unlikely that a specialized APC or thymic stromal compartment is involved in Treg-development. Thymic negative selection of precursors interacting too strongly with thymic DC avoids development of Treg that would overly dampen useful immune-responses. The contribution of multiple and functionally quite distinct APCs to selection of Treg precursors is expected to select cells with a wide variety of specificities. Thus, a Treg-repertoire is shaped in a way that is exquisitely adapted for prevention of (auto)immune-pathology without paralyzing the immune system.

IV. Thymic Commitment to the Treg-Lineage

Cell-fate determination of developing thymocytes is dictated by sequential interactions of hematopoietic precursors with the different components of the thymic microenvironment. Lineage commitment of the distinct T cell subpopulations (e.g., CD4 vs. CD8 and CD8αβ vs. CD8αα) was, in the past, often viewed as an "instructional" or "stochastic" process, that is, directed or not by the TCR's specificity respectively.[114] Similar arguments have also been applied to Tconv/Treg fate-decision. In the "instructional" model, T-cell precursors with high affinity receptors for self MHC/peptide ligands are directed to the Treg lineage as an alternative to negative selection. The "stochastic" model postulates that factors other than the TCR's specificity determine Treg lineage fate. Presently, most reports support the instructional model for Treg-lineage choice, but alternative interpretations should be taken into account. Moreover, some experimental data are rather difficult to reconcile with an instructive model. We will here summarize and discuss these reports.

A. A Role for the TCR in Treg Lineage Choice?

Support for an instructive role of TCR signaling in Treg lineage commitment comes from studies using TCR transgenic animals. In these mice, Treg did not develop in a RAG-deficient background in which only the transgenic TCRαβ heterodimer was expressed. By contrast, Tconv developed in these mice. These observations clearly demonstrated that Tconv versus Treg selection requires distinct MHC/peptide ligands. Importantly, in TCR-transgenic RAG-deficient mice, transgenic expression of agonist ligand allowed for development of high proportions of Treg. This observation was interpreted as proof for an instructional model for Treg development.[58–65] Also, in hematopoietic chimeras in which superantigens were expressed on thymic epithelium but not on hematopoietic cells, increased number of superantigen-specific Treg developed in the thymus.[67] However, the increase in absolute numbers of Treg in all these studies was actually rather modest and well below the number of cells that could be expected from a purely instructional model. Only 1 to 17% of antigen-specific precursors developed into Treg in the presence of antigen.[66] It remained therefore difficult to distinguish between the potential explanations stating that agonist ligand (1) instructs Treg-lineage choice versus (2) allows for positive selection of precommitted precursors. However, for one of these processes, elevated levels of TCR signaling is clearly involved. Additional data supporting this conclusion came from studies on the role of negative regulators of TCR signaling, such as SHIP and CD5, on Treg development: In mice deficient in SHIP or CD5, increased numbers of Treg developed.[115,116]

Whereas up to two-thirds of developing T cell precursors undergo negative selection upon interaction with DC,[4] only approximately 4% of $CD4^+CD8^-$ thymocytes express Foxp3. Treg lineage-choice clearly therefore is not simply an alternative to negative selection and other thus far unidentified factors must be involved. However, autospecific Treg-precursors are sensitive to negative selection induced upon recognition of antigen presented by thymic DC.[38] Treg lineage-commitment may therefore be induced whenever autospecific precursors recognize antigen presented by TEC. These radioresistant cells are known to induce clonal anergy of autospecific thymocytes.[83] Interestingly, Treg have a similar "anergic" phenotype *in vitro*, that is, upon TCR-mediated activation they do not proliferate unless large concentrations of IL-2 are added to the culture. It remained therefore possible that recognition of self-antigen presented by TEC induces Treg, rather than anergy. However, in our studies on bone marrow chimeric animals in which superantigen was presented exclusively by TEC, only a small proportion of developing superantigen-specific T cells (i.e., 10–17%) expressed Foxp3.[67] As expected, the others were functionally anergic, that is, upon TCR-mediated activation they did not proliferate *in vitro* (our unpublished observations). In conclusion, Treg lineage-choice requires "something more" than simply recognition of agonist ligand expressed by TEC.

CD4/CD8 lineage choice in the thymus has been extensively studied using mice transgenic for TCR derived from CD4 and CD8 T cell clones.[114] In these transgenic mice, only $CD4^+CD8^-$ and $CD4^-CD8^+$ thymocytes developed, respectively. To get a better insight into Treg lineage-choice, several laboratory have generated mice expressing transgenic TCR derived from Treg. Surprisingly, no or only very limited numbers of Treg developed in these animals.[117–119] In some (but not all) cases, substantial negative selection of thymocytes occurred, showing that the availability of selecting ligand was not (always) limiting. Chimera-experiments in which limited numbers of Treg TCR-transgenic thymocytes developed, suggested that more Treg differentiated under such conditions, but appropriate controls using polyclonal irrelevant TCR were lacking.[118,119] The tentative explanation for these results stated that the niche for development of individual Treg-clones is of very limited size. Again, the precursor's TCR-specificity is clearly insufficient for commitment to the Treg lineage.

B. The Requirement for Costimulatory and Adhesion Molecules for Optimal Treg-Development

In addition to TCR-mediated signaling ("signal 1"), productive T cell activation necessitates a second signal ("signal 2") provided by the engagement of costimulatory molecules.[120] The most prominent costimulatory pathway is mediated by CD28 expressed by the T cell and CD80/86 expressed by the APC. Paradoxically, CD28 or CD80/86 deficiency leads to strongly exacerbated development of spontaneous autoimmune type I diabetes in NOD mice. This observation was due to strongly reduced proportions of peripheral Treg.[121] Later, it was shown that this co-stimulatory pathway is required for development of normal numbers of Treg in the thymus.[122] However, the numbers of Treg were reduced by 80% in CD28-deficient mice and, whereas this co-stimulatory molecule strongly enhances Treg development, it is not strictly required. CD28 also plays a major role in thymic negative selection of autospecific precursors.[123] Given the parallel between negative selection of autospecific precursors and Treg development demonstrated in for example, the TCR/ligand doubly transgenic mice, it will be important to assess if other co-stimulatory molecules involved in thymic negative selection (e.g., CD43)[124] play redundant roles in Treg differentiation. Interestingly, a detailed analysis of the different signaling regions of the CD28 cytoplasmic tail indicated that the tyrosine kinase p56lck-binding domain is necessary for thymic Treg generation and *in vitro* induction of Foxp3.[60] In contrast, the PI3K and the Itk (IL-2 inducible T cell kinase) binding domains did not play a detectable role. Recently, it was shown that the p56lck binding domain of CD28 is required for stabilization of Foxp3 mRNA in thymic Treg precursors.[125] Importantly, it was also found that CD28 engagement regulates CD25 expression. Other cell-surface markers may also be involved in thymic Treg-development. CD40, a molecule thought to have costimulatory activity,[126–128] has been shown to somewhat affect thymic Treg development.[129] Similarly, deficiency of the adhesion molecule LFA-1 leads to diminished Treg numbers.[130] In contrast, CD5, known to dampen TCR-signaling, appears to inhibit Treg development.[116] Therefore, several cell-surface molecules other than the TCR contribute to development of Treg. These molecules have in common the fact that they modulate the levels of "TCR-derived" signals, potentially explaining their role in Treg differentiation. However, as for the TCR, it remains unclear if they are involved in Treg lineage-choice or selection of committed precursors.

C. Cytokines

One of the earliest identified markers for Treg is the IL-2-receptor α chain CD25,[20] raising the intriguing possibility that IL-2 is critically involved in the dampening of immune responses. This cytokine was mostly known for its involvement in T cell proliferation. It was therefore a very unexpected observation that IL-2 deficient mice had a disorder characterized by lymphoproliferation, autoimmune pathology, and chronic enteric inflammation.[131,132] Mice deficient in the IL-2Rα and β chains were affected by a similar pathology.[133,134] Whereas initially it was thought that the lymphoproliferative disorder was due to defective IL-2-dependent activation-induced cell-death (AICD), it soon became evident that defects in dominant tolerance mechanisms were responsible.[135–137] Transgenic mice in which IL-2Rβ-expression was limited to the thymus were largely healthy.[138] These data showed that a perturbed intrathymic process was responsible for the immune-pathology in IL-2Rβ deficient mice. Soon thereafter, defective Treg development in IL-2Rβ-deficient mice and its reconstitution in mice expressing the IL-2Rβ only in the thymus were shown to be responsible for these observations.[139] Also IL-2 deficient mice have strongly reduced numbers of $CD25^+$ Treg in the thymus.[140] Collectively, these data clearly indicate a very important role for IL-2 in the intrathymic differentiation of Treg. However, while subsequent studies in which novel experimental tools were used largely confirmed this conclusion, they also showed that IL-2 does not play a nonredundant role in thymic development of Foxp3-expressing Treg.[64,141]

Whereas the IL-2Rα chain (CD25) is only part of the receptor for IL-2, the β chain (CD122) is also a component of the IL-15R. IL-2 or IL-2Rα-deficiency reduces thymic $Foxp3^+$ Treg levels to approximately half of that found in wt animals, indicating an important but nonredundant role of IL-2 in Treg development.[141–143] Deficiency in the IL-2/15Rβ-chain causes a more severe defect in thymic Foxp3$^+$ Treg development than that found in IL-2 or IL-2Rα-deficient mice.[142,143] This observation indicates an important role for IL-15, confirmed by the observation that mice deficient in this cytokine have slightly reduced thymic Foxp3$^+$ Treg number. Moreover, mice doubly deficient in IL-2 and IL-15 have even more drastically reduced thymic Treg levels.[142] Mice deficient in the IL-2Rγ chain CD132 (also a component of several other cytokine receptors, including IL-4, 7, 9, 15, and 21) have a very profound defect in thymic Foxp3$^+$ Treg numbers.[141,142] This effect is largely due to defective responsiveness to IL-2 and IL-15. On the other hand, IL-4Rα and IL-7Rα deficient mice have normal proportions of Treg.[142,144] Also IL-7 deficient mice have normal numbers of fully functional Foxp3$^+$ Treg.[145] Collectively, these reports indicate an important role for IL-2 and IL-15 in Treg-development, but other still-to-be-identified cytokines using γc-containing receptors must also play a role.

Thymic stromal lymphopoietin (TSLP) is a cytokine produced in the thymus by Hassall's corpuscles. In humans, it was shown that TSLP can activate thymic DC to express high levels of CD80 and CD86. These DC then induced differentiation of Foxp3$^+$ Treg *in vitro*.[110] In the human thymus, Treg were found in close association with TSLP-producing Hassall's corpuscles. It was therefore concluded that TSLP would play an important role in thymic Treg development. However, whereas DC can induce Treg differentiation *in vitro* as well as *in vivo*,[111,112] interaction with MHC class II/peptide complexes expressed by these cells is not required for efficient Treg generation.[38] Moreover, in the thymi of mice deficient in the receptor for TSLP, normal numbers of Foxp3$^+$ Treg developed.[146] It can therefore not be excluded that TSLP plays some redundant role in thymic Treg development, but its importance remains uncertain.

The immunomodulatory cytokine transforming growth factor β (TGF-β) can induce peripheral conversion of Tconv into Treg.[147] However, TGF-β1 deficient mice have normal numbers of thymic CD25^{high} Treg.[148] Among thymocytes incapable of responding to this cytokine only moderately reduced numbers of apparently functional Foxp3$^+$ Treg were found.[149,150] These data suggested that TGF-β does not play a major role in the intrathymic differentiation of Treg. However, conditional ablation of the TGF-βRI led to substantially reduced numbers of thymic Treg just after birth.[151] IL-2 mediated expansion normalized Treg numbers soon thereafter. As a consequence, thymic Treg levels in adult TGF-βRI deficient mice were normal or even higher than those found in wt animals. However, the TCR-repertoire of these cells is probably very different from the one found in wt animals, but this hypothesis remains to be verified. Hence, TGF-β probably plays a very important role in intrathymic differentiation of a normally diverse repertoire of Treg.

It is therefore clear that IL-2, IL-15, and TGF-β play important roles in the intrathymic differentiation of Treg. Less clear is the exact mechanism by which these cytokines act in this process. An interesting approach to address this question was recently reported by Zheng and colleagues.[152] These authors analyzed the respective contributions of distinct enhancer elements in the promoter of the *Foxp3*-gene to Treg differentiation. Deletion of the conserved noncoding DNA-sequence (CNS) 3 strongly reduced thymic Treg generation. Moreover, CNS3 was found to be in an active chromatin configuration in Treg precursors. This enhancer element binds the NF-κB family member c-Rel, which is activated by for example, a TCR-dependent pathway. These observations are consistent with the reported role of the IKK complex, required for activation of c-Rel,[153] in thymic development of Treg (but not Tconv).[154] TGF-β activates Smad3, and this transcription factor binds to CNS1. IL-2 and IL-15 activate STAT5, which is known to play an important role in thymic Treg development.[142] This transcription factor binds to CNS2. Interestingly, deletion of CNS1 and CNS2 in the *Foxp3*-promoter did not disturb thymic Treg development. CNS1 was found to be involved in peripheral induction of the regulatory phenotype and CNS2 in the maintenance of Foxp3 expression. Combined, these observations suggest an important role for signals emanating from the TCR and the co-stimulatory molecule CD28 in early Treg development, maybe in Treg lineage choice. They would also render it rather unlikely that TGF-β and interleukins are involved in commitment to the Treg lineage. However, more work will need to be done to thoroughly test this model.

D. Something Else?

Signals emanating from the TCR and/or receptors for cytokines may therefore be involved in Treg lineage choice. However, as discussed above, in experimental models in which large numbers of T cell-precursors should commit to the Treg lineage, disappointingly small numbers of cells actually successfully developed into Treg. This observation may be due to limited availability of required cytokines, but experimental *in vivo* evidence for this hypothesis is lacking. The "stochastic" model for Treg lineage commitment provides an alternative explanation. In this model, any factor other than the TCR's specificity directs precursors to the Treg-lineage. Subsequent selection processes would assure enrichment of autospecific cells. A subpopulation of very immature DN2 thymocytes (DN2-S), that do not yet express the TCR, appear to preferentially develop into Treg. This observation suggests that Treg lineage commitment takes place well before positive selection.[155] The observation that Treg differentiation is strongly influenced by the genetic background of the strain analyzed[28,156,157] is difficult to reconcile with an instructional model for Treg lineage commitment. However, genetic factors and mechanisms responsible for this phenomenon remain to be identified.

A stochastic (or "noninstructional") model for Treg lineage commitment appears at odds with the currently probably most popular model for CD4 versus CD8 lineage commitment.[114] In this "kinetic" model, immature $CD4^{+}CD8^{+}$ thymocytes expressing appropriate TCRs will interact with MHC class I or II molecules and next downregulate the CD8 coreceptor. $CD4^{+}CD8^{low}$ thymocytes that had received a TCR-mediated signal via interaction with MHC class II will, given the unaltered expression of CD4, continue to receive signals. Consequently, these cells will proceed in their differentiation and become $CD4^{+}CD8^{-}$ T cells, consistent with the TCR's MHC class II restriction. In contrast, $CD4^{+}CD8^{low}$ thymocytes specific to MHC class I will, given the reduced CD8 expression, receive a reduced TCR-mediated signal and "as a consequence" turn off CD4 expression and restore CD8 expression. However, this model is rather difficult to adapt for the process of Treg generation. How would a $CD4^{+}$ thymocyte with high avidity for thymic stroma choose between being negatively selected and differentiating into Treg? A stochastic model for Treg development would provide an answer to this question: Precommitted Treg-precursors are simply less sensitive to negative selection. This is not an unprecedented situation: $CD4^{+}CD8^{+}$ thymocytes with high affinity MHC class I restricted TCR can differentiate into CD8αα expressing cells which have very particular biological properties.[158,159] How such precursors choose between being negatively selected and differentiating into CD8αα T lymphocytes was revealed by an interesting study in which preselection precursors for CD8αα T cells were identified.[159] At the $CD4^{+}CD8^{+}$ stage, prior to any selection, a small subpopulation expressed the CD8αα homodimer (most $CD8^{+}$ T cells only express the αβ heterodimer). These cells were resistant to deletion and selectively developed into CD8αα T cells. A similar scenario may be valid for autospecific precursors that ultimately differentiate into Treg. A precursor population potentially committed to the Treg-lineage has indeed been identified.[155] However, more work will need to be done to firmly confirm the biological properties of this population.

V. Concluding Remarks

Since the identification of CD25 as a marker for T cells with immunoregulatory properties, very substantial progress has been made in understanding their TCR-repertoire and the thymic processes involved in selection of their precursors. However, it remains unclear how precursors commit to the Treg lineage. The lineage choice may be made independently of the precursor's specificity,[155] similar to what was found for development of CD8αα T cells.[159] Subsequent selection would then enrich the Treg repertoire in autospecific cells by positive selection of high affinity/avidity cells and relative resistance to negative selection. Precommitment could allow Tconv and Treg precursors to have distinct TCR-signaling thresholds.[50,51] In this "stochastic" model, the committed precursor population needs to be unequivocally identified. Alternatively, lineage choice could be concomitant with selection, that is, high affinity/avidity interactions would direct precursors to the Treg lineage. In this "instructional" model, it remains to be revealed how a common high affinity T cell precursor would choose between being negatively selected and being directed to the Treg lineage. Consistent with both models, positive selection of Treg precursors can take place upon interaction with a variety of thymic stromal cells. The emerging notion that stromal cells other than TEC can positively select appears to be restricted to quantitatively minor T cell subsets with particular biological functions.[160–164] Maybe identification of the gene(s) responsible for the documented genetic differences in Treg development[28,156,157] will shed light on the mechanisms involved in Treg lineage commitment.

References

1. Zerrahn J, Held W, Raulet DH. The MHC reactivity of the T cell repertoire prior to positive and negative selection. *Cell* 1997;**88**:627–36.
2. Egerton M, Scollay R, Shortman K. Kinetics of mature T-cell development in the thymus. *Proc Natl Acad Sci USA* 1990;**87**:2579–82.
3. von Boehmer H, Melchers F. Checkpoints in lymphocyte development and autoimmune disease. *Nat Immunol* 2010;**11**:14–20.
4. van Meerwijk JPM, Marguerat S, Lees RK, Germain RN, Fowlkes BJ, MacDonald HR. Quantitative impact of thymic clonal deletion on the T cell repertoire. *J Exp Med* 1997;**185**:377–83.
5. Bouneaud C, Kourilsky P, Bousso P. Impact of negative selection on the T cell repertoire reactive to a self-peptide: a large fraction of T cell clones escapes clonal deletion. *Immunity* 2000;**13**:829–40.
6. Mueller DL. Mechanisms maintaining peripheral tolerance. *Nat Immunol* 2010;**11**:21–7.
7. Wing K, Sakaguchi S. Regulatory T cells exert checks and balances on self tolerance and autoimmunity. *Nat Immunol* 2010;**11**:7–13.
8. Aluvihare VR, Betz AG. The role of regulatory T cells in alloantigen tolerance. *Immunol Rev* 2006;**212**:330–43.
9. Belkaid Y, Tarbell K. Regulatory T cells in the control of host-microorganism interactions. *Annu Rev Immunol* 2009;**27**:551–89.
10. Izcue A, Coombes JL, Powrie F. Regulatory lymphocytes and intestinal inflammation. *Annu Rev Immunol* 2009;**27**:313–38.
11. Bendelac A. Nondeletional pathways for the development of autoreactive thymocytes. *Nat Immunol* 2004;**5**:557–8.
12. Pomie C, Menager-Marcq I, van Meerwijk JP. Murine CD8(+) regulatory T lymphocytes: the new era. *Hum Immunol* 2008;**69**:708–14.
13. Rocha B. The extrathymic T-cell differentiation in the murine gut. *Immunol Rev* 2007;**215**:166–77.
14. Ishikawa H, Naito T, Iwanaga T, Takahashi-Iwanaga H, Suematsu M, Hibi T, et al. Curriculum vitae of intestinal intraepithelial T cells: their developmental and behavioral characteristics. *Immunol Rev* 2007;**215**:154–65.

15. Lambolez F, Kronenberg M, Cheroutre H. Thymic differentiation of TCR alpha beta(+) CD8 alpha alpha(+) IELs. *Immunol Rev* 2007;**215**:178–88.
16. Curotto de Lafaille MA, Lafaille JJ. Natural and adaptive foxp3$^+$ regulatory T cells: more of the same or a division of labor? *Immunity* 2009;**30**:626–35.
17. Floess S, Freyer J, Siewert C, Baron U, Olek S, Polansky J, et al. Epigenetic control of the foxp3 locus in regulatory T cells. *PLoS Biol* 2007;**5**:e38.
18. Nishizuka Y, Sakakura T. Thymus and reproduction: sex-linked dysgenesia of the gonad after neonatal thymectomy in mice. *Science* 1969;**166**:753–5.
19. Sakaguchi S, Takahashi T, Nishizuka Y. Study on cellular events in post-thymectomy autoimmune oophoritis in mice. II. Requirement of Lyt-1 cells in normal female mice for the prevention of oophoritis. *J Exp Med* 1982;**156**:1577–86.
20. Sakaguchi S, Sakaguchi N, Asano M, Itoh M, Toda M. Immunologic self-tolerance maintained by activated T cells expressing IL-2 receptor alpha-chains (CD25). Breakdown of a single mechanism of self-tolerance causes various autoimmune diseases. *J Immunol* 1995;**155**:1151–64.
21. Asano M, Toda M, Sakaguchi N, Sakaguchi S. Autoimmune disease as a consequence of developmental abnormality of a T cell subpopulation. *J Exp Med* 1996;**184**:387–96.
22. Fontenot JD, Dooley JL, Farr AG, Rudensky AY. Developmental regulation of Foxp3 expression during ontogeny. *J Exp Med* 2005;**202**:901–6.
23. Le Douarin N, Corbel C, Bandeira A, Thomas-Vaslin V, Modigliani Y, Coutinho A, et al. Evidence for a thymus-dependent form of tolerance that is not based on elimination or anergy of reactive T cells. *Immunol Rev* 1996;**149**:35–53.
24. Saoudi A, Seddon B, Heath V, Fowell D, Mason D. The physiological role of regulatory T cells in the prevention of autoimmunity: the function of the thymus in the generation of the regulatory T cell subset. *Immunol Rev* 1996;**149**:195–216.
25. Itoh M, Takahashi T, Sakaguchi N, Kuniyasu Y, Shimizu J, Otsuka F, et al. Thymus and autoimmunity: production of CD25$^+$CD4$^+$ naturally anergic and suppressive T cells as a key function of the thymus in maintaining immunologic self-tolerance. *J Immunol* 1999;**162**:5317–26.
26. Agus DB, Surh CD, Sprent J. Re-entry of T cells to the adult thymus is restricted to activated T cells. *J Exp Med* 1991;**173**:1039–46.
27. Bosco N, Agenes F, Rolink AG, Ceredig R. Peripheral T cell lymphopenia and concomitant enrichment in naturally arising regulatory T cells: the case of the pre-Talpha gene-deleted mouse. *J Immunol* 2006;**177**:5014–23.
28. Feuerer M, Jiang W, Holler PD, Satpathy A, Campbell C, Bogue M, et al. Enhanced thymic selection of FoxP3$^+$ regulatory T cells in the NOD mouse model of autoimmune diabetes. *Proc Natl Acad Sci USA* 2007;**104**:18181–6.
29. Tung KS, Smith S, Teuscher C, Cook C, Anderson RE. Murine autoimmune oophoritis, epididymoorchitis, and gastritis induced by day 3 thymectomy. Immunopathology. *Am J Pathol* 1987;**126**:293–302.
30. Powrie F, Mason D. OX-22high CD4+ T cells induce wasting disease with multiple organ pathology: prevention by the OX-22low subset. *J Exp Med* 1990;**172**:1701–8.
31. Bennett CL, Christie J, Ramsdell F, Brunkow ME, Ferguson PJ, Whitesell L, et al. The immune dysregulation, polyendocrinopathy, enteropathy, X-linked syndrome (IPEX) is caused by mutations of FOXP3. *Nat Genet* 2001;**27**:20–1.
32. Brunkow ME, Jeffery EW, Hjerrild KA, Paeper B, Clark LB, Yasayko SA, et al. Disruption of a new forkhead/winged-helix protein, scurfin, results in the fatal lymphoproliferative disorder of the scurfy mouse. *Nat Genet* 2001;**27**:68–73.
33. Fontenot JD, Gavin MA, Rudensky AY. Foxp3 programs the development and function of CD4(+)CD25(+) regulatory T cells. *Nat Immunol* 2003;**3**:3.

34. Khattri R, Cox T, Yasayko SA, Ramsdell F. An essential role for Scurfin in CD4(+)CD25(+) T regulatory cells. *Nat Immunol* 2003;**3**:3.
35. Hori S, Nomura T, Sakaguchi S. Control of regulatory T cell development by the transcription factor Foxp3. *Science* 2003;**299**:1057–61.
36. Gavin MA, Rasmussen JP, Fontenot JD, Vasta V, Manganiello VC, Beavo JA, et al. Foxp3-dependent programme of regulatory T-cell differentiation. *Nature* 2007;**445**:771–5.
37. Lu LF, Rudensky A. Molecular orchestration of differentiation and function of regulatory T cells. *Genes Dev* 2009;**23**:1270–82.
38. Romagnoli P, Hudrisier D, van Meerwijk JPM. Preferential recognition of self-antigens despite normal thymic deletion of $CD4^{+}CD25^{+}$ regulatory T cells. *J Immunol* 2002;**168**:1644–8.
39. Fisson S, Darrasse-Jeze G, Litvinova E, Septier F, Klatzmann D, Liblau R, et al. Continuous activation of autoreactive $CD4^{+}$ $CD25^{+}$ regulatory T cells in the steady state. *J Exp Med* 2003;**198**:737–46.
40. Belkaid Y, Piccirillo CA, Mendez S, Shevach EM, Sacks DL. $CD4^{+}CD25^{+}$ regulatory T cells control Leishmania major persistence and immunity. *Nature* 2002;**420**:502–7.
41. Suvas S, Kumaraguru U, Pack CD, Lee S, Rouse BT. $CD4^{+}$ $CD25^{+}$ T cells regulate virus-specific primary and memory $CD8^{+}$ T cell responses. *J Exp Med* 2003;**198**:889–901.
42. Aluvihare VR, Kallikourdis M, Betz AG. Regulatory T cells mediate maternal tolerance to the fetus. *Nat Immunol* 2004;**5**:266–71.
43. Joffre O, Santolaria T, Calise D, Al Saati T, Hudrisier D, Romagnoli P, et al. Prevention of acute and chronic allograft rejection with $CD4^{+}$ $CD25^{+}$ $Foxp3^{+}$ regulatory T lymphocytes. *Nat Med* 2008;**14**:88–92.
44. Joffre O, Gorsse N, Romagnoli P, Hudrisier D, van Meerwijk JPM. Induction of antigen-specific tolerance to bone marrow allografts with $CD4^{+}$ $CD25^{+}$ T lymphocytes. *Blood* 2004;**103**:4216–21.
45. Hsieh CS, Liang Y, Tyznik AJ, Self SG, Liggitt D, Rudensky AY. Recognition of the peripheral self by naturally arising $CD25^{+}$ $CD4^{+}$ T cell receptors. *Immunity* 2004;**21**:267–77.
46. Pacholczyk R, Ignatowicz H, Kraj P, Ignatowicz L. Origin and T cell receptor diversity of $Foxp3^{+}$ $CD4^{+}$ $CD25^{+}$ T cells. *Immunity* 2006;**25**:249–59.
47. Wong J, Obst R, Correia-Neves M, Losyev G, Mathis D, Benoist C. Adaptation of TCR repertoires to self-peptides in regulatory and nonregulatory $CD4^{+}$ T cells. *J Immunol* 2007;**178**:7032–41.
48. Stepankova R, Powrie F, Kofronova O, Kozakova H, Hudcovic T, Hrncir T, et al. Segmented filamentous bacteria in a defined bacterial cocktail induce intestinal inflammation in SCID mice reconstituted with CD45RB high $CD4^{+}$ T cells. *Inflamm Bowel Dis* 2007;**13**:1202–11.
49. Pacholczyk R, Kern J, Singh N, Iwashima M, Kraj P, Ignatowicz L. Nonself-antigens are the cognate specificities of $Foxp3^{+}$ regulatory T cells. *Immunity* 2007;**27**:493–504.
50. Li QJ, Chau J, Ebert PJ, Sylvester G, Min H, Liu G, et al. miR-181a is an intrinsic modulator of T cell sensitivity and selection. *Cell* 2007;**129**:147–61.
51. Ebert PJ, Jiang S, Xie J, Li QJ, Davis MM. An endogenous positively selecting peptide enhances mature T cell responses and becomes an autoantigen in the absence of microRNA miR-181a. *Nat Immunol* 2009;**10**:1162–9.
52. Ferreira C, Singh Y, Furmanski AL, Wong FS, Garden OA, Dyson J. Non-obese diabetic mice select a low-diversity repertoire of natural regulatory T cells. *Proc Natl Acad Sci USA* 2009;**106**:8320–5.
53. Bevan MJ. In a radiation chimaera, host H-2 antigens determine immune responsiveness of donor cytotoxic cells. *Nature* 1977;**269**:417–8.

54. Fink PJ, Bevan MJ. H-2 antigens of the thymus determine lymphocyte specificity. *J Exp Med* 1978;**148**:766–75.
55. Coutinho A, Salaun J, Corbel C, Bandeira A, Le Douarin N. The role of thymic epithelium in the establishment of transplantation tolerance. *Immunol Rev* 1993;**133**:225–40.
56. Modigliani Y, Coutinho A, Pereira P, Le Douarin N, Thomas-Vaslin V, Burlen-Defranoux O, et al. Establishment of tissue-specific tolerance is driven by regulatory T cells selected by thymic epithelium. *Eur J Immunol* 1996;**26**:1807–15.
57. Modigliani Y, Bandeira A, Coutinho A. A model for developmentally acquired thymus-dependent tolerance to central and peripheral antigens. *Immunol Rev* 1996;**149**: 155–174.
58. Jordan MS, Boesteanu A, Reed AJ, Petrone AL, Holenbeck AE, Lerman MA, et al. Thymic selection of $CD4^+$ $CD25^+$ regulatory T cells induced by an agonist self-peptide. *Nat Immunol* 2001;**2**:301–6.
59. Apostolou I, Sarukhan A, Klein L, von Boehmer H. Origin of regulatory T cells with known specificity for antigen. *Nat Immunol* 2002;**3**:756–63.
60. Tai X, Cowan M, Feigenbaum L, Singer A. CD28 costimulation of developing thymocytes induces Foxp3 expression and regulatory T cell differentiation independently of interleukin 2. *Nat Immunol* 2005;**6**:152–62.
61. van Santen H-M, Benoist C, Mathis D. Number of Treg cells that differentiate does not increase upon encounter of agonist ligand on thymic epithelial cells. *J Exp Med* 2004;**200**:1221–30.
62. Aschenbrenner K, D'Cruz LM, Vollmann EH, Hinterberger M, Emmerich J, Swee LK, et al. Selection of Foxp3(+) regulatory T cells specific for self antigen expressed and presented by Aire(+) medullary thymic epithelial cells. *Nat Immunol* 2007;**8**:351–8.
63. Cabarrocas J, Cassan C, Magnusson F, Piaggio E, Mars L, Derbinski J, et al. $Foxp3^+$ $CD25^+$ regulatory T cells specific for a neo-self-antigen develop at the double-positive thymic stage. *Proc Natl Acad Sci USA* 2006;**103**:8453–8.
64. D'Cruz LM, Klein L. Development and function of agonist-induced $CD25^+$ $Foxp3^+$ regulatory T cells in the absence of interleukin 2 signaling. *Nat Immunol* 2005;**6**:1152–9.
65. Kawahata K, Misaki Y, Yamauchi M, Tsunekawa S, Setoguchi K, Miyazaki J-i, et al. Generation of $CD4^+$ $CD25^+$ regulatory T cells from autoreactive T cells simultaneously with their negative selection in the thymus and from nonautoreactive T cells by endogenous TCR expression. *J Immunol* 2002;**168**:4399–405.
66. Romagnoli P, Ribot J, Tellier J, van Meerwijk JPM. Thymic and peripheral generation of $CD4^+$ $Foxp3^+$ regulatory T cells. In: Jiang S, editor. *Regulatory T cells and clinical application*. NewYork, USA: Springer Science+Business Media; 2008. p. 29–55.
67. Ribot J, Romagnoli P, van Meerwijk JPM. Agonist ligands expressed by thymic epithelium enhance positive selection of regulatory T lymphocytes from precursors with a normally diverse TCR repertoire. *J Immunol* 2006;**177**:1101–7.
68. Cosgrove D, Chan SH, Waltzinger C, Benoist C, Mathis D. The thymic compartment responsible for positive selection of $CD4^+$ T cells. *Int Immunol* 1992;**4**:707–10.
69. Laufer TM, DeKoning J, Markowitz JS, Lo D, Glimcher LH. Unopposed positive selection and autoreactivity in mice expressing class II MHC only on thymic cortex. *Nature* 1996;**383**:81–5.
70. Capone M, Romagnoli P, Beermann F, MacDonald HR, van Meerwijk JPM. Dissociation of thymic positive and negative selection in transgenic mice expressing major histocompatibility complex class I molecules exclusively on thymic cortical epithelial cells. *Blood* 2001;**97**:1336–42.
71. MacDonald HR, Lees RK, Schneider R, Zinkernagel RM, Hengartner H. Positive selection of $CD4^+$ thymocytes controlled by MHC class II gene products. *Nature* 1988;**336**:471–3.

72. Benoist C, Mathis D. Positive selection of the T cell repertoire: where and when does it occur? *Cell* 1989;**58**:1027–33.
73. Nakagawa T, Roth W, Wong P, Nelson A, Farr A, Deussing J, et al. Cathepsin L: critical role in Ii degradation and CD4 T cell selection in the thymus. *Science* 1998;**280**:450–3.
74. Bowlus CL, Ahn J, Chu T, Gruen JR. Cloning of a novel MHC-encoded serine peptidase highly expressed by cortical epithelial cells of the thymus. *Cell Immunol* 1999;**196**:80–6.
75. Carrier A, Nguyen C, Victorero G, Granjeaud S, Rocha D, Bernard K, et al. Differential gene expression in CD3epsilon- and RAG1-deficient thymuses: definition of a set of genes potentially involved in thymocyte maturation. *Immunogenetics* 1999;**50**:255–70.
76. Murata S, Sasaki K, Kishimoto T, Niwa S, Hayashi H, Takahama Y, et al. Regulation of $CD8^+$ T cell development by thymus-specific proteasomes. *Science* 2007;**316**:1349–53.
77. Nitta T, Murata S, Sasaki K, Fujii H, Ripen AM, Ishimaru N, et al. Thymoproteasome shapes immunocompetent repertoire of CD8(+) T cells. *Immunity* 2010;**32**:29–40.
78. Nedjic J, Aichinger M, Emmerich J, Mizushima N, Klein L. Autophagy in thymic epithelium shapes the T-cell repertoire and is essential for tolerance. *Nature* 2008;**455**:396–400.
79. Bensinger SJ, Bandeira A, Jordan MS, Caton AJ, Laufer TM. Major histocompatibility complex class II-positive cortical epithelium mediates the selection of $CD4^+25^+$ immunoregulatory T cells. *J Exp Med* 2001;**194**:427–38.
80. Ribot J, Enault G, Pilipenko S, Huchenq A, Calise M, Hudrisier D, et al. Shaping of the autoreactive regulatory T cell repertoire by thymic cortical positive selection. *J Immunol* 2007;**179**:6741–8.
81. Liston A, Nutsch KM, Farr AG, Lund JM, Rasmussen JP, Koni PA, et al. Differentiation of regulatory $Foxp3^+$ T cells in the thymic cortex. *Proc Natl Acad Sci USA* 2008;**105**:11903–8.
82. Le Borgne M, Ladi E, Dzhagalov I, Herzmark P, Liao YF, Chakraborty AK, et al. The impact of negative selection on thymocyte migration in the medulla. *Nat Immunol* 2009;**10**:823–30.
83. Ramsdell F, Fowlkes BJ. Clonal deletion versus clonal anergy: the role of the thymus in inducing self tolerance. *Science* 1990;**248**:1342–8.
84. Akiyama T, Maeda S, Yamane S, Ogino K, Kasai M, Kajiura F, et al. Dependence of self-tolerance on TRAF6-directed development of thymic stroma. *Science* 2005;**308**:248–51.
85. Kurobe H, Liu C, Ueno T, Saito F, Ohigashi I, Seach N, et al. CCR7-dependent cortex-to-medulla migration of positively selected thymocytes is essential for establishing central tolerance. *Immunity* 2006;**24**:165–77.
86. Davalos-Misslitz AC, Rieckenberg J, Willenzon S, Worbs T, Kremmer E, Bernhardt G, et al. Generalized multi-organ autoimmunity in CCR7-deficient mice. *Eur J Immunol* 2007;**37**:613–22.
87. Laufer TM, Fan L, Glimcher LH. Self-reactive T cells selected on thymic cortical epithelium are polyclonal and are pathogenic in vivo. *J Immunol* 1999;**162**:5078–84.
88. Kyewski B, Klein L. A central role for central tolerance. *Annu Rev Immunol* 2006;**24**:571–606.
89. Mathis D, Benoist C. Aire. *Annu Rev Immunol* 2009;**27**:287–312.
90. Nagamine K, Peterson P, Scott HS, Kudoh J, Minoshima S, Heino M, et al. Positional cloning of the APECED gene. *Nat Genet* 1997;**17**:393–8.
91. The Finnish-German APECED Consortium . An autoimmune disease, APECED, caused by mutations in a novel gene featuring two PHD-type zinc-finger domains. *Nat Genet* 1997;**17**:399–403.
92. Ramsey C, Winqvist O, Puhakka L, Halonen M, Moro A, Kampe O, et al. Aire deficient mice develop multiple features of APECED phenotype and show altered immune response. *Hum Mol Genet* 2002;**11**:397–409.
93. Anderson MS, Venanzi ES, Klein L, Chen Z, Berzins S, Turley SJ, et al. Projection of an immunological self-shadow within the thymus by the Aire protein. *Science* 2002;**298**:1395–401.

94. Kuroda N, Mitani T, Takeda N, Ishimaru N, Arakaki R, Hayashi Y, et al. Development of autoimmunity against transcriptionally unrepressed target antigen in the thymus of Aire-deficient mice. *J Immunol* 2005;**174**:1862–70.
95. Hubert FX, Kinkel SA, Crewther PE, Cannon PZ, Webster KE, Link M, et al. Aire-deficient C57BL/6 mice mimicking the common human 13-base pair deletion mutation present with only a mild autoimmune phenotype. *J Immunol* 2009;**182**:3902–18.
96. Pontynen N, Miettinen A, Arstila TP, Kampe O, Alimohammadi M, Vaarala O, et al. Aire deficient mice do not develop the same profile of tissue-specific autoantibodies as APECED patients. *J Autoimmun* 2006;**27**:96–104.
97. Liston A, Lesage S, Wilson J, Peltonen L, Goodnow CC. Aire regulates negative selection of organ-specific T cells. *Nat Immunol* 2003;**4**:350–4.
98. Liston A, Gray DH, Lesage S, Fletcher AL, Wilson J, Webster KE, et al. Gene dosage-limiting role of aire in thymic expression, clonal deletion, and organ-specific autoimmunity. *J Exp Med* 2004;**200**:1015–26.
99. Anderson MS, Venanzi ES, Chen Z, Berzins SP, Benoist C, Mathis D. The cellular mechanism of AIRE control of T cell tolerance. *Immunity* 2005;**23**:227–39.
100. Romagnoli P, Hudrisier D, van Meerwijk JPM. Molecular signature of recent thymic selection events on effector and regulatory $CD4^+$ T lymphocytes. *J Immunol* 2005;**175**:5751–8.
101. Proietto AI, Lahoud MH, Wu L. Distinct functional capacities of mouse thymic and splenic dendritic cell populations. *Immunol Cell Biol* 2008;**86**:700–8.
102. Bonasio R, Scimone ML, Schaerli P, Grabie N, Lichtman AH, von Andrian UH. Clonal deletion of thymocytes by circulating dendritic cells homing to the thymus. *Nat Immunol* 2006;**7**:1092–100.
103. Li J, Park J, Foss D, Goldschneider I. Thymus-homing peripheral dendritic cells constitute two of the three major subsets of dendritic cells in the steady-state thymus. *J Exp Med* 2009;**206**:607–22.
104. Humblet C, Rudensky A, Kyewski B. Presentation and intercellular transfer of self antigen within the thymic microenvironment: expression of the E alpha peptide-I-Ab complex by isolated thymic stromal cells. *Int Immunol* 1994;**6**:1949–58.
105. Koble C, Kyewski B. The thymic medulla: a unique microenvironment for intercellular self-antigen transfer. *J Exp Med* 2009;**206**:1505–13.
106. Gallegos AM, Bevan MJ. Central tolerance to tissue-specific antigens mediated by direct and indirect antigen presentation. *J Exp Med* 2004;**200**:1039–49.
107. Matzinger P, Guerder S. Does T-cell tolerance require a dedicated antigen-presenting cell? *Nature* 1989;**338**:74–6.
108. Owen RD. Immunogenic consequences of vascular anastomoses between bovine twins. *Science* 1945;**102**:400–1.
109. Pacholczyk R, Kraj P, Ignatowicz L. Peptide specificity of thymic selection of $CD4^+CD25^+$ T cells. *J Immunol* 2002;**168**:613–20.
110. Watanabe N, Wang YH, Lee HK, Ito T, Wang YH, Cao W, et al. Hassall's corpuscles instruct dendritic cells to induce $CD4^+CD25^+$ regulatory T cells in human thymus. *Nature* 2005;**436**:1181–5.
111. Wirnsberger G, Mair F, Klein L. Regulatory T cells differentiation of thymocytes does not require a dedicated antigen-presenting cell but is under T cell-intrinsic developmental control. *Proc Natl Acad Sci USA* 2009;doi:10.1073/pnas.0901877106.
112. Proietto AI, van Dommelen S, Zhou P, Rizzitelli A, D'Amico A, Steptoe RJ, et al. Dendritic cells in the thymus contribute to T-regulatory cell induction. *Proc Natl Acad Sci USA* 2008;**105**:19869–74.
113. Atibalentja DF, Byersdorfer CA, Unanue ER. Thymus-blood protein interactions are highly effective in negative selection and regulatory T cell induction. *J Immunol* 2009;**183**:7909–18.

114. Singer A, Adoro S, Park JH. Lineage fate and intense debate: myths, models and mechanisms of CD4- versus CD8-lineage choice. *Nat Rev Immunol* 2008;**8**:788–801.
115. Carter JD, Calabrese GM, Naganuma M, Lorenz U. Deficiency of the Src homology region 2 domain-containing phosphatase 1 (SHP-1) causes enrichment of $CD4^{+}CD25^{+}$ regulatory T cells. *J Immunol* 2005;**174**:6627–38.
116. Ordonez-Rueda D, Lozano F, Sarukhan A, Raman C, Garcia-Zepeda EA, Soldevila G. Increased numbers of thymic and peripheral $CD4^{+}$ $CD25^{+}Foxp3^{+}$ cells in the absence of CD5 signaling. *Eur J Immunol* 2009;**39**:2233–47.
117. DiPaolo RJ, Shevach EM. $CD4^{+}$ T-cell development in a mouse expressing a transgenic TCR derived from a Treg. *Eur J Immunol* 2009;**39**:234–40.
118. Leung MW, Shen S, Lafaille JJ. TCR-dependent differentiation of thymic $Foxp3^{+}$ cells is limited to small clonal sizes. *J Exp Med* 2009;**206**:2121–30.
119. Bautista JL, Lio CW, Lathrop SK, Forbush K, Liang Y, Luo J, et al. Intraclonal competition limits the fate determination of regulatory T cells in the thymus. *Nat Immunol* 2009;**10**:610–7.
120. Mueller DL, Jenkins MK, Schwartz RH. Clonal expansion versus functional clonal inactivation: a costimulatory signalling pathway determines the outcome of T cell antigen receptor occupancy. *Annu Rev Immunol* 1989;**7**:445–80.
121. Salomon B, Lenschow DJ, Rhee L, Ashourian N, Singh B, Sharpe A, et al. B7/CD28 costimulation is essential for the homeostasis of the $CD4^{+}CD25^{+}$ immunoregulatory T cells that control autoimmune diabetes. *Immunity* 2000;**12**:431–40.
122. Tang Q, Henriksen KJ, Boden EK, Tooley AJ, Ye J, Subudhi SK, et al. Cutting edge: CD28 controls peripheral homeostasis of $CD4^{+}CD25^{+}$ regulatory T cells. *J Immunol* 2003;**171**:3348–52.
123. Degermann S, Surh CD, Glimcher LH, Sprent J, Lo D. B7 expression on thymic medullary epithelium correlates with epithelium-mediated deletion of V beta 5+ thymocytes. *J Immunol* 1994;**152**:3254–63.
124. Kishimoto H, Sprent J. Several different cell surface molecules control negative selection of medullary thymocytes. *J Exp Med* 1999;**190**:65–73.
125. Nazarov-Stoica C, Surls J, Bona C, Casares S, Brumeanu TD. CD28 signaling in T regulatory precursors requires p56lck and rafts integrity to stabilize the Foxp3 message. *J Immunol* 2009;**182**:102–10.
126. Blotta MH, Marshall JD, DeKruyff RH, Umetsu DT. Cross-linking of the CD40 ligand on human $CD4^{+}$ T lymphocytes generates a costimulatory signal that up-regulates IL-4 synthesis. *J Immunol* 1996;**156**:3133–40.
127. Munroe ME, Bishop GA. A costimulatory function for T cell CD40. *J Immunol* 2007;**178**:671–82.
128. van Essen D, Kikutani H, Gray D. CD40 ligand-transduced co-stimulation of T cells in the development of helper function. *Nature* 1995;**378**:620–3.
129. Guiducci C, Valzasina B, Dislich H, Colombo MP. CD40/CD40L interaction regulates $CD4^{+}CD25^{+}$ T reg homeostasis through dendritic cell-produced IL-2. *Eur J Immunol* 2005;**35**:557–67.
130. Marski M, Kandula S, Turner JR, Abraham C. CD18 is required for optimal development and function of $CD4^{+}CD25^{+}$ T regulatory cells. *J Immunol* 2005;**175**:7889–97.
131. Sadlack B, Lohler J, Schorle H, Klebb G, Haber H, Sickel E, et al. Generalized autoimmune disease in interleukin-2-deficient mice is triggered by an uncontrolled activation and proliferation of $CD4^{+}$ T cells. *Eur J Immunol* 1995;**25**:3053–9.
132. Sadlack B, Merz H, Schorle H, Schimpl A, Feller AC, Horak I. Ulcerative colitis-like disease in mice with a disrupted interleukin-2 gene. *Cell* 1993;**75**:253–61.

133. Willerford DM, Chen J, Ferry JA, Davidson L, Ma A, Alt FW. Interleukin-2 receptor alpha chain regulates the size and content of the peripheral lymphoid compartment. *Immunity* 1995;**3**:521–30.
134. Suzuki H, Kundig TM, Furlonger C, Wakeham A, Timms E, Matsuyama T, et al. Deregulated T cell activation and autoimmunity in mice lacking interleukin-2 receptor beta. *Science* 1995;**268**:1472–6.
135. Kramer S, Schimpl A, Hunig T. Immunopathology of interleukin (IL) 2-deficient mice: thymus dependence and suppression by thymus-dependent cells with an intact IL-2 gene. *J Exp Med* 1995;**182**:1769–76.
136. Wolf M, Schimpl A, Hunig T. Control of T cell hyperactivation in IL-2-deficient mice by CD4 (+)CD25(−) and CD4(+)CD25(+) T cells: evidence for two distinct regulatory mechanisms. *Eur J Immunol* 2001;**31**:1637–45.
137. Almeida AR, Legrand N, Papiernik M, Freitas AA. Homeostasis of peripheral $CD4^+$ T cells: IL-2R alpha and IL-2 shape a population of regulatory cells that controls $CD4^+$ T cell numbers. *J Immunol* 2002;**169**:4850–60.
138. Malek TR, Porter BO, Codias EK, Scibelli P, Yu A. Normal lymphoid homeostasis and lack of lethal autoimmunity in mice containing mature T cells with severely impaired IL-2 receptors. *J Immunol* 2000;**164**:2905–14.
139. Malek TR, Yu A, Vincek V, Scibelli P, Kong L. CD4 regulatory T cells prevent lethal autoimmunity in IL-2R beta-deficient mice. Implications for the nonredundant function of IL-2. *Immunity* 2002;**17**:167–78.
140. Papiernik M, de Moraes ML, Pontoux C, Vasseur F, Penit C. Regulatory CD4 T cells: expression of IL-2R alpha chain, resistance to clonal deletion and IL-2 dependency. *Int Immunol* 1998;**10**:371–8.
141. Fontenot JD, Rasmussen JP, Gavin MA, Rudensky AY. A function for interleukin 2 in Foxp3-expressing regulatory T cells. *Nat Immunol* 2005;.
142. Burchill MA, Yang J, Vogtenhuber C, Blazar BR, Farrar MA. IL-2 receptor beta-dependent STAT5 activation is required for the development of $Foxp3^+$ regulatory T cells. *J Immunol* 2007;**178**:280–90.
143. Soper DM, Kasprowicz DJ, Ziegler SF. IL-2R beta links IL-2R signaling with Foxp3 expression. *Eur J Immunol* 2007;**37**:1817–26.
144. Vang KB, Yang J, Mahmud SA, Burchill MA, Vegoe AL, Farrar MA. IL-2, -7, and -15, but not thymic stromal lymphopoeitin, redundantly govern $CD4^+Foxp3^+$ regulatory T cell development. *J Immunol* 2008;**181**:3285–90.
145. Peffault de Latour R, Dujardin HC, Mishellany F, Burlen-Defranoux O, Zuber J, Marques R, et al. Ontogeny, function, and peripheral homeostasis of regulatory T cells in the absence of interleukin-7. *Blood* 2006;**108**:2300–6.
146. Mazzucchelli R, Hixon JA, Spolski R, Chen X, Li WQ, Hall VL, et al. Development of regulatory T cells requires IL-7Ralpha stimulation by IL-7 or TSLP. *Blood* 2008;**112**:3283–92.
147. Kretschmer K, Apostolou I, Hawiger D, Khazaie K, Nussenzweig MC, von Boehmer H. Inducing and expanding regulatory T cell populations by foreign antigen. *Nat Immunol* 2005;**6**:1219–27.
148. Marie JC, Letterio JJ, Gavin M, Rudensky AY. TGF-beta1 maintains suppressor function and Foxp3 expression in $CD4^+CD25^+$ regulatory T cells. *J Exp Med* 2005;**201**:1061–7.
149. Fahlen L, Read S, Gorelik L, Hurst SD, Coffman RL, Flavell RA, et al. T cells that cannot respond to TGF-beta escape control by CD4(+)CD25(+) regulatory T cells. *J Exp Med* 2005;**201**:737–46.
150. Marie JC, Liggitt D, Rudensky AY. Cellular mechanisms of fatal early-onset autoimmunity in mice with the T cell-specific targeting of transforming growth factor-beta receptor. *Immunity* 2006;**25**:441–54.

151. Liu Y, Zhang P, Li J, Kulkarni AB, Perruche S, Chen W. A critical function for TGF-beta signaling in the development of natural $CD4^+CD25^+Foxp3^+$ regulatory T cells. *Nat Immunol* 2008;**9**:632–40.
152. Zheng Y, Josefowicz S, Chaudhry A, Peng XP, Forbush K, Rudensky AY. Role of conserved non-coding DNA elements in the Foxp3 gene in regulatory T-cell fate. *Nature* 2010;**463**:808–12.
153. Vallabhapurapu S, Karin M. Regulation and function of NF-kappaB transcription factors in the immune system. *Annu Rev Immunol* 2009;**27**:693–733.
154. Schmidt-Supprian M, Courtois G, Tian J, Coyle AJ, Israel A, Rajewsky K, et al. Mature T cells depend on signaling through the IKK complex. *Immunity* 2003;**19**:377–89.
155. Pennington DJ, Silva-Santos B, Silberzahn T, Escorcio-Correia M, Woodward MJ, Roberts SJ, et al. Early events in the thymus affect the balance of effector and regulatory T cells. *Nature* 2006;**444**:1073–7.
156. Romagnoli P, Tellier J, van Meerwijk JPM. Genetic control of thymic development of $CD4^+CD25^+FoxP3^+$ regulatory T lymphocytes. *Eur J Immunol* 2005;**35**:3525–32.
157. Tellier J, van Meerwijk JP, Romagnoli P. An MHC-linked locus modulates thymic differentiation of $CD4^+CD25^+Foxp3^+$ regulatory T lymphocytes. *Int Immunol* 2006;**18**:1509–19.
158. Yamagata T, Mathis D, Benoist C. Self-reactivity in thymic double-positive cells commits cells to a CD8 alpha alpha lineage with characteristics of innate immune cells. *Nat Immunol* 2004;**5**:597–605.
159. Gangadharan D, Lambolez F, Attinger A, Wang-Zhu Y, Sullivan BA, Cheroutre H. Identification of pre- and postselection TCRalphabeta+intraepithelial lymphocyte precursors in the thymus. *Immunity* 2006;**25**:631–41.
160. Bendelac A, Savage PB, Teyton L. The Biology of NKT cells. *Annu Rev Immunol* 2007;**25**:297–336.
161. Li W, Kim MG, Gourley TS, McCarthy BP, Sant'Angelo DB, Chang CH. An alternate pathway for CD4 T cell development: thymocyte-expressed MHC class II selects a distinct T cell population. *Immunity* 2005;**23**:375–86.
162. Choi EY, Jung KC, Park HJ, Chung DH, Song JS, Yang SD, et al. Thymocyte-thymocyte interaction for efficient positive selection and maturation of CD4 T cells. *Immunity* 2005;**23**:387–96.
163. Li W, Sofi MH, Yeh N, Sehra S, McCarthy BP, Patel DR, et al. Thymic selection pathway regulates the effector function of CD4 T cells. *J Exp Med* 2007;**204**:2145–57.
164. Lee YJ, Jeon YK, Kang BH, Chung DH, Park CG, Shin HY, et al. Generation of $PLZF^+CD4^+$ T cells via MHC class II-dependent thymocyte-thymocyte interaction is a physiological process in humans. *J Exp Med* 2010;**207**:237–46, S231–7.

Molecular Mechanisms of Regulatory T Cell Development and Suppressive Function

Jeong M. Kim

Genentech, 1 DNA Way, South San Francisco, California, USA

The requirement for regulatory T cells (Treg) to maintain tolerance to self-tissues is evidenced by fatal autoimmune disease that results from genetic deficiencies in Treg cell development or Treg cell depletion *in vivo*. These observations revealed that a normal T cell repertoire harbors self-reactive T cells

Progress in Molecular Biology and Translational Science, Vol. 92
DOI: 10.1016/S1877-1173(10)92011-5

that are kept dormant by Treg cells. In order to prevent auto-reactive T cell activation, Treg cells disarm antigen-presenting cells (APC) through multiple suppressive mechanisms including B7 signaling and sequestration, ATP catabolism, cytolysis, and immunosuppressive cytokine secretion. In addition to APCs, multiple leukocyte subsets are subjected to Treg cell mediated suppression. The acquisition of suppressive activity occurs concomitantly with Treg cell lineage commitment. The identification of molecular cues that guide differentiation of Treg cells versus auto-reactive cells or other $CD4^+$ T cell subsets have been aided by the differential expression of the transcription factor Foxp3 by Treg cells. Foxp3 is the most faithful marker for Treg cells and in its absence, Treg cell development is abrogated. Utilizing Foxp3 expression as a surrogate for Treg cell commitment, factors that promote Foxp3 transcription have provided new insights to Treg cell development at a molecular level.

The recent resurgence of Treg research has been instigated by the identification of CD25 as a marker for Tregs.[1] Prior to this, the very existence of Tregs was questioned due to the relative impurity of fractions containing suppressive activity. Based on CD25 expression, it is now known that Tregs constitute approximately 10% of $CD4^+$ T cells in mice and 5% of the human $CD4^+$ T cell subset. Studies that removed Tregs from the T cell repertoire have revealed a previously unappreciated pool of self-reactive T cells harbored by a normal immune system. Manipulating Treg numbers unleashed pathogenic T cell mediated gastritis, thyroiditis, glomerulonephritis, sialoadenitis, adrenalitis, arthritis, insulitis, and oophoritis. The breadth of target organs and the severity of autoimmunity induced by Treg depletion emphasizes the incomplete nature of clonal deletion of self-reactive T cells during thymic development. Conversely, Treg transfer ameliorates or provides complete protection in mouse models of type 1 diabetes mellitus, inflammatory bowel disease, systemic lupus erythematosus, and multiple sclerosis.[2–5] Although perturbations in Treg numbers or function have been implicated in nearly all autoimmune diseases, single nucleotide polymorphisms in genes that influence Treg dynamics have been detected in type 1 diabetes mellitus and autoimmune thyroiditis, suggesting that Treg impairment can be a primary cause of human disease development.[6–9] The most prominent human disease linked to Treg dysfunction is immune dysregulation polyendocrinopathy, enteropathy X-linked (IPEX) syndrome, which is characterized by early onset of multiorgan autoimmunity.

Positional cloning of the gene responsible for causing the rare autoimmune disorder IPEX identified loss of function mutations in *Foxp3*.[10–12] In parallel, the causative mutation of the mouse analog of IPEX was found in the *Foxp3* gene, providing a mouse model of Treg deficiency.[13] In mice and IPEX patients with a nonfunctional *Foxp3* allele, Treg cell development is completely blocked. [14–16] Treg cell deficiency is primary for autoimmunity development in *foxp3*$^{-}$ mice because transferring wild-type Tregs prevents multiorgan autoimmunity and lymphoproliferative disorders.[17] The *Foxp3* gene encodes a transcription factor that is selectively expressed by Tregs. In contrast to other Treg cell markers such as CD25, non-Tregs do not express Foxp3 upon standard mitogenic stimulation. Furthermore, ectopic Foxp3 expression in effector T cells partially transfers suppressive activity, and continual Foxp3 expression is required to maintain suppressive activity because conditional Foxp3 deletion in peripheral Tregs abrogates suppressive function.[15,17] Thus, Foxp3 is necessary for Treg cell development, sufficient to initiate suppressive function, and serves as the most reliable marker for Tregs.

Based on the importance of Foxp3 in Treg cell development and its effector function, recent Treg research has focused on identifying Foxp3 target genes, signaling pathways and transcription factors that regulate Foxp3 expression, and the mechanism of Foxp3 activity. This chapter will focus on the developmental origins and suppressive function of Tregs.

I. Tracing Thymic Treg Development with Foxp3

The developmental origins of Treg cells were examined using Foxp3 expression as a surrogate marker for Treg lineage commitment. Within the thymus, approximately 90% of Foxp3^{+} cells reside within the CD4 single positive (SP) subset.[18,19] More than 80% of Foxp3^{+} CD4 SP cells express low levels of CD24 or HSA, a marker for immature SP thymocytes. Other thymocyte subsets comprise a minor proportion of Foxp3^{+} cells, with CD8 SP constituting 4% and CD4 CD8 double positive (DP) 2% of Foxp3^{+} thymocytes. The remaining unaccounted populations of Foxp3 expressing cells display lower levels of CD4 and/or CD8, suggesting these cells are not bona fide CD4 CD8 double negative thymocytes. Among DP cells, Foxp3^{+} cells express CD69, lower levels of HSA, and higher levels of TCRβ, indicating that Treg lineage commitment is either coincident or occurs after positive selection of thymocytes that recognize self-peptide MHC complexes.[19] Thus, the majority of Treg differentiation occurs late in thymocyte development.

Consistent with Foxp3 expression occurring during the later stages of T cell development, the majority of Foxp3^{+} thymocytes are located within the thymic medulla.[18] In human Treg cell development, medulla-resident cells, including reticular epithelial cells and dendritic cells (DCs), collaborate to promote Foxp3 expression.[20] The medullary determinants that guide Foxp3 expression in CD4 SP cells have not yet been identified. High levels of B7 expression by medullary DCs may contribute to costimulating Treg cell precursors, thus providing sustained TCR signaling. Consistent with this hypothesis, deficiency of B7.1 and B7.2 or their ligand CD28 causes a dramatic reduction in Treg numbers.[21,22] Although the highest concentrations of thymic Treg cells are detected in the medulla, Treg lineage commitment also occurs in the cortex.[23,24] The contribution of the cortical Foxp3$^+$ cells to the medullary pool of Foxp3$^+$ cells is not known, making the precursor–progeny relationship between Foxp3$^+$ DP and SP ambiguous. It is possible that only a small fraction of Foxp3$^+$ CD4 SP cells are derived from Foxp3$^+$ DP cells may be a possibility. Differences in the nature of peptides presented by cortical and medullary APCs are predicted to generate distinct Treg repertoires, underscoring the importance of the site of Treg cell lineage commitment.

Applying the signal strength model for thymocyte selection, TCR avidities for self-antigens that lie between positive and negative selection are thought to promote Treg cell differentiation.[25–29] Provision of cognate antigens to TCR transgenic thymocytes in "double transgenic" systems induces negative selection of self-reactive T cells and increases both Treg frequency and absolute number. Thus, TCR avidities that minimize clonal deletion while maintaining the quality of TCR signaling are thought to induce Treg cell development. The molecular factors that differentiate self-reactive cells that are culled from the peripheral T cell repertoire from the Treg cells are largely unknown. However, it is hypothesized that Foxp3 expression confers selective survival of thymocytes that react to self-antigens.[30]

In addition to signals derived from the TCR complex, IL-2 receptor signal transduction is required for optimal Treg generation. TCR and IL-2 signaling are thought to occur sequentially as transcription of the IL-2 receptor-α or CD25, the high affinity subunit, is controlled by TCR signaling. Once CD25 is upregulated in a TCR-dependent manner, antigen-independent activation of the IL-2 receptor completes Treg cell differentiation. Kinetic analysis of Treg development in neonates and adult mice has demonstrated that Foxp3$^-$CD4$^+$CD25$^+$ SP cells precede Foxp3$^+$CD4$^+$CD25$^+$ SP cells.[14,19] Interestingly, a sharp reduction in Foxp3$^-$CD4$^+$CD25$^+$ SP cells immediately precedes a wave of Foxp3$^+$ cell production, suggesting that Foxp3$^-$CD4$^+$CD25$^+$ SP cells serve as Treg precursors. In support of this "two-step" model of Treg differentiation, Foxp3$^-$CD4$^+$CD25$^+$ SP thymocytes are more capable of inducing Foxp3 transcription than their CD25$^-$ counterparts *in vivo* and *in vitro*. Furthermore, TCR sequences derived from Foxp3$^-$CD4$^+$CD25$^+$ SP thymocytes largely overlap with Foxp3$^+$ TCRs; this is consistent with the notion that a majority of Foxp3$^-$CD4$^+$CD25$^+$ SP cells are poised to express Foxp3 and represent a transitional stage between a non-Treg and Treg cell.[31] Genetic deficiencies in the alpha, beta, or gamma subunits of the IL-2 receptor complex, or Stat5, the transducer for IL-2 signaling, profoundly impair Treg development.[32,33] Conversely, expression of a constitutively active Stat5 transgene enhances Treg cell development.[34]

A. Genes Regulated by Foxp3

Genome-wide expression profiling of Treg cells with multiple permutations of Foxp3 status has been conducted. Microarray experiments using Treg cells isolated from Foxp3 reporter mice, T effector cells ectopically expressing Foxp3, Treg cells that no longer express Foxp3, and Treg cells that cannot express Foxp3 were designed to identify genes differentially expressed in Tregs and Foxp3 target genes. Combined with Foxp3 chromatin immunoprecipitation (ChIP) to identify genes that associate with Foxp3 protein, a list of approximately 300 Foxp3 gene products has been generated.[35,36] Although differences in methodology and fold-change cutoff yield variable results, reports from multiple laboratories have reached the consensus that Foxp3 serves as a transcriptional activator in three times the number of genes in which it serves as a repressor.

Based on gain-of-function activity by ectopic expression, Foxp3 has been dubbed as a "master regulator" of Treg development.[15,17] However, more recent analyses of Treg cells expressing a functionally null (T_{FN}) Foxp3 and a closer examination of Foxp3 target genes have challenged this notion.[37–39] T_{FN} were engineered by replacing Foxp3 with GFP expression, permitting tracking and isolation of these cells. Surprisingly, approximately 50% of the Treg gene expression signature is maintained in T_{FN}. Corroborating these results, ectopic Foxp3 expression in non-Treg cells reproduces only 40% of the Treg cell transcriptional signature.[38] Although the Foxp3-dependent transcriptional program is essential for Treg effector function, Foxp3 does not serve as a bona fide master regulator in the manner that MyoD and Eyeless determine muscle and eye differentiation respectively.

B. Foxp3 Binding Partners

The identification of Foxp3 binding partners provided new insights into how Foxp3 mediates transcriptional activation and silencing. The transcription factor Runx1 normally transactivates IFN-γ and IL-2 in effector T cells. However in Tregs, Foxp3 physically interacts with Runx1, preventing Runx1 transcriptional activity at the IL-2 and IFN-γ loci.[40] Similarly, Foxp3 binds to the transcription factors NFAT and ROR-γt to silence their target gene expression.[41,42] In effector T cells, NFAT activates IL-2 transcription and ROR-γt promotes IL-17 expression. In Tregs, NFAT or ROR-γt, when complexed to Foxp3, fails to promote IL-2 and IL-17 transcription, respectively. How Foxp3 binding suppresses NFAT, ROR-γt, or Runx1 transcriptional activity is largely unknown.

Besides binding to and inhibiting gene-specific transcriptional activators, Foxp3 binds to the transcriptional repressors Eos and C-terminal binding protein (CtBP).[43] Eos is a member of the Ikaros family of transcription factors that associates with CtBP to mediate gene silencing. Silencing Eos expression in Treg cells induces the upregulation of normally suppressed Foxp3 target genes including

Ifng and *Il2*. Eos was demonstrated to suppress a total of 52 Foxp3 gene products. It remains to be determined if Eos is recruited to NFAT:Foxp3 protein complexes to suppress *Il2* transcription. Extending the Eos:Foxp3 model for target gene silencing, additional transcriptional repressors may associate with Foxp3 to inhibit the expression of genes unaccounted for by the Eos:Foxp3 complexes.

II. Acquisition of Foxp3 Expression by Peripheral $CD4^+$ T Cells

Naïve $CD4^+$ T cells differentiate into specialized effector cells under the guidance of cytokines during T cell activation. The discovery that TGF-β administration during T cell activation induces Foxp3 expression and suppressive activity raised the possibility that Treg cells can be extra-thymically generated from naïve $CD4^+$ T cells.[44] As an extension of the T_H1–T_H2 model of helper T cell differentiation, TGF-β serves as the instructional cytokine that activates Smad transcription factors. Licensed Smad family members translocate to the nucleus to bind to DNA elements in the Foxp3 regulatory loci and enhance Foxp3 transcription. Experimentally, activating naïve $CD4^+$ cells in the presence of TGF-β yields T cell cultures comprising up to 90% $Foxp3^+$ cells.[45] TGF-β induced Treg (iTreg) cells share many characteristics with the natural regulatory T cells including *in vitro* anergy, and comparable CTLA-4 and CD45RB expression levels.[46] Additionally, iTregs produce lower levels of IFN-γ and IL-4, confirming previous reports that TGF-β attenuates T_H1 and T_H2 differentiation, respectively. However, the suppressive quality and the stability of these iTreg cells remain controversial. In some studies, iTregs were as suppressive as thymically derived natural Tregs (nTregs) in inhibiting $CD4^+$ cell proliferation *in vitro* and *in vivo*. Other reports provided evidence for partial acquisition of suppressor activity and unstable expression of Foxp3.

In *in vitro* culture and *in vivo* transfer systems, TGF-β responsiveness is critical for iTreg differentiation. In tissue culture, TGF-β induces expression of the Treg markers Foxp3 and CD25 in a dose dependent manner. *In vivo* conversion of TCR-transgenic $Foxp3^-CD4^+$ effector T cells to Treg is severely compromised in cells expressing a dominant negative TGF-β receptor.[47] Therefore, it is hypothesized that TGF-β responsiveness is essential for peripheral acquisition of suppressive activity.

There are two overlapping mechanisms of TGF-β action during peripheral $CD4^+$ T cell stimulation. First, TGF-β directly promotes Foxp3 transcription through Smad3–Foxp3 enhancer interactions.[48] Additionally, TGF-β indirectly promotes iTreg differentiation by the dampening of TCR signal transduction and the ensuing mitogenesis, which are conditions favorable for Foxp3 induction.[47]

A. iTreg Cell Generation Is Linked to Suboptimal and Sterile T Cell Stimulation

While nTreg cell development in the thymus is associated with high affinity TCR–MHC peptide interactions, iTreg cell differentiation in the periphery is induced under sub-immunogenic conditions.[47,49,50] In studies that utilized T cell transfer models, the largest induction of Foxp3 in peripheral T cells occurred when the TCR transgenic T cells were primed with low doses of their cognate ligand. *In vitro* stimulation of T cells with TGF-β and low concentrations of anti-CD3 antibodies produced the highest frequency of Foxp3 expressing cells.[51,52] Conversely, increasing TCR-derived signals through deficiencies in negative regulators of TCR signaling such as CTLA-4 and Cbl impaired iTreg generation.[53,54] Suboptimal TCR signaling impairs proliferation and accordingly, Foxp3 expression is highly enriched in peripheral T cells that have undergone the fewest rounds of cell divisions.[47,50] As TGF-β is known to inhibit TCR-dependent expansion, TGF-β partly promotes Treg cell differentiation by limiting T cell proliferation.

Optimal iTreg generation is associated with nonimmunogenic antigen delivery methods such as oral or intravenous injections, peptide pumps, or antibody-mediated DC targeting in the absence of adjuvants.[47,50] Administering LPS with antigen abrogates iTreg cell differentiation, suggesting that inflammatory cytokine environments are incompatible with iTreg differentiation.[50] *In vitro*, the addition of the T_H1 or T_H2 instructive cytokines, IL-12 or IL-4, respectively, inhibits TGF-β dependent iTreg generation.[55] Conversely, TGF-β antagonizes IL-12 or IL-4 activity in promoting the T_H1 and T_H2 cell lineage specification.[56,57] Mechanistically, TGF-β impairs T_H1 and T_H2 development by reducing the expression of T-bet and Gata-3, transcription factors that are required for T_H1 and T_H2 differentiation respectively.[58–60] Thus, TGF-β induces iTreg generation while simultaneously repressing differentiation of alternate lineages.

The physiological correlate of experimentally derived iTreg cells has not been identified. Teleogically, iTreg cells may arise in response to self-antigens that are not expressed in the thymus in TGF-β-rich environments. One potential anatomical location of iTreg generation is the gut-associated lymphoid tissue. Bacterial antigens derived from commensal microflora and antigens administered orally are tolerated by our immune system through undefined mechanisms. Interestingly, the highest frequencies of $CD4^+$ $Foxp3^{GFP-}$ cells that had acquired Foxp3 expression in lymphopenic recipients were detected in the lamina propria.[61] The APC subset that is responsible for the enhanced Foxp3 acquisition in gut-associated lymphoid tissue was identified as $CD103^+$DCs.[61,62] CD103 is a transcriptional target of TGF-β signaling, suggesting that these DCs were exposed to TGF-β. Not only are $CD103^+$ DCs recipients of TGF-β signals, but they are also expressors of elevated levels of *Tgfb2* and other genes required for TGF-β processing, as revealed by transcriptional profiling experiments. [61] In addition to TGF-β, CD103^{+} DCs produce the vitamin A metabolite, retinoic acid, which augments the ability of TGF-β to induce Foxp3 expression.[51,61,62] Although, the physiological relevance of retinoic acid in iTreg generation is controversial, it is widely accepted that iTreg cell differentiation is potentiated by a subset of naturally occurring DCs that are enriched in gastrointestinal tissue.[63,64]

B. TGF-β Signaling Directly Influences Foxp3 Transcription

TGF-β signal transduction results in the activation and nuclear translocation of the transcription factor, Smad3. Among the multiple transcriptional target genes, active Smad3 binds to a conserved noncoding sequence (CNS) in the *Foxp3* locus that displays enhancer activity in reporter gene assays. Smad3 occupancy at the TGF-β responsive element in the *Foxp3* locus is correlated with Foxp3 transcription and demethylated cytosines within the *Foxp3* promoter. In contrast, effector T cells stimulated in the absence of TGF-β, and therefore Foxp3 negative, show methylated CpG dinucleotides in the *Foxp3* promoter. TGF-β may antagonize the function of a DNA methylase at the *Foxp3* locus or indirectly promote CpG demethylation by recruiting transcription factors to occupy the CpG islands.

At the transcriptional level, TGF-β signaling induces and represses the expression of most Foxp3 dependent transcriptional targets that define the "natural" Treg cell signature. In addition to the canonical Treg cell genes, TGF-β influences the transcription of nearly 2000 gene products, only 10% of which are encompassed within the Treg cell transcriptional signature. Therefore, TGF-β induces the expression of genes beyond the Treg transcriptome, and iTregs are not equivalent to nTregs by gene expression analysis.

III. Bottom-Up Approach to Analyzing Treg Cell Differentiation

Reliable Foxp3 expression in Treg cells suggests that identifying transcription factors that induce Foxp3 expression will provide new insights into Treg cell differentiation. CNSs within the Foxp3 promoter and within the first intron that could serve as enhancer elements were tested for transcriptional activity.[65] Consensus binding sites for basal transcription factors, including the CAAT, GC, and TATA box were identified −217, −138, and −44 base pairs (bp) upstream of the transcriptional start site, respectively, in the human Foxp3 promoter designated as CNS1.[66] Upstream of these consensus motifs for core transcription factors, evolutionarily conserved nucleotides that conform to AP-1, NFAT, NFκB, and Runx binding sites were identified within CNS1 (Fig. 1). Given the importance of TCR signal transduction in Treg cell lineage

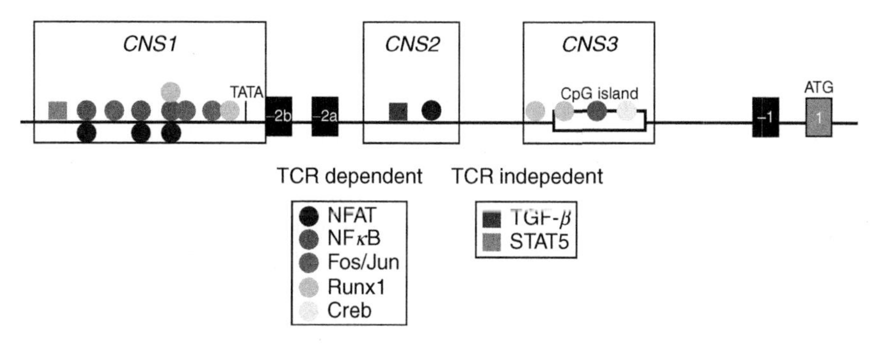

Fig. 1. Schematic map of *Foxp3* CNS 1–3 and its associated transcription factors. The relative location the TCR dependent and cytokine dependent transcription factors are represented as circles and squares, respectively. The CpG methylation island located in CNS3 is designated as a white rectangle. (See Color Insert.)

commitment, it is hypothesized that AP-1, NFAT, and NFκB comprise the TCR response elements within the Foxp3 promoter, similar to the convergence of these transcription factors on the IL-2 promoter to mediate TCR-dependent mitogenesis. In addition to these TCR-derived signals, the Foxp3 promoter is also sensitive to IL-2 administration and harbors a consensus STAT5 binding motif. Although IL-2 and its downstream signaling component STAT5 were known to be required for Treg cell competitive fitness, this finding implicated IL-2 signaling in directly influencing Foxp3 expression.

Two evolutionarily conserved sites found within the first Foxp3 intron were designated CNS2 and CNS3.[65] When sequences derived from either CNS2 or CNS3 were added to the Foxp3 promoter in luciferase assays, reporter activity significantly increased over the Foxp3 promoter alone.[48] As with CNS1, the ability of either CNS2 or CNS3 to augment Foxp3 reporter activity was dependent on TCR engagement. In accordance with this notion, NFAT and NFκB were shown to occupy CNS2, and CREB/ATF and NFκB occupied CNS3 in a TCR-dependent manner. CNS3 also contains a SMAD binding site that is responsible for Foxp3 induction in response to exogenous TGF-β treatment and TCR crosslinking. The evidence and relative contribution of each transcription factor for stable Foxp3 expression are individually discussed below.

A. NFAT and AP-1

The NFAT family of transcription factors has been shown to bind to the Foxp3 promoter and a TCR-responsive enhancer element in CNS2.[48,66] Three NFAT binding sites reside within 500 bp upstream of the Foxp3 transcriptional start site.[66] Mutations in any of the NFAT binding sites within the Foxp3 promoter result in a twofold decrease in luciferase reporter activity.

Of the four NFAT transcription factors, only NFATc1 binds to CNS2.[48] In contrast, only NFATc2 was shown to bind to the Foxp3 promoter through chromatin IP or nucleotide pull-down assays.[66,67] However, the ability of other NFAT family members to bind to the Foxp3 promoter has not been tested.

In T cells, regulation of NFAT activity is mediated by TCR-induced calcium flux. TCR clustering initiates the activation of tyrosine kinases that eventually phosphorylate LAT, a plasma membrane protein that serves as a key platform for recruiting cytosolic signaling molecules to the cell surface. One key step in the initiation of calcium signaling is recruitment of PLC-γ1 to LAT. PLC-γ1 binding to LAT is regulated by the phosphorylation of tyrosine 136 on LAT. Gene targeted mutagenesis of LAT Y136 to abrogate PLC-γ1 recruitment results in impaired Treg cell development.[68] Once PLC is recruited to LAT and activated, it generates second messengers for calcium release from the endoplasmic reticulum. Calcium release from the intracellular endoplasmic store induces the activation of STIM1 and STIM2, which are essential components that sustain calcium signaling for NFAT activation. T cell specific elimination of STIM1 and STIM2 causes a near complete cell-intrinsic block in Treg cell development.[69] Cytosolic calcium binds to Calcineurin, the target of the immunosuppressant drug Cyclosporine A. Upon calcium binding, Calcineurin licenses NFAT family members for nuclear translocation and transcriptional activity. Inhibiting Calcineurin, and thus NFAT activity, through Cyclosporine A treatment in neonatal mice induces multiorgan autoimmunity that is similar in severity and kinetics to that in neonatally thymectomized mice.[70,71] In hindsight, impaired Treg cell development in the presence of Cyclosporine A may be caused by reduced NFAT activation and thus, NFAT-dependent Foxp3 expression. Additionally, pharmacological inhibition of NFAT activation through Cyclosporine A treatment impairs TCR-induced Foxp3 expression in human $CD4^{+}CD25^{-}$ effector T cells.[66]

B. Creb/ATF2

CNS3 contains a CpG island and the Creb binding site, 5′-TGACGTCA-3′, which matches the consensus Creb binding motif, 5′-TGAnnTCA-3′.[65] In luciferase reporter assays, disruption of the Creb binding site *in cis* or expression of a dominant negative Creb construct *in trans* abrogates TCR-induced enhancer function.

Creb transcriptional activity is regulated by serine 216 phosphorylation. Multiple kinases including PKA, CamK, and Rsk are capable of phosphorylating and activating Creb, thus invoking the cyclic AMP (cAMP)-, calcium-, and Ras-dependent signal transduction pathways. In T cells, Creb is primarily activated by Rsk2, an upstream kinase in the MAPK pathway, suggesting that Creb activity in Treg cells is governed by TCR-induced signaling via the RAS-MAPK pathway.[72]

C. CpG Methylation by Dnmt1

Pharmacological inhibition of DNA methyltransferase activity by 5-azacytidine treatment permits Foxp3 expression in non-Treg cells.[65,73] This result was independently confirmed using shRNA mediated reduction of Dnmt1, the major DNA methyltransferase active during cell division and in Dnmt1 deficient T cells.[65,74] Thus, repression of Foxp3 expression in non-Treg cells is mediated by DNA methylation.

Impairments in DNA methylation can influence the expression of multiple gene products that indirectly promote Foxp3 expression. Alternatively, it may directly affect Foxp3 transcription via methylated CpG islands within the Foxp3 promoter and enhancer regions.[65,74] Within the Foxp3 locus, methylated CpG islands have been detected in CNS3 and in the Foxp3 promoter. As expected, these CpG islands are completely demethylated in natural Treg cells. Conversely, the CNS3 CpG island in $CD4^+CD25^-$ effector T cells is completely methylated whereas the promoter is partially methylated. Artificially methylating the CNS3 CpG island abolishes luciferase reporter activity, suggesting that a demethylated CNS3 is essential for Foxp3 transcription.

D. Runt Domain Containing Transcription Factors—Runx

Runx transcription factors were initially implicated in Treg cell effector function as Foxp3 interacting partners.[40] The Runx transcription factor family is comprised of Runx1 (AML1), Runx2 (AML3), and Runx3 (AML2). For transcriptional functionality, Runx transcription factors have to heterodimerize with their binding partner CBFβ. Since overexpressed Foxp3 can co-immunoprecipitate with all three Runx proteins, definitive analysis of the requirement for Runx family members in Treg cell biology was assessed in mice expressing Cre in T cells or Tregs, and a *CBF*β floxed allele (referred to as *CBF*$\beta^{F/F}$ mice). Unexpectedly, Tregs isolated from *CBF*$\beta^{F/F}$ mice expressed lower amounts of Foxp3 protein on a per cell basis, which correlated with a decrease in Foxp3 transcript levels.[75–78] The conditional deletion of Runx1 or RNA silencing in Treg cells phenocopied the decrease in Foxp3 expression observed in *CBF*$\beta^{F/F}$ Treg cells, indicating that, of the three Runx transcription factors, Runx1 plays a nonredundant role. In contrast, Runx3 deficient Treg cells expressed wild-type Foxp3 levels, while a requirement for Runx2 in Treg cells has not been determined.[75,78]

ChIP with anti-CBFb antibodies revealed that the Runx complex bound to the Foxp3 promoter (CNS1) as well as CNS2 and CNS3.[75–77] Within these Foxp3 regulatory regions, two Runx-binding sites were identified in both CNS1 and CNS2. Mutations in putative Runx-binding sites in either CNS1 or CNS3 did not reduce reporter gene activity in unstimulated cells, but CNS1 mutants showed impairment of PMA and ionomycin-induced increase in transcription. Within CNS1, the predicted 5′ Runx-binding site overlaps with a NFAT binding site, which is required for optimal Foxp3 expression. Therefore the relative contribution of Runx-binding to CNS1 for Foxp3 transcription is not clear.

Given the limitations of episomal reporter constructs to monitor gene transcription in its native chromosomal context and the dramatic reduction in Foxp3 expression compared to only marginal differences in reporter gene activities, the Runx transcription factor complex was hypothesized to mediate Foxp3 transcription by influencing chromatin dynamics. In support of this notion, ChIP with anti-H3-K9me3, a histone modification associated with gene silencing precipitated DNA segments distributed near the Foxp3 promoter in $CBF\beta^{F/F}$ Treg cells at a higher frequency compared to $CBF\beta^{WT}$ Treg cells.[76] Thus, the Runx–CBF complex is thought to recruit chromatin modifying molecules to the Foxp3 locus in Tregs.

E. NFκB/c-Rel

Mice engineered with a genetic deficiency in components of the TCR-dependent NFκB signal transduction pathway, including BCL10, Carma1, TAK1 and IKK2, show a reduction in Treg absolute number and frequency.[79,80] Genetic complementation of Carma1 or TAK1 deficiency with constitutively active IKK-β transgene (IKKEE) restores Treg cell numbers in the thymus and periphery, further validating a role for NFκB in Treg cell development.[81] Additionally, enforcement of NFκB activation during thymic development elevates the frequency of $Foxp3^{+}$ cells among CD4 SP thymocytes in IKKEE transgenic mice on a wild-type background. In aggregate, these data are compatible with the hypothesis that transcriptional targets of NFκB are necessary for Treg cell differentiation. However, the identification of consensus NFκB binding sites within the Foxp3 promoter and CNS3 suggest that impaired Treg development in mice with attenuated NFκB signaling is due to reduced Foxp3 expression.[81]

Two putative NFκB binding sites at −382 and −327 bp upstream of the Foxp3 transcriptional start site were confirmed through luciferase reporter gene analysis.[67] The relative contribution of each NFκB binding site to Foxp3 expression is not known as comparisons of individual NFκB-binding mutants have not been conducted. However, mutating the 5′ NFκB binding site is sufficient to abrogate c-Rel mediated Foxp3 expression. As both the NFAT and NFκB proteins employ a Rel homology domain for DNA binding, the c-Rel binding domain in the Foxp3 promoter overlaps with the NFATc2 site.[67] Thus, it is plausible that the consensus sequence, TTCC, accommodates both NFAT and NFκB. Indeed, nucleotide pull-down assays demonstrated that Foxp3 promoter sequences are capable of precipitating both NFATc2 and NFκB. Furthermore, complexes containing both NFATc2 and NFκB were detected in sequential chromatin IP using antibodies to both transcription factors.

From sequence analysis, it is observed that CNS3 contains three potential NFκB docking sites. Although these three sites are separated by approximately 500 bp, c-Rel chromatin IP primarily pulled down the second NFκB binding site that resides within the CpG island.[81] NFκB occupancy at this site correlates with demethylation of the CpG island.[67] Whether c-Rel binding recruits chromatin remodeling enzymes or prevents DNA methyltransferase binding such as Dnmt1 has not been determined.

Among the NFκB transcription factors, c-Rel mediated the largest induction of Foxp3 reporter activity in a transient overexpression system, whereas, p65 yielded a threefold lower signal than c-Rel in PMA and ionomycin treated cells.[81] However, RelB or p50 overexpression does not enhance reporter activity, suggesting that the NFκB binding sequences within the Foxp3 regulatory region display specificity for c-Rel and p65. In agreement with this notion, c-Rel deficiency dramatically impairs Treg cell differentiation.[67] Total thymic cellularity and Foxp3^{-}CD4^{+} SP cellularity were reduced only by 20% in *c-Rel*$^{-/-}$ mice whereas Foxp3^{+}CD4^{+} SP was reduced by 90%, suggesting that Treg cell development is more sensitive to c-Rel deficiency than CD4^{+} effector cell development.

F. TGF-β/SMAD3

Optimal thymic and extrathymic Treg cell differentiation is dependent on TGF-β signaling.[82] Experimentally, TGF-β is best known for its ability to induce Foxp3 expression by peripheral CD4^{+} T cells when combined with reagents that mimic TCR engagement.[44] Through unclear mechanisms, TGF-β attenuates TCR-induced activation and proliferation. As the acquisition of Foxp3 expression by peripheral T cells is correlated with suboptimal TCR signaling and fewer cell divisions, it was thought that TGF-β indirectly influences Foxp3 induction as a T cell immunosuppressant.[47,52]

A TGF-β response element was identified in CNS2, providing evidence for direct involvement of TGF-β signaling in Foxp3 transcription.[48] As with Stat signaling, TGF-β binding induces phosphorylation and nuclear translocation of the Smad family of transcripton factors. Smad3, but not Smad2 or Smad4, binds to the sequence 5′-AGACTGTCT-3′ in CNS2 that matches the consensus binding motif for the Smad proteins. In the presence of TGF-β, Smad3 binding to CNS2 gradually diminishes over time ($<$ 24 h), but imprints Foxp3 expression for longer time periods as acetylated histone H4 is detected at CNS3 and CNS1, 48 h after stimulation. Consistent with this, TGF-β treated T cells have reduced methylated DNA at the Foxp3 promoter.[65] The relative contribution of Smad3 versus other TCR-induced factors in opening the Foxp3 locus is not known because TGF-β administration alone is not sufficient to induce Foxp3 expression. Interestingly, the CpG island within *Foxp3* CNS3 is completely demethylated in thymically derived Treg cells, but nearly completely methylated in TGF-β induced iTreg cells.[65] It is thought that this difference in the CNS3 methylation status is linked to the unstable nature of Foxp3 expression in iTreg cells. As proposed by Tone and colleagues, it is hypothesized that TGF-β works in concert with TCR-derived signals to enable full CNS2 enhancer activity.[48]

During natural Treg cell development in the thymus, TGF-β signaling is necessary only during the first 5 days of a neonatal mouse's life.[82] Afterwards, Treg cellularity equilibrates to wild-type levels in an adult mouse. Thus, it is envisioned that other signaling arms that promote Foxp3 transcription compensate for this TGF-β RI deficiency. Interestingly, the neonatal thymic architecture provides a suboptimal environment for Treg cell development compared to an adult thymus.[14] Therefore, it can be envisioned that TGF-β signaling and Smad3 signaling are required for Foxp3 transcription when the other Foxp3 promoter occupants are limiting in the neonatal thymus.

G. STAT5

Six evolutionarily conserved sites that permit STAT5 binding are located in the *Foxp3* regulatory sites, three of which reside in the Foxp3 promoter and three in CNS3.[34,73] Through chromatin IP analysis, bona fide STAT5 binding was detected in the Foxp3 promoter, but not in the CNS3 enhancer region. Consistent with a role for STAT5 in Foxp3 transcription, T cell specific deletion of STAT5 resulted in a threefold decrease in $Foxp3^{+}$ cells in the thymus and spleen.[34]

STAT5 serves as a signaling unit and transcription factor for cytokine receptors that utilize the common gamma chain (γc). These γc-dependent cytokines include IL-2, IL-7, and IL-15. Although IL-2 has been extensively characterized as a survival factor for Treg cells, it is also implicated in Treg cell development, as Treg specific elimination of the IL-2 receptor reduces thymic Treg cellularity twofold.[32,83,84] Compound elimination of both IL-2 and IL-15 further reduces Treg cell development to levels detected in STAT5 deficient mice.[34] Furthermore, a Foxp3 transgene rescues $IL2Rb^{-/-}$ mice from lymphoproliferative disorders. Thus, reconstitution of Foxp3 expression is sufficient to restore Treg development on an $IL2Rb^{-/-}$ background suggesting that a major function of IL-2 signaling is to promote *Foxp3* transcription.

In developing thymocytes, STAT5 is thought to participate in Foxp3 induction in $CD4^{+}CD25^{+}Foxp3^{-}$ SP precursors in the aforementioned "two-step" model for Treg development.[31] In the first step, $CD25^{-}CD4^{+}$ SP Treg cell precursors that express a TCR specificity conducive to Treg cell differentiation are instructed to express CD25 via TCR-derived signals. After acquiring high-affinity IL-2 receptor expression, the $CD25^+CD4^+$ SP Treg cell precursors gain competence to transduce IL-2 signals that are chiefly mediated by STAT5. Therefore, STAT5 is thought to participate in Foxp3 expression in cells that have the TCR derived components, such as NFkB, NFAT, and CREB, assembled at the Foxp3 promoter.

IV. Cellular Targets of Treg Cells

A. $CD4^+$ T Cells

Treg effector function is assessed by two major assays. First, Treg suppressive activity can be tested by cotransferring Treg cells with wild-type $CD4^+$ effector T cells into lymphopenic recipient mice.[1,85] Transferring T cell preparations devoid of Tregs or reduced Treg: T effector cell ratios induces generalized lymphadenopathy, wasting disease, and colitis development in the recipient mice. In this system, pathogenic T cells are the primary mediators of the autoimmunity, suggesting that $CD4^+$ T cells are direct targets of suppression.

Second, suppression of effector T cell proliferation can be assessed *in vitro* by adding Tregs to standard T cell stimulation cultures. As Tregs are anergic in response to TCR crosslinking, and APCs are irradiated prior to culture, measurements of total proliferation reflect effector T cell divisions. *In vitro*, Tregs can directly suppress $CD4^+$ effector T cell proliferation in the absence of APCs.[86,87] Based on these two assays for Treg suppressive function, it was assumed that $CD4^+$ T cells serve as direct targets of Treg suppression. However, the inclusion of APCs in *in vitro* Treg- mediated suppression cultures significantly reduces the ratio of Tregs to T effector cells required to inhibit T effector cell proliferation, suggesting that APCs catalyze $CD4^+$ T cell suppression.[88] Irrespective of the presence or absence of APCs, suppression of effector T cell proliferation is contact dependent and soluble factor independent. However, intravital imaging studies revealed that Tregs rarely come in contact with pathogenic effector T cells, and thus fail to provide supportive evidence that Tregs directly suppress $CD4^+$ effector T cells in the draining lymph nodes.[89,90] Additionally, discordant conclusions reached from *in vivo* versus *in vitro* suppression assays for cytokine involvement reveal the limitations of such tissue culture assays.

B. Dendritic Cells

An emerging body of evidence suggests that Tregs maintain tolerance to self-tissues by suppressing DC function, numbers, and maturation. *In vitro* observations of DC maturation and cytokine production suggest that Tregs can directly influence DC function. The culture of immature bone marrow derived DCs or primary $CD11b^+$ DCs with Tregs reduces the expression of the costimulatory molecules, B7.1 and B7.2.[91–93] Additionally, Tregs inhibited the upregulation of B7.1 and B7.2 in response to LPS signaling.[92] This suppression of costimulatory molecule expression was specific to Treg cells because coculturing effector T cells failed to reduce B7.1 and B7.2 expression. Among primary DCs, the $CD11b^+$, but not the plasmacytoid subset was sensitive to Treg cell mediated suppression. Treg cell conditioned DCs were less potent stimulators of conventional $CD4^+$ T cell proliferation than DCs exposed to effector $CD4^+$ T cells, suggesting that Treg cell mediated B7.1 and B7.2 downregulation may contribute to impaired T cell priming.

Similar coculture systems revealed that Treg cells directly influence the DC cytokine profile. DCs modestly upregulated the expression of the proinflammatory cytokine IL-6 and produced low levels of IL-10 in the presence of naïve effector T cells.[93,94] However, IL-10 production is significantly increased when DCs are cocultured with Treg cells. Although Treg cells are known to produce IL-10, DCs were the primary source of IL-10 because $IL\text{-}10^{-/-}$ DCs and IL-10 sufficient Treg cells failed to reproduce these results. Furthermore, autocrine IL-10 signaling in DCs has been demonstrated to suppress inflammatory cytokine production. Adding Treg cells to DC cultures suppressed LPS-induced production of IL-12p40, TNF-α, and IL-6.[92] DC-derived IL-10 plays an integral role in suppressing inflammatory cytokine production since the administration of anti-IL-10R antibodies reversed Treg cell mediated suppression of IL-12p40, TNF-α, and IL-6 expression.

These results are not unique to Treg and DCs in isolation, as cultures containing unstimulated DCs, effector, and Treg cells were indistinguishable from cultures comprising DCs and Treg cells, suggesting that Treg cells can oppose the DC stimulatory effect of effector T cells in a competitive situation (Fig. 2A). Furthermore, Treg cells suppressed IL-6 production by DCs to levels below those detected in cultures containing DC and effector T cells. Thus, Treg cells were dominant over effector T cells in influencing the cytokine profile of unstimulated DCs. However, IL-6 levels produced by DCs exposed to LPS and effector T cells were not suppressed by the addition of Treg cells, suggesting that Treg cells are not able to suppress DCs in the presence of strong inflammatory stimulation.

In aggregate, these *in vitro* findings indicate that Tregs are capable of impairing the quality of the DC antigen presentation function. *In vivo*, inducible elimination of Tregs in mice engineered to express a toxin receptor in Treg cells corroborated these findings.[95] In these studies, Treg cell ablation increased the absolute numbers of DCs. Similar to germline $Foxp3^-$ mice, inducible Treg cell elimination induces lymphadenopathy and splenomegaly increasing the cellularity of all leukocyte subsets[95]. However, the relative

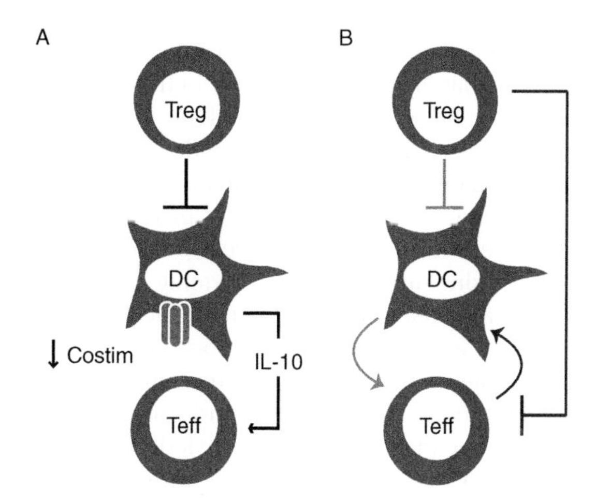

FIG. 2. Celullar interactions involved in Treg mediated suppression. (A) Direct interactions between Treg and dendritic cells induce reduced surface expression of costimulatory molecules and promote inhibitory cytokine production. (B) Dendritic cells integrate suppressive and stimulatory signals from Treg and effector T cells respectively. Two nonmutually exclusive models are depicted as black or gray connectors in the flow chart. In gray, Treg cells directly suppress DCs to indirectly inhibit T effector cell activation. In black, Treg cells directly suppress effector T cells to prevent the expression of DC stimulatory molecules such as GM-CSF and CD40L. The integration of both models results in the Treg mediated suppression of a feed-forward loop between DC and T effector cells.

frequency of DCs was also increased approximately fivefold, indicating that the elevated DC numbers were not solely due to an increase in all leukocyte subsets. Among leukocytes, myeloid cells experienced the greatest expansion. Enlargement of the DC compartment was associated with increased proliferation of committed DC precursors, but not the common myeloid precursor.[96] Treg cell control over DC numbers is dependent on the presence of effector T cells as DC numbers are not elevated in pan T cell deficient mice such as the *TCRb*$^{-/-}$ or *Rag*$^{-/-}$ mice. Furthermore, combining CD4 depletion with Treg cell elimination abrogates DC expansion, suggesting that effector CD4^{+} T cells oppose Treg cell suppression of DCs.[95] Indeed, effector CD4^{+} T cells generated upon Treg cell ablation produce DC modulating factors including GM-CSF, IL-13, and CD40L. Together, these observations evoke a model comprising dendritic, Treg, and T effector cells, with DCs serving as a sensor to integrate the inhibitory signals transmitted by Treg cells and the stimulatory cues transmitted by the effector T cells (Fig. 2B).

Intravital imaging studies that observed the interaction of DCs, Treg cells, and effector T cells support the notion that Treg cells suppress the ability of DCs to prime effector T cells. In a study that compared the migratory behavior of diabetogenic CD4^{+} T cells in Treg cell sufficient or insufficient mice, the authors observed that the transferred effector T cells tended to cluster around DCs in the pancreatic draining lymph node only in the absence of Treg cells.[90] These descriptive differences were supported by significant reductions in effector T cell displacement and velocity in Treg deficient mice. The physiological consequence of increased effector T cell and DC interactions include a more robust proliferation and increased IFN-γ production by pathogenic T cells. Together, these results imply that Treg cells indirectly affect self-reactive T cell priming and cytokine production by disarming DCs.

An independent study that visualized the dynamics of pathogenic $CD4^+$ T cells in an experimental autoimmune encephalomyelitis (EAE) model largely corroborated the findings from the diabetes prone mice.[89] While the antigen that elicited the above results was unknown and was therefore derived from endogenous sources in the diabetes model, knowledge of the cognate peptide for the pathogenic $CD4^+$ T cells enabled the authors to directly compare cognate peptide presenting or irrelevant DCs in the draining popliteal lymph node. In this analysis, pathogenic T cell velocity and displacement decreased only in the presence of the cognate peptide, suggesting that DCs discriminately suppress effector T cell priming in an antigen-dependent manner.

C. $CD8^+$ T Cells

Most of the consequences of coculturing $CD8^+$ T cell with Treg cells are similar to those observed with $CD4^+$ effector T cells. These include *in vitro* suppression of $CD8^+$ T cell proliferation and cytokine production that occur in a cell contact-dependent manner.[97] *In vivo*, suppression of $CD8^+$ T cells has been examined in tumor immunosurveillance and viral infection models.

Immunity to transplanted tumors is significantly augmented in either Treg cell-free conditions or Treg cell depleted environments.[98–100] Enhanced tumor immunosurveillance was associated with increased $CD8^+$ T cell maturation, IFN-γ production, and tumor cell cytolytic activity. The suppression of tumor cell killing is hypothesized to require TGF-β, based on studies that neutralized TGF-β with specific antibodies or impaired TGF-β signaling through dominant negative TGF-β receptor expression.[101,102] TGF-β limits $CD8^+$ T cell cytolytic potential by repressing target cell killing, not the expression of core granule components such as perforin, and granzyme A and B.[101] Thus, it is possible that Treg cells provide a TGF-β environment that prohibits CTL differentiation. However, results obtained from experiments that utilized DN-TGFβR expressing cells or TGF-β neutralization highlight the importance of TGF-β responsiveness but do not describe whether Treg cells provide relevant TGF-β to mediate these effects. In this regard, tumor cells are known to establish an immunosuppressive microenvironment by producing TGF-β.[103]

Depleting Tregs has been demonstrated to be beneficial for boosting $CD8^+$ T cell responses against acute and persistent viral infections. In a HSV-1 infection model, a single injection of CD25 depleting antibodies prior to infection enhanced $CD8^+$ T cell proliferation, IFN-γ production, and cytotoxicity.[104] Interestingly, heightened $CD8^+$ T cell responses were detected at both the acute and memory phases of the anti-HSV response. In another study, acutely depleting Treg cells during secondary infection with *Listeria monocytogenes* impaired $CD8^+$ memory T cell expansion and cytokine production, indicating that Tregs limit the $CD8^+$ recall memory T cell responses.[105] Furthermore, polyclonal $CD8^+$ memory T cells from aged mice proliferated more robustly when transferred in the absence of Treg cells compared to cotransfers with Treg cells, suggesting that Treg cells are capable of reducing steady state levels of an established $CD8^+$ T cell memory population.[106] Thus, Treg cells may impinge on $CD8^+$ memory T cell generation, maintenance, and memory cell reactivation.

D. NK Cells

In cancer patients and in multiple tumor models in mice, tumor growth has been associated with Treg cell mediated suppression of NK cell activity.[98,107] In mice, Treg cell depletion enhances NK cell mediated clearance of transplanted tumors. In cancer patients, Treg cell numbers are inversely proportional to NK cell activity. Specifically, the absolute number of Tregs was shown to be elevated in gastrointestinal stromal tumor bearing (GIST) patients with comprised NK cell activity compared to GIST patients with normal NK cell function.[108] As in $CD8^+$ T cells, NK cell proliferation, IFN-γ production, and cytotoxic potential are impaired in the presence of Treg cells in tissue culture, indicating that NK cells are direct targets of Treg cell suppressor function.[107,109] Consistent with this, transferring Tregs into T cell deficient but NK cell sufficient mice abrogates NK cell cytotoxicity against tumor cell targets.[110] Furthermore, decreased cytolytic function in transferred mice was associated with reduced expression of NKG2D, an NK cell stimulatory receptor, by the suppressed NK cells. Tumors expressing NKG2D ligands were able to escape NK cell mediated tumor immunosurveillance more efficiently than tumors that did not express the ligands, suggesting that Tregs suppressed NK cell activity in a NKG2D-dependent fashion. In this system, TGF-β deficient Tregs were unable to suppress NK cell cytolytic activity, indicating that Treg derived TGF-β may directly suppress NK tumoricidal activity. In other systems, tumor derived TGF-β down-regulates NKG2D surface expression on NK cells.[110] Therefore, it is possible that regulatory T cell derived TGF-β mediates NK cell suppression by reducing NKG2D expression levels.

E. B Cells

Genetically or surgically modified mice with a reduced or no Treg compartment exhibit multiple signs of humoral dysregulation. Generalized hypergammaglobulinemia as well as organ specific antibodies are detected in these mice. Additionally, splenic B cells isolated from Foxp3 deficient mice display higher levels of costimulatory molecules, suggesting that B cells possess enhanced antigen presentation function in the absence of Tregs.[111] As Tregs are known to suppress T_H cell activity, B cells may be the indirect cellular targets of Treg cell mediated suppression. However, *in vitro* culture experiments that analyzed B cell responses in the absence of T_H cells suggest that Tregs may directly suppress B cells *in vivo*.[112] To bypass this requirement of T_H cells, human B cells were stimulated with CD40 crosslinking antibodies and T_H derived cytokines, in the presence or absence of Treg cells. The addition of Treg cells to B cell cultures reduced the production of IgG and IgA isotypes from IgD^+ precursors in a dose-dependent manner. The reduced capacity to produce isotype-switched antibodies in the presence of Treg cells was associated with decreased activation-induced cytosine deaminase expression in response to CD40 and IL-4 signaling. Therefore, Tregs are capable of inhibiting T-dependent antibody production by impairing class switch recombination.

Tregs have been shown to directly suppress T-independent B cell activation. LPS-induced B cell proliferation is markedly impaired in the presence of an increasing number of Treg cells.[113] This suppression of B cell proliferation is not unique to T-independent mitogenic signaling as proliferative responses to CD40 and IgM crosslinking are also suppressed by Treg cells. Detection of reduced B cell proliferation was correlated with an increase in B cell death in Treg cell containing cultures.[114] Similar to monocytes, DCs, and T_H cells, B cells were shown to be susceptible to granzyme-dependent cell death induction by Tregs. Consistent with the idea that granzyme B mediates target cell killing by murine Tregs, administering DCI, a specific granzyme B inhibitor decreased B cell death. In contrast to another report, perforin-deficient Treg cells displayed reduced cytolytic activity compared to wild-type counterparts. Perhaps perforin is differentially required for B cell versus T_H cell cytolysis.

Intrasplenic Treg and B cell colocalization supports the notion that Tregs are capable of directly controlling B cell numbers *in vivo*. Human Tregs are recruited to B cell zones through the expression of the germinal center associated chemokine receptor CXCR5, after TCR stimulation. In mice, B–Treg cell encounters are facilitated by B cell derived CCL4, which recruit CCR5 expressing Tregs.[113] Indeed, CCL4 neutralization *in vivo*, which is predicted to disrupt B–Treg cell interactions, elevated serum autoantibody titers. Once a B–Treg cell interaction is established, it is hypothesized that Tregs preferentially kill antigen presenting B cells compared to nonspecific B cells. This idea is supported by results from *in vitro* experiments in which TCR transgenic Tregs were mixed with two populations of allelically marked B cells that were either loaded with the cognate peptide or left untreated.[114] In this competitive situation, antigen presenting B cells were four times more frequently killed than nonpulsed B cells, suggesting that conditions that favor conjugate formation increase the likelihood of target cell lysis. It remains to be determined whether cognate antigen-presenting dendritic and myeloid cells are also preferentially lysed by Treg cells.

V. Molecular Mechanisms of Treg Cell Suppression

A. TGF-β

The significance of TGF-β in maintaining immunological tolerance is evidenced by the striking phenotype of TGF-β deficient mice. TGF-β_1 deficient mice develop fatal lymphoproliferative disease by 3 months of age. Transgenic mice that express a dominant TGF-β receptor under the proximal lck promoter largely recapitulate the autoimmune phenotype of germline $TGF\beta_1^{-/-}$ mice, revealing that T cells are a major target of TGF-β mediated immunosuppression.[115] Multiple hematopoeitic and nonhematopoeitic cells produce TGF-β, but T cells are an essential source of TGF-β as T cell specific TGF-β knockout mice succumb to autoimmunity with the kinetics and severity seen in germline TGF-β deficient mice.[116] Among T cells, Treg cells express high levels to TGF-β. The role of TGF-β as a Treg suppressor molecule remains controversial.[86,87,117] As negative results obtained from cytokine neutralization experiments are generally inconclusive, *in vitro* suppression assays were repeated with T cells isolated from mice with genetic deficiencies in either TGF-β production or in responsiveness to resolve the discrepant results.[118] In support of TGF-β having a nonessential role in Treg function, responder T cell proliferation was equally suppressed by $Tgf\beta_1^{-/-}$ and wild-type Tregs. Although exogenous TGF-β is sufficient to impair proliferation of unseparated T cells, responder T cells that are insensitive to TGF-β signaling, caused either by a Smad3 deficiency or by the expression of a dominant negative TGF-β receptor, were responsive to Treg suppressor function. Thus, Treg mediated inhibition of conventional T cell proliferation *in vitro* occurs independent of TGF-β activity.

In contrast to suppression of *in vitro* proliferation, a requirement for TGF-β has been demonstrated in the aforementioned T cell transfer model of colitis. In this system, effector T cells that expressed a dominant negative TGF-βRII receptor were resistant to wild-type Tregs, indicating that TGF-β responsiveness was necessary to prevent development of colitis. However, $Tgf\beta_1^{-/-}$ Tregs were equally competent as wild-type regulatory T cells in their ability to suppress wild-type effector T cells and colitis. Together, these experiments revealed that although TGF-β plays an essential role in colitis prevention, TGF-β from non-Treg sources could compensate for TGF-β deficiency in Tregs.

B. IL-10

IL-10 is an immunosuppressive cytokine that inhibits DC maturation and inflammatory cytokine production. Germline *Il10* knockout mice develop severe colitis by three months of age, but are otherwise unremarkable.[119] T cell or Treg specific *Il10* deletion reproduces the kinetics and severity of colitis development seen in germline $Il10^{-/-}$ mice indicating that Treg-derived IL-10 is essential to maintain gastrointestinal homeostasis.[120,121] Besides gut associated lymphoid tissues, generalized lymphadenopathy and T cell infiltration in target organs affected by manipulating Treg numbers are not detected in $Il10^{-/-}$ mice. As predicted by the lack of widespread immunopathologies in $Il10^{-/-}$ mice, suppression of effector $CD4^{+}$ T cell proliferation is unaffected in the presence of IL-10 neutralizing antibodies.[86]

Consistent with a role for IL-10 in gastrointestinal tolerance, studies that employed the T cell transfer model for colitis confirmed that Treg cell-derived IL-10 is necessary to suppress immunopathology *in vivo.*[122] The severity of colitis induced by wild-type $CD4^{+}$ effector T cells in the absence of Tregs can be ameliorated by exogenous IL-10 administration. Furthermore, colitis is prevented by equipping transferred effector $CD4^{+}$ T cells with an *Il10* transgene.[123] Treg cells were demonstrated to be the physiologic IL-10 source *in vivo* because protection from colitis development was abrograted when wild-type effector T cells were cotransferred with *Il10* deficient Treg cells.

Although Treg derived IL-10 is not essential for tolerance induction at most anatomical locations, Tregs promote a tolerogenic state by programming other immune cells to produce IL-10. In a T cell dependent model of asthma, IL-10 plays an essential role in suppressing airway inflammation. The sources of protective IL-10 in this asthma model are the $CD4^{+}$ effector cells and not the Treg cells. However, Tregs are required to induce IL-10 expression in $CD4^{+}$ effector cells to prevent airway hyperreactivity.[124] Similar to the role of Tregs in influencing IL-10 production by DCs, Tregs suppress asthma development by redirecting effector T cells to produce to IL-10 and suppressing inflammatory cytokine production.[92,94,125] In summary, Treg derived IL-10 contributes to the maintenance of mucosal homeostasis. In addition to this Treg instrinsic role for IL-10, Tregs instruct other immune cell types to adopt an anti-inflammatory cytokine profile that includes IL-10.

C. CTLA-4

The immunosuppressive function of CTLA-4 was originally thought to occur in a cell autonomous manner. However, in mixed bone marrow chimeras, the presence of wild-type T cells prevents the activation of CTLA-4 deficient T cells and lymphoproliferative disease development that was detected in CTLA-4 knockout mice.[126] Based on this observation and high CTLA-4 expression levels in Tregs, the role of CTLA-4 in maintaining dominant tolerance was further investigated through a series of CTLA-4 blocking experiments. In the adoptive T cell transfer model of experimental colitis, CTLA-4 blockade did not impair disease development in mice that received only $CD4^+$ effector cells.[127] However, the protective effect of Treg cells in this system was abrogated when CTLA-4 specific antibodies were administered to the recipient mice, suggesting that Treg cells are the primary targets of CTLA-4 inhibition *in vivo*. Importantly, the CTLA-4 specific antibody utilized in these studies does not deplete Tregs.[128] In support of these findings, enhanced effector T cell proliferation was also detected in *in vitro* suppression assays performed in the presence of a CTLA-4 blockade, reversing the anti-proliferative effect of Treg cells. As in *in vivo* transfer studies, CTLA-4 inhibition was specific for CTLA-4 expressed on Treg cells because cultures containing $Ctla4^{-/-}$ effector T cells and CTLA4-sufficient Tregs were sensitive to the addition of anti-CTLA4 antibodies.

As with effector $CD4^+$ T cells, CTLA-4 may function in Treg cells by directly transmitting signals that mediate immunosuppression. In this regard, the cross-linking of CD3, CD28, and CTLA-4 induces the production of the immunosuppressive cytokine TGF-β by unsorted $CD4^+$ T cells.[117] Thus, the suppressor function CTLA-4 on Treg cells may be partially mediated by TGF-β.

Alternatively, constitutive CTLA-4 expression by Tregs may passively suppress effector T cells by outcompeting for B7 binding, as B7 binds ten to twenty times more avidly to CTLA-4 than to CD28.[129,130] In this model, B7 sequestration by CTLA-4 is predicted to limit the provision of costimulatory signals to effector T cells. Therefore, CTLA-4 signaling in Treg cells would not be essential for maintaining its immunosuppressive properties. In support of this hypothesis, the addition of CTLA-4-Ig fusion protein is sufficient to prevent hyperproliferative T cell responses *in vitro*.[131] Furthermore, CTLA-4-Ig administered *in vivo* protects CTLA-4 deficient neonates from lethal autoimmunity and lymphoproliferative disorder.[132] Although these studies suggest that blocking costimulation may be achievable through CTLA-4-Ig administration, predictions based on surface CTLA4 versus CD28 expression levels suggest that cellular CTLA-4 is incapable of saturating B7 molecules.[133]

In addition to functioning as a costimulatory ligand for T cells, an emerging body of evidence suggests that B7 engagement by the CD28 family members initiates a "reverse signaling" cascade in the APCs. In a process that requires autocrine IFN-γ and STAT-1 signaling, B7 engagement induces indoleamine 2,3-dioxygenase (IDO) expression in macrophages and DCs.[134,135] IDO effects immunosuppression by catalyzing the degradation of the essential amino acid tryptophan into kynurenine. Microenvironments that are depleted of tryptophan have been demonstrated to impair T cell proliferation *in vitro*[136,137] and to limit T cell mediated skin inflammation *in vivo*[134] (Fig. 3A). In contrast to CTLA-4, B7 ligation by CD28 was not capable of inducing IDO expression in APCs,

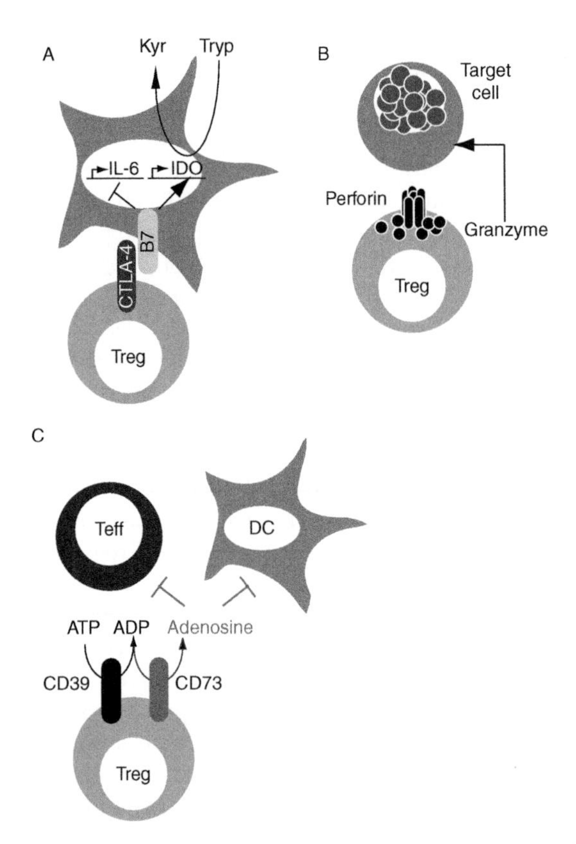

FIG. 3. Tregs inhibit effector T cell activation through CTLA-4, Granzyme, and ATP depletion. (A) CTLA-4 dependent metabolic regulation of effector T cell activation. Treg cells constitutively express high levels of CTLA-4. B7 engagement by CTLA-4 induces the expression of IDO that catabolizes tryptophan to kynurenine. Tryptophan depleted environments impair effector T cell activation. IL-6 production is inhibited by B7 binding to CTLA-4. (B) Treg cells induce target cell apoptosis in a perforin and granzyme dependent manner. (C) Extracellular ATP is converted to adenosine by sequential activities of CD39 and CD73 expressed by Treg cells. Adenosine induces IL-10 production by DCs and attenuates TCR signaling through cAMP generation.

suggesting that IDO expression is dependent on high-affinity B7 interactions.[138] The inability of CD28 to substitute for CTLA-4 in promoting IDO expression was traced to the production of IL-6 in APCs upon stimulation by CD28 but not CTLA-4. Neutralizing IL-6 upon CD28 treatment was sufficient to promote IDO expression, suggesting that IL-6 can overcome the immunosuppressive barrier imposed by CTLA-4 expressing Tregs. Interestingly, IL-6 was demonstrated to make effector T cells refractory to Treg suppression.[139] It is plausible that IL-6 makes effector cells resistant to Tregs by preventing IDO expression.

As discussed in Section IV.B of *Cellular Targets of Treg Cell Immunosuppression*, DCs integrate signals derived from contacts with either CTLA-4 bearing Tregs or CD28 expressing T effector cells. As visualized in intravital imaging studies, Treg cells preferentially cluster around DCs in the presence of effector T cells. This phenomenon is dependent on CTLA-4 expression as *Ctla4*$^{-/-}$ Tregs fail to oust effector T cells when competing for DC access,and provision of CTLA-4 antibodies is sufficient to break Treg-DC aggregates *in vitro*.[140] Based on the higher affinity of CTLA-4: B7 compared to CD28: B7 interactions, and preferential DC access by Treg cells over effector T cells, it is hypothesized that CTLA-4 mediated IDO expression prevails over competing CD28 signals. The importance of CTLA-4 in Treg cell effector function is underscored by the observed phenotype of Treg cell-specific CTLA-4 knockout mice (*Ctla4*$^{F/F}$). These mice succumb to fatal multiorgan autoimmunity by 7 weeks of age. Autoimmunity is correlated with elevated DC expression of B7 family members, indicating that CTLA-4 expression on Tregs plays a nonredundant role in controlling DC activation.[141] Although essentially all activated CD4^{+} effector cells in *Ctla4*$^{F/F}$ produce high levels of CTLA-4, they are incapable of suppressing DC maturation. Therefore, B7 engagement by CTLA-4 cannot be substituted by non-Tregs in DC homeostasis.

Although B7 family members are primarily expressed by professional APCs, it is known that activated T cells also express B7.1 and B7.2.[142–144] To explore the possibility that reverse B7 signaling in T cells may play a role in T cell homeostasis, wild-type Tregs were tested for their ability to suppress B7 deficient effector CD4^{+} T cells.[145] In this system, *B7.1*$^{-/-}$*B7.2*$^{-/-}$ effector T cells were insensitive to Treg mediated suppression and induced severe gastritis and colitis in the lymphopenic recipients. Additionally, *in vitro* proliferation of *B7.1*$^{-/-}$*B7.2*$^{-/-}$ effector T cells was not suppressed by the addition of wild-type Tregs. Reconstituting *B7.1*$^{-/-}$*B7.2*$^{-/-}$ effector T cells with a B7 truncation mutant lacking the cytoplasmic domain failed to rescue lymphoproliferation, indicating that B7 signaling, and not B7 occupancy, is essential for the suppression. Therefore, the CTLA-4 expressed by Tregs may exert immunosuppressive activity in a B7-dependent manner on both effector CD4^{+} T cells and APCs. These experiments also provide the best evidence to support the fact that Tregs may directly suppress both effector CD4^{+} T cells *in vivo*.

D. Fibroleukin

The immunosuppressive properties of fibroleukin were originally identified in unsorted peripheral blood T cells. Further dissection of human T cell subsets revealed that the $CD45RO^+$ T cell population, which comprises memory and Tregs, primarily expresses fibroleukin.[146] Microarray analyses of genes that are differentially expressed in Tregs confirmed elevated fibroleukin expression levels in Tregs compared to naïve and activated $CD4^+$ T helper cells.[18,147]

T cells express fibroleukin as a soluble secreted molecule that binds to DC and T cells.[148] Exogenous fibroleukin treatment impairs T cell proliferation in APC-containing T cell cultures, suggesting that APCs may also be susceptible to fibroleukin-mediated immunosuppression. Indeed, monocyte-derived DCs exposed to fibroleukin express lower levels of surface B7.1, CD40, and MHC II. Additionally, fibroleukin-conditioned DCs exhibited a decreased capacity to support the proliferation of untreated T cells; whereas a more modest reduction in T cell proliferation was detected when T cells, but not DCs, were pretreated with fibroleukin. However, exogenous fibroleukin administration weakly impairs T cell proliferation in response to concavalin A or plate-bound CD3 and CD28 antibody stimulation, indicating that T cells are also direct targets of fibroleukin immunosuppression, albeit less sensitive to fibroleukin than DCs. Furthermore, fibroleukin-deficient mice show increased number and enhanced maturation of DCs, and this is correlated to increased activity of effector T cells.[149] The clinical outcome of fibroleukin deficiency is restricted to the development of severe glomerulonephritis. Increased protein deposition in the glomeruli is dependent on hematopoietic cells and is correlated to decreased Treg activity, but the requirement of Treg- derived fibroleukin to prevent kidney damage is unclear. These studies demonstrate that fibroleukin is capable of suppressing T cell activation both by reducing T cell priming capacity of APCs and by directly inhibiting the proliferative capacity of effector T cells.

E. Granzyme Dependent Cytotoxicity

Treg cell mediated inhibition of *in vitro* effector T cell proliferation was demonstrated to require cell-to-cell contact. Although the molecular basis for contact-mediated suppression is largely unknown, recent reports have revealed that Tregs also require cellular contact for target cell killing via the granule exocytosis pathway.[150,151] Granule-mediated cytoxicity is dependent on granzymes, granule resident proteases, which initiate a cascade of apoptosis-promoting cleavage events. As in effector T cells, granzyme expression is induced in Tregs in response to T cell receptor signaling. While granzyme A is primarily expressed by activated human Tregs,[151] granzyme B is the predominant granzyme induced in murine Tregs.[150] Granzyme A and B differ in substrate specificity and the kinetics of cell death induction, but activated murine and human Tregs comparably induce effector T cell death at 1:1 ratio of regulatory to effector T cells. *In vitro* cytotoxicity was dependent on granzyme function, as suppression of effector T cell proliferation was severely compromised in cultures containing granzyme B deficient Tregs.[150]

Cytolytic granules also contain perforin, which is essential for target cell lysis in $CD8^+$ CTLs and NK cells. The deposited perforin polymerizes on the target cell plasma membrane in a calcium dependent manner and generates holes that were hypothesized to serve as granzyme conduits into the target cell. However, accurate measurements of pores formed by perforin suggest that the diameter of polyperforin channels do not accommodate granzyme passage.[152] Although the exact function of perforin remains unknown, phenotypic similarities in mice deficient in either perforin or granzyme B provide evidence that perforin plays a nonredundant role in targeted cytolysis by lymphocytes. In support of this idea, inhibiting perforin by either EDTA or concanamycin A treatment abrogates target cell killing by human Tregs. In contrast to these findings, perforin deficient murine Tregs were equally suppressive as its wild-type counterparts *in vitro*, suggesting that perforin is not essential for granzyme B dependent target cell lysis in murine Treg cells. These discrepant results may reflect the usage of different granzymes for target cell killing in mouse versus human Tregs. In this regard, granzyme A may be more dependent on perforin for killing that granzyme B. Alternatively, calicium chelators or concanamycin A may not be specific for perforin inhibition, affecting target cell cytolysis independent of perforin function. Human Tregs, additionally, have been demonstrated to kill monocytes, DCs, and activated $CD8^+$ T cells[151] (Fig. 3B). Murine Tregs are also capable of killing B cells *in vitro*.[114]

F. Extracellular ATP Depletion

Tissue damage and ensuing cell death generate extracellular ATP, which serves as an inflammatory mediator. Tregs express cell surface proteins, CD39 and CD73, which sequentially convert inflammation-inducing ATP to immunosuppressant adenosine. CD39 catalyzes the first step of ATP breakdown to 5′ adenosine monophosphophate (AMP). The ecto-5′ nucleotidase of CD73 cleaves the last phosphate in AMP to generate adenosine. Multiple immune cell subtypes are suppressed by adenosine, most notably T cells and DCs. Adenosine exposure during T cell stimulation attenuates proliferation, cytokine secretion, and cytotoxic activity[153,154] (Fig. 3C). In T cells, adenosine binds to its receptor A_{2a} and exerts its suppressive activity by elevating cAMP concentrations in cells. The second messenger, cAMP, serves as a cofactor for protein kinase A (PKA), which in turn suppresses T cell activation. In DCs, adenosine decreases antigen presentation and increases the expression of the immunosuppressive cytokine IL-10.[155,156] CD39 deficiency significantly diminishes Treg mediated survival of allograft skin transplants. However, $CD39^{-/-}$ Tregs are minimally impaired in suppressing effector T cell proliferation and $CD39^{-/-}$ mice do not display any overt signs of autoimmunity.[153] The CD39-CD73 axis of adenosine generation is not essential for maintaining immune homeostasis in unchallenged mice, but may play an important role in diseases with significant tissue destruction and hypoxia.

VI. Concluding Remarks

New technologies in gene expression profiling, cell sorting, and mouse engineering have aided the identification of gene products and signaling pathways that are involved in Treg development and suppressive function. Lists of genes differentially expressed by Tregs and Foxp3 target genes have been generated as a result of these advances, but one of the major challenges of Treg research is to assign functional data to these gene products. Most molecules that comprise the Treg suppressive artillery are not unique to Tregs. Thus, the context in which each candidate suppressor molecule exerts its activity needs to be determined. Likewise, hypotheses on Treg development presume that the TCR and cytokine signaling that are common to all T cells are also required for Treg differentiation. Therefore, the unique interplay of common signaling pathways and not novel signaling modules are thought to underlie Treg development. Understanding how common signaling pathways and transcription factors integrate to specify the Treg lineage are likely to provide new insights into how Tregs diverge from effector cells.

References

1. Sakaguchi S, Sakaguchi N, Asano M, Itoh M, Toda M. Immunologic self-tolerance maintained by activated T cells expressing IL-2 receptor alpha-chains (CD25). Breakdown of a single mechanism of self-tolerance causes various autoimmune diseases. *J Immunol* 1995;**155**:1151–64.
2. Tang Q, Henriksen KJ, Bi M, Finger EB, Szot G, Ye J, et al. In vitro-expanded antigen-specific regulatory T cells suppress autoimmune diabetes. *J Exp Med* 2004;**199**:1455–65.
3. Mottet C, Uhlig HH, Powrie F. Cutting edge: cure of colitis by $CD4^+CD25^+$ regulatory T cells. *J Immunol* 2003;**170**:3939–43.
4. Kohm AP, Carpentier PA, Anger HA, Miller SD. Cutting edge: $CD4^+CD25^+$ regulatory T cells suppress antigen-specific autoreactive immune responses and central nervous system inflammation during active experimental autoimmune encephalomyelitis. *J Immunol* 2002;**169**:4712–6.
5. Scalapino KJ, Tang Q, Bluestone JA, Bonyhadi ML, Daikh DI. Suppression of disease in New Zealand Black/New Zealand White lupus-prone mice by adoptive transfer of ex vivo expanded regulatory T cells. *J Immunol* 2006;**177**:1451–9.
6. Yanagawa T, Hidaka Y, Guimaraes V, Soliman M, DeGroot LJ. CTLA-4 gene polymorphism associated with Graves' disease in a Caucasian population. *J Clin Endocrinol Metab* 1995;**80**:41–5.
7. Atabani SF, Thio CL, Divanovic S, Trompette A, Belkaid Y, Thomas DL, et al. Association of CTLA4 polymorphism with regulatory T cell frequency. *Eur J Immunol* 2005;**35**:2157–62.

8. Nistico L, Buzzetti R, Pritchard LE, Van der Auwera B, Giovannini C, Bosi E, et al. The CTLA-4 gene region of chromosome 2q33 is linked to, and associated with, type 1 diabetes. Belgian Diabetes Registry. *Hum Mol Genet* 1996;**5**:1075–80.
9. Vella A, Cooper JD, Lowe CE, Walker N, Nutland S, Widmer B, et al. Localization of a type 1 diabetes locus in the IL2RA/CD25 region by use of tag single-nucleotide polymorphisms. *Am J Hum Genet* 2005;**76**:773–9.
10. Bennett CL, Christie J, Ramsdell F, Brunkow ME, Ferguson PJ, Whitesell L, et al. The immune dysregulation, polyendocrinopathy, enteropathy, X-linked syndrome (IPEX) is caused by mutations of FOXP3. *Nat Genet* 2001;**27**:20–1.
11. Chatila TA, Blaeser F, Ho N, Lederman HM, Voulgaropoulos C, Helms C, et al. JM2, encoding a fork head-related protein, is mutated in X-linked autoimmunity-allergic disregulation syndrome. *J Clin Invest* 2000;**106**:R75–81.
12. Wildin RS, Ramsdell F, Peake J, Faravelli F, Casanova JL, Buist N, et al. X-linked neonatal diabetes mellitus, enteropathy and endocrinopathy syndrome is the human equivalent of mouse scurfy. *Nat Genet* 2001;**27**:18–20.
13. Brunkow ME, Jeffery EW, Hjerrild KA, Paeper B, Clark LB, et al. Disruption of a new forkhead/winged-helix protein, scurfin, results in the fatal lymphoproliferative disorder of the scurfy mouse. *Nat Genet* 2001;**27**:68–73.
14. Fontenot JD, Dooley JL, Farr AG, Rudensky AY. Developmental regulation of Foxp3 expression during ontogeny. *J Exp Med* 2005;**202**:901–6.
15. Hori S, Nomura T, Sakaguchi S. Control of regulatory T cell development by the transcription factor Foxp3. *Science* 2003;**299**:1057–61.
16. Khattri R, Cox T, Yasayko SA, Ramsdell F. An essential role for Scurfin in $CD4^{+}CD25^{+}$ T regulatory cells. *Nat Immunol* 2003;**4**:337–42.
17. Fontenot JD, Gavin MA, Rudensky AY. Foxp3 programs the development and function of $CD4^{+}CD25^{+}$ regulatory T cells. *Nat Immunol* 2003;**4**:330–6.
18. Fontenot JD, et al. Regulatory T cell lineage specification by the forkhead transcription factor foxp3. *Immunity* 2005;**22**:329–41.
19. Lee HM, Hsieh CS. Rare development of $Foxp3^{+}$ thymocytes in the $CD4^{+}CD8^{+}$ subset. *J Immunol* 2009;**183**:2261–6.
20. Watanabe N, Wang YH, Lee HK, Ito T, Wang YH, Cao W, et al. Hassall's corpuscles instruct dendritic cells to induce $CD4^{+}CD25^{+}$ regulatory T cells in human thymus. *Nature* 2005;**436**:1181–5.
21. Salomon B, et al. B7/CD28 costimulation is essential for the homeostasis of the $CD4^{+}CD25^{+}$ immunoregulatory T cells that control autoimmune diabetes. *Immunity* 2000;**12**:431–40.
22. Tai X, Cowan M, Feigenbaum L, Singer A. CD28 costimulation of developing thymocytes induces Foxp3 expression and regulatory T cell differentiation independently of interleukin 2. *Nat Immunol* 2005;**6**:152–62.
23. Bensinger SJ, Bandeira A, Jordan MS, Caton AJ, Laufer TM. Major histocompatibility complex class II-positive cortical epithelium mediates the selection of CD4(+)25(+) immunoregulatory T cells. *J Exp Med* 2001;**194**:427–38.
24. Liston A, Nutsch KM, Farr AG, Lund JM, Rasmussen JP, Koni PA, et al. Differentiation of regulatory Foxp3+ T cells in the thymic cortex. *Proc Natl Acad Sci USA* 2008;**105**:11903–8.
25. Apostolou I, Sarukhan A, Klein L, von Boehmer H. Origin of regulatory T cells with known specificity for antigen. *Nat Immunol* 2002;**3**:756–63.
26. Caton AJ, Cozzo C, Larkin 3rd J, Lerman MA, Boesteanu A, Jordan MS, et al. CD4(+) CD25 (+) regulatory T cell selection. *Ann NY Acad Sci* 2004;**1029**:101–14.
27. Jordan MS, Boesteanu A, Reed AJ, Petrone AL, Holenbeck AE, Lerman MA, et al. Thymic selection of $CD4^{+}CD25^{+}$ regulatory T cells induced by an agonist self-peptide. *Nat Immunol* 2001;**2**:301–6.

28. Kawahata K, Misaki Y, Yamauchi M, Tsunekawa S, Setoguchi K, Miyazaki J, et al. Generation of CD4(+)CD25(+) regulatory T cells from autoreactive T cells simultaneously with their negative selection in the thymus and from nonautoreactive T cells by endogenous TCR expression. *J Immunol* 2002;**168**:4399–405.
29. Walker LS, Chodos A, Eggena M, Dooms H, Abbas AK. Antigen-dependent proliferation of $CD4^+CD25^+$ regulatory T cells in vivo. *J Exp Med* 2003;**198**:249–58.
30. van Santen HM, Benoist C, Mathis D. Number of T reg cells that differentiate does not increase upon encounter of agonist ligand on thymic epithelial cells. *J Exp Med* 2004;**200**:1221–30.
31. Lio CW, Hsieh CS. A two-step process for thymic regulatory T cell development. *Immunity* 2008;**28**:100–11.
32. Fontenot JD, Rasmussen JP, Gavin MA, Rudensky AY. A function for interleukin 2 in Foxp3-expressing regulatory T cells. *Nat Immunol* 2005;**6**:1142–51.
33. Malek TR, Yu A, Vincek V, Scibelli P, Kong L. CD4 regulatory T cells prevent lethal autoimmunity in IL-2Rbeta-deficient mice. Implications for the nonredundant function of IL-2. *Immunity* 2002;**17**:167–78.
34. Burchill MA, Yang J, Vogtenhuber C, Blazar BR, Farrar MA. IL-2 receptor beta-dependent STAT5 activation is required for the development of $Foxp3^+$ regulatory T cells. *J Immunol* 2007;**178**:280–90.
35. Marson A, Kretschmer K, Frampton GM, Jacobsen ES, Polansky JK, MacIsaac KD, et al. Foxp3 occupancy and regulation of key target genes during T-cell stimulation. *Nature* 2007;**445**:931–5.
36. Zheng Y, Josefowicz SZ, Kas A, Chu TT, Gavin MA, Rudensky AY. Genome-wide analysis of Foxp3 target genes in developing and mature regulatory T cells. *Nature* 2007;**445**:936–40.
37. Gavin MA, Rasmussen JP, Fontenot JD, Vasta V, Manganiello VC, Beavo JA, et al. Foxp3-dependent programme of regulatory T-cell differentiation. *Nature* 2007;**445**:771–5.
38. Hill JA, Feuerer M, Tash K, Haxhinasto S, Perez J, Melamed R, et al. Foxp3 transcription-factor-dependent and -independent regulation of the regulatory T cell transcriptional signature. *Immunity* 2007;**27**:786–800.
39. Lin W, Haribhai D, Relland LM, Truong N, Carlson MR, Williams CB, et al. Regulatory T cell development in the absence of functional Foxp3. *Nat Immunol* 2007;**8**:359–68.
40. Ono M, Borde M, Heissmeyer V, Feuerer M, Lapan AD, Stroud JC, et al. Foxp3 controls regulatory T-cell function by interacting with AML1/Runx1. *Nature* 2007;**446**:685–9.
41. Wu Y, Borde M, Heissmeyer V, Feuerer M, Lapan AD, Stroud JC, et al. FOXP3 controls regulatory T cell function through cooperation with NFAT. *Cell* 2006;**126**:375–87.
42. Zhou L, Lopes JE, Chong MM, Ivanov, I, Min R, Victora GD, et al. TGF-beta-induced Foxp3 inhibits T(H)17 cell differentiation by antagonizing RORgammat function. *Nature* 2008;**453**:236–40.
43. Pan F, Yu H, Dang EV, Barbi J, Pan X, Grosso JF, et al. Eos mediates Foxp3-dependent gene silencing in $CD4^+$ regulatory T cells. *Science* 2009;**325**:1142–6.
44. Chen W, Jin W, Hardegen N, Lei KJ, Li L, Marinos N, et al. Conversion of peripheral $CD4^+CD25^-$ naive T cells to $CD4^+CD25^+$ regulatory T cells by TGF-beta induction of transcription factor Foxp3. *J Exp Med* 2003;**198**:1875–86.
45. Wan YY, Flavell RA. Identifying Foxp3-expressing suppressor T cells with a bicistronic reporter. *Proc Natl Acad Sci USA* 2005;**102**:5126–31.
46. Fu S, Zhang N, Yopp AC, Chen D, Mao M, Chen D, et al. TGF-beta induces $Foxp3^+$ T-regulatory cells from $CD4^+CD25^-$ precursors. *Am J Transplant* 2004;**4**:1614–27.
47. Kretschmer K, Apostolou I, Hawiger D, Khazaie K, Nussenzweig MC, von Boehmer H, et al. Inducing and expanding regulatory T cell populations by foreign antigen. *Nat Immunol* 2005;**6**:1219–27.

48. Tone Y, Furuuchi K, Kojima Y, Tykocinski ML, Greene MI, Tone M. Smad3 and NFAT cooperate to induce Foxp3 expression through its enhancer. *Nat Immunol* 2008;**9**:194–202.
49. Apostolou I, von Boehmer H. In vivo instruction of suppressor commitment in naive T cells. *J Exp Med* 2004;**199**:1401–8.
50. Thorstenson KM, Khoruts A. Generation of anergic and potentially immunoregulatory $CD25^{+}CD4$ T cells in vivo after induction of peripheral tolerance with intravenous or oral antigen. *J Immunol* 2001;**167**:188–95.
51. Benson MJ, Pino-Lagos K, Rosemblatt M, Noelle RJ. All-trans retinoic acid mediates enhanced T reg cell growth, differentiation, and gut homing in the face of high levels of co-stimulation. *J Exp Med* 2007;**204**:1765–74.
52. Kim JM, Rudensky A. The role of the transcription factor Foxp3 in the development of regulatory T cells. *Immunol Rev* 2006;**212**:86–98.
53. Wohlfert EA, Gorelik L, Mittler R, Flavell RA, Clark RB. Cutting edge: deficiency in the E3 ubiquitin ligase Cbl-b results in a multifunctional defect in T cell TGF-beta sensitivity in vitro and in vivo. *J Immunol* 2006;**176**:1316–20.
54. Zheng SG, Wang JH, Stohl W, Kim KS, Gray JD, Horwitz DA. TGF-beta requires CTLA-4 early after T cell activation to induce FoxP3 and generate adaptive $CD4^{+}CD25^{+}$ regulatory cells. *J Immunol* 2006;**176**:3321–9.
55. Wei J, Duramad O, Perng OA, Reiner SL, Liu YJ, Qin FX. Antagonistic nature of T helper 1/2 developmental programs in opposing peripheral induction of $Foxp3^{+}$ regulatory T cells. *Proc Natl Acad Sci USA* 2007;**104**:18169–74.
56. Sad S, Mosmann TR. Single IL-2-secreting precursor CD4 T cell can develop into either Th1 or Th2 cytokine secretion phenotype. *J Immunol* 1994;**153**:3514–22.
57. Swain SL, Huston G, Tonkonogy S, Weinberg A. Transforming growth factor-beta and IL-4 cause helper T cell precursors to develop into distinct effector helper cells that differ in lymphokine secretion pattern and cell surface phenotype. *J Immunol* 1991;**147**:2991–3000.
58. Heath VL, Murphy EE, Crain C, Tomlinson MG, O'Garra A. TGF-beta1 down-regulates Th2 development and results in decreased IL-4-induced STAT6 activation and GATA-3 expression. *Eur J Immunol* 2000;**30**:2639–49.
59. Gorelik L, Constant S, Flavell RA. Mechanism of transforming growth factor beta-induced inhibition of T helper type 1 differentiation. *J Exp Med* 2002;**195**:1499–505.
60. Gorelik L, Fields PE, Flavell RA. Cutting edge: TGF-beta inhibits Th type 2 development through inhibition of GATA-3 expression. *J Immunol* 2000;**165**:4773–7.
61. Sun CM, Hall JA, Blank RB, Bouladoux N, Oukka M, Mora JR, et al. Small intestine lamina propria dendritic cells promote de novo generation of Foxp3 T reg cells via retinoic acid. *J Exp Med* 2007;**204**:1775–85.
62. Coombes JL, Siddiqui KR, Arancibia-Carcamo CV, Hall J, Sun CM, Belkaid Y, et al. A functionally specialized population of mucosal CD103+ DCs induces Foxp3+ regulatory T cells via a TGF-beta and retinoic acid-dependent mechanism. *J Exp Med* 2007;**204**:1757–64.
63. Hill JA, Hall JA, Sun CM, Cai Q, Ghyselinck N, Chambon P, et al. Retinoic acid enhances Foxp3 induction indirectly by relieving inhibition from $CD4^{+}$ CD44hi Cells. *Immunity* 2008;**29**:758–70.
64. Mucida D, Pino-Lagos K, Kim G, Nowak E, Benson MJ, Kronenberg M, et al. Retinoic acid can directly promote TGF-beta-mediated Foxp3(+) Treg cell conversion of naive T cells. *Immunity* 2009;**30**:471–2 author reply 472–473.
65. Kim HP, Leonard WJ. CREB/ATF-dependent T cell receptor-induced FoxP3 gene expression: a role for DNA methylation. *J Exp Med* 2007;**204**:1543–51.
66. Mantel PY, et al. Molecular mechanisms underlying FOXP3 induction in human T cells. *J Immunol* 2006;**176**:3593–602.

67. Ruan Q, Kameswaran V, Tone Y, Li L, Liou HC, Greene MI, et al. Development of Foxp3(+) regulatory t cells is driven by the c-Rel enhanceosome. *Immunity* 2009;**31**:932–40.
68. Koonpaew S, Shen S, Flowers L, Zhang W. LAT-mediated signaling in $CD4^{+}CD25^{+}$ regulatory T cell development. *J Exp Med* 2006;**203**:119–29.
69. Oh-Hora M, Yamashita M, Hogan PG, Sharma S, Lamperti E, Chung W, et al. Dual functions for the endoplasmic reticulum calcium sensors STIM1 and STIM2 in T cell activation and tolerance. *Nat Immunol* 2008;**9**:432–43.
70. Sakaguchi S, Sakaguchi N. Thymus and autoimmunity. Transplantation of the thymus from cyclosporin A-treated mice causes organ-specific autoimmune disease in athymic nude mice. *J Exp Med* 1988;**167**:1479–85.
71. Sakaguchi S, Sakaguchi N. Organ-specific autoimmune disease induced in mice by elimination of T cell subsets. V. Neonatal administration of cyclosporin A causes autoimmune disease. *J Immunol* 1989;**142**:471–80.
72. Muthusamy N, Leiden JM. A protein kinase C-, Ras-, and RSK2-dependent signal transduction pathway activates the cAMP-responsive element-binding protein transcription factor following T cell receptor engagement. *J Biol Chem* 1998;**273**:22841–7.
73. Zorn E, Nelson EA, Mohseni M, Porcheray F, Kim H, Litsa D, et al. IL-2 regulates FOXP3 expression in human $CD4^{+}CD25^{+}$ regulatory T cells through a STAT-dependent mechanism and induces the expansion of these cells in vivo. *Blood* 2006;**108**:1571–9.
74. Josefowicz SZ, Wilson CB, Rudensky AY. Cutting edge: TCR stimulation is sufficient for induction of Foxp3 expression in the absence of DNA methyltransferase 1. *J Immunol* 2009;**182**:6648–52.
75. Kitoh A, Ono M, Naoe Y, Ohkura N, Yamaguchi T, Yaguchi H, et al. Indispensable role of the Runx1-Cbfbeta transcription complex for in vivo-suppressive function of $FoxP3^{+}$ regulatory T cells. *Immunity* 2009;**31**:609–20.
76. Rudra D, Egawa T, Chong MM, Treuting P, Littman DR, Rudensky AY. Runx-CBFbeta complexes control expression of the transcription factor Foxp3 in regulatory T cells. *Nat Immunol* 2009;**10**:1170–7.
77. Bruno L, Mazzarella L, Hoogenkamp M, Hertweck A, Cobb BS, Sauer S, et al. Runx proteins regulate Foxp3 expression. *J Exp Med* 2009;**206**:2329–37.
78. Klunker S, Chong MM, Mantel PY, Palomares O, Bassin C, Ziegler M, et al. Transcription factors RUNX1 and RUNX3 in the induction and suppressive function of $Foxp3^{+}$ inducible regulatory T cells. *J Exp Med* 2009;**206**:2701–15.
79. Schmidt-Supprian M, Tian J, Grant EP, Pasparakis M, Maehr R, Ovaa H, et al. Differential dependence of $CD4^{+}CD25^{+}$ regulatory and natural killer-like T cells on signals leading to NF-kappaB activation. *Proc Natl Acad Sci USA* 2004;**101**:4566–71.
80. Wan YY, Chi H, Xie M, Schneider MD, Flavell RA. The kinase TAK1 integrates antigen and cytokine receptor signaling for T cell development, survival and function. *Nat Immunol* 2006;**7**:851–8.
81. Long M, Park SG, Strickland I, Hayden MS, Ghosh S. Nuclear factor-kappaB modulates regulatory T cell development by directly regulating expression of Foxp3 transcription factor. *Immunity* 2009;**31**:921–31.
82. Liu Y, Zhang P, Li J, Kulkarni AB, Perruche S, Chen W. A critical function for TGF-beta signaling in the development of natural $CD4^{+}CD25^{+}Foxp3^{+}$ regulatory T cells. *Nat Immunol* 2008;**9**:632–40.
83. D'Cruz LM, Klein L. Development and function of agonist-induced $CD25^{+}Foxp3^{+}$ regulatory T cells in the absence of interleukin 2 signaling. *Nat Immunol* 2005;**6**:1152–9.
84. Liston A, Siggs OM, Goodnow CC. Tracing the action of IL-2 in tolerance to islet-specific antigen. *Immunol Cell Biol* 2007;**85**:338–42.

85. Suri-Payer E, Amar AZ, Thornton AM, Shevach EM. CD4$^+$CD25$^+$ T cells inhibit both the induction and effector function of autoreactive T cells and represent a unique lineage of immunoregulatory cells. *J Immunol* 1998;**160**:1212–8.
86. Takahashi T, Kuniyasu Y, Toda M, Sakaguchi N, Itoh M, Iwata M, et al. Immunologic self-tolerance maintained by CD25$^+$CD4+ naturally anergic and suppressive T cells: induction of autoimmune disease by breaking their anergic/suppressive state. *Int Immunol* 1998;**10**:1969–80.
87. Thornton AM, Shevach EM. CD4$^+$CD25$^+$ immunoregulatory T cells suppress polyclonal T cell activation in vitro by inhibiting interleukin 2 production. *J Exp Med* 1998;**188**:287–96.
88. Shevach EM. Mechanisms of Foxp3$^+$ T regulatory cell-mediated suppression. *Immunity* 2009;**30**:636–45.
89. Tadokoro CE, Shakhar G, Shen S, Ding Y, Lino AC, Maraver A, et al. Regulatory T cells inhibit stable contacts between CD4$^+$ T cells and dendritic cells in vivo. *J Exp Med* 2006;**203**:505–11.
90. Tang Q, Adams JY, Tooley AJ, Bi M, Fife BT, Serra P, et al. Visualizing regulatory T cell control of autoimmune responses in nonobese diabetic mice. *Nat Immunol* 2006;**7**:83–92.
91. Cederbom L, Hall H, Ivars F. CD4$^+$CD25$^+$ regulatory T cells down-regulate co-stimulatory molecules on antigen-presenting cells. *Eur J Immunol* 2000;**30**:1538–43.
92. Houot R, Perrot I, Garcia E, Durand I, Lebecque S. Human CD4$^+$CD25high regulatory T cells modulate myeloid but not plasmacytoid dendritic cells activation. *J Immunol* 2006;**176**:5293–8.
93. Misra N, Bayry J, Lacroix-Desmazes S, Kazatchkine MD, Kaveri SV. Cutting edge: human CD4$^+$CD25$^+$ T cells restrain the maturation and antigen-presenting function of dendritic cells. *J Immunol* 2004;**172**:4676–80.
94. Veldhoen M, Moncrieffe H, Hocking RJ, Atkins CJ, Stockinger B. Modulation of dendritic cell function by naive and regulatory CD4+ T cells. *J Immunol* 2006;**176**:6202–10.
95. Kim JM, Rasmussen JP, Rudensky AY. Regulatory T cells prevent catastrophic autoimmunity throughout the lifespan of mice. *Nat Immunol* 2007;**8**:191–7.
96. Liu K, Victora GD, Schwickert TA, Guermonprez P, Meredith MM, Yao K, et al. In vivo analysis of dendritic cell development and homeostasis. *Science* 2009;**324**:392–7.
97. Piccirillo CA, Shevach EM. Cutting edge: control of CD8$^+$ T cell activation by CD4$^+$CD25$^+$ immunoregulatory cells. *J Immunol* 2001;**167**:1137–40.
98. Shimizu J, Yamazaki S, Sakaguchi S. Induction of tumor immunity by removing CD25$^+$CD4+ T cells: a common basis between tumor immunity and autoimmunity. *J Immunol* 1999;**163**:5211–8.
99. Onizuka S, Tawara I, Shimizu J, Sakaguchi S, Fujita T, Nakayama E. Tumor rejection by in vivo administration of anti-CD25 (interleukin-2 receptor alpha) monoclonal antibody. *Cancer Res* 1999;**59**:3128–33.
100. Tanaka H, Tanaka J, Kjaergaard J, Shu S. Depletion of CD4$^+$CD25$^+$ regulatory cells augments the generation of specific immune T cells in tumor-draining lymph nodes. *J Immunother* 2002;**25**:207–17.
101. Mempel TR, Pittet MJ, Khazaie K, Weninger W, Weissleder R, von Boehmer H, et al. Regulatory T cells reversibly suppress cytotoxic T cell function independent of effector differentiation. *Immunity* 2006;**25**:129–41.
102. Chen ML, Pittet MJ, Gorelik L, Flavell RA, Weissleder R, von Boehmer H, et al. Regulatory T cells suppress tumor-specific CD8 T cell cytotoxicity through TGF-beta signals in vivo. *Proc Natl Acad Sci USA* 2005;**102**:419–24.
103. Pasche B. Role of transforming growth factor beta in cancer. *J Cell Physiol* 2001;**186**:153–68.
104. Suvas S, Kumaraguru U, Pack CD, Lee S, Rouse BT. CD4$^+$CD25$^+$ T cells regulate virus-specific primary and memory CD8$^+$ T cell responses. *J Exp Med* 2003;**198**:889–901.

105. Kursar M, Bonhagen K, Fensterle J, Kohler A, Hurwitz R, Kamradt T, et al. Regulatory $CD4^+CD25^+$ T cells restrict memory $CD8^+$ T cell responses. *J Exp Med* 2002;**196**:1585–92.
106. Murakami M, Sakamoto A, Bender J, Kappler J, Marrack P. $CD25^+CD4^+$ T cells contribute to the control of memory $CD8^+$ T cells. *Proc Natl Acad Sci USA* 2002;**99**:8832–7.
107. Wolf AM, Wolf D, Steurer M, Gastl G, Gunsilius E, Grubeck-Loebenstein B. Increase of regulatory T cells in the peripheral blood of cancer patients. *Clin Cancer Res* 2003;**9**:606–12.
108. Ghiringhelli F, Menard C, Terme M, Flament C, Taieb J, Chaput N, et al. $CD4^+CD25^+$ regulatory T cells inhibit natural killer cell functions in a transforming growth factor-beta-dependent manner. *J Exp Med* 2005;**202**:1075–85.
109. Trzonkowski P, Szmit E, Mysliwska J, Mysliwski A. $CD4^+CD25^+$ T regulatory cells inhibit cytotoxic activity of CTL and NK cells in humans-impact of immunosenescence. *Clin Immunol* 2006;**119**:307–16.
110. Smyth MJ, Teng MW, Swann J, Kyparissoudis K, Godfrey DI, Hayakawa Y. $CD4^+CD25^+$ T regulatory cells suppress NK cell-mediated immunotherapy of cancer. *J Immunol* 2006;**176**:1582–7.
111. Clark LB, Appleby MW, Brunkow ME, Wilkinson JE, Ziegler SF, Ramsdell F. Cellular and molecular characterization of the scurfy mouse mutant. *J Immunol* 1999;**162**:2546–54.
112. Lim HW, Hillsamer P, Banham AH, Kim CH. Cutting edge: direct suppression of B cells by $CD4^+CD25^+$ regulatory T cells. *J Immunol* 2005;**175**:4180–3.
113. Bystry RS, Aluvihare V, Welch KA, Kallikourdis M, Betz AG. B cells and professional APCs recruit regulatory T cells via CCL4. *Nat Immunol* 2001;**2**:1126–32.
114. Zhao DM, Thornton AM, DiPaolo RJ, Shevach EM. Activated $CD4^+CD25^+$ T cells selectively kill B lymphocytes. *Blood* 2006;**107**:3925–32.
115. Gorelik L, Flavell RA. Abrogation of TGFbeta signaling in T cells leads to spontaneous T cell differentiation and autoimmune disease. *Immunity* 2000;**12**:171–81.
116. Li MO, Wan YY, Flavell RA. T cell-produced transforming growth factor-beta1 controls T cell tolerance and regulates Th1- and Th17-cell differentiation. *Immunity* 2007;**26**:579–91.
117. Nakamura K, Kitani A, Strober W. Cell contact-dependent immunosuppression by CD4(+) CD25(+) regulatory T cells is mediated by cell surface-bound transforming growth factor beta. *J Exp Med* 2001;**194**:629–44.
118. Piccirillo CA, Letterio JJ, Thornton AM, McHugh RS, Mamura M, Mizuhara H, et al. CD4(+) CD25(+) regulatory T cells can mediate suppressor function in the absence of transforming growth factor beta1 production and responsiveness. *J Exp Med* 2002;**196**:237–46.
119. Kuhn R, Lohler J, Rennick D, Rajewsky K, Muller W. Interleukin-10-deficient mice develop chronic enterocolitis. *Cell* 1993;**75**:263–74.
120. Roers A, Siewe L, Strittmatter E, Deckert M, Schluter D, Stenzel W, et al. T cell-specific inactivation of the interleukin 10 gene in mice results in enhanced T cell responses but normal innate responses to lipopolysaccharide or skin irritation. *J Exp Med* 2004;**200**:1289–97.
121. Rubtsov YP, Rasmussen JP, Chi EY, Fontenot J, Castelli L, Ye X, et al. Regulatory T cell-derived interleukin-10 limits inflammation at environmental interfaces. *Immunity* 2008;**28**:546–58.
122. Asseman C, Mauze S, Leach MW, Coffman RL, Powrie F. An essential role for interleukin 10 in the function of regulatory T cells that inhibit intestinal inflammation. *J Exp Med* 1999;**190**:995–1004.
123. Hagenbaugh A, Sharma S, Dubinett SM, Wei SH, Aranda R, Cheroutre H, et al. Altered immune responses in interleukin 10 transgenic mice. *J Exp Med* 1997;**185**:2101–10.
124. Kearley J, Barker JE, Robinson DS, Lloyd CM. Resolution of airway inflammation and hyperreactivity after in vivo transfer of $CD4^+CD25^+$ regulatory T cells is interleukin 10 dependent. *J Exp Med* 2005;**202**:1539–47.
125. Kryczek I, Wei S, Zou L, Zhu G, Mottram P, Xu H, et al. Cutting edge: induction of B7-H4 on APCs through IL-10: novel suppressive mode for regulatory T cells. *J Immunol* 2006;**177**:40–4.

126. Bachmann MF, Kohler G, Ecabert B, Mak TW, Kopf M. Cutting edge: lymphoproliferative disease in the absence of CTLA-4 is not T cell autonomous. *J Immunol* 1999;**163**:1128–31.
127. Read S, Malmstrom V, Powrie F. Cytotoxic T lymphocyte-associated antigen 4 plays an essential role in the function of CD25(+)CD4(+) regulatory cells that control intestinal inflammation. *J Exp Med* 2000;**192**:295–302.
128. Takahashi T, Tagami T, Yamazaki S, Uede T, Shimizu J, Sakaguchi N, et al. Immunologic self-tolerance maintained by CD25(+)CD4(+) regulatory T cells constitutively expressing cytotoxic T lymphocyte-associated antigen 4. *J Exp Med* 2000;**192**:303–10.
129. Linsley PS, Brady W, Urnes M, Grosmaire LS, Damle NK, Ledbetter JA. CTLA-4 is a second receptor for the B cell activation antigen B7. *J Exp Med* 1991;**174**:561–9.
130. Peach RJ, Bajorath J, Brady W, Leytze G, Greene J, Naemura J, et al. Complementarity determining region 1 (CDR1)- and CDR3-analogous regions in CTLA-4 and CD28 determine the binding to B7-1. *J Exp Med* 1994;**180**:2049–58.
131. Walunas TL, Bakker CY, Bluestone JA. CTLA-4 ligation blocks CD28-dependent T cell activation. *J Exp Med* 1996;**183**:2541–50.
132. Tivol EA, Boyd SD, McKeon S, Borriello F, Nickerson P, Strom TB, et al. CTLA4Ig prevents lymphoproliferation and fatal multiorgan tissue destruction in CTLA-4-deficient mice. *J Immunol* 1997;**158**:5091–4.
133. Thompson CB, Allison JP. The emerging role of CTLA-4 as an immune attenuator. *Immunity* 1997;**7**:445–50.
134. Fallarino F, Grohmann U, Hwang KW, Orabona C, Vacca C, Bianchi R, et al. Modulation of tryptophan catabolism by regulatory T cells. *Nat Immunol* 2003;**4**:1206–12.
135. Grohmann U, Orabona C, Fallarino F, Vacca C, Calcinaro F, Falorni A, et al. CTLA-4-Ig regulates tryptophan catabolism in vivo. *Nat Immunol* 2002;**3**:1097–101.
136. Hwu P, Du MX, Lapointe R, Do M, Taylor MW, Young HA. Indoleamine 2, 3-dioxygenase production by human dendritic cells results in the inhibition of T cell proliferation. *J Immunol* 2000;**164**:3596–9.
137. Munn DH, Shafizadeh E, Attwood JT, Bondarev I, Pashine A, Mellor AL. Inhibition of T cell proliferation by macrophage tryptophan catabolism. *J Exp Med* 1999;**189**:1363–72.
138. Orabona C, Grohmann U, Belladonna ML, Fallarino F, Vacca C, Bianchi R. CD28 induces immunostimulatory signals in dendritic cells via CD80 and CD86. *Nat Immunol* 2004;**5**:1134–42.
139. Pasare C, Medzhitov R. Toll pathway-dependent blockade of $CD4^{+}CD25^{+}$ T cell-mediated suppression by dendritic cells. *Science* 2003;**299**:1033–6.
140. Onishi Y, Fehervari Z, Yamaguchi T, Sakaguchi S. $Foxp3^{+}$ natural regulatory T cells preferentially form aggregates on dendritic cells in vitro and actively inhibit their maturation. *Proc Natl Acad Sci USA* 2008;**105**:10113–8.
141. Wing K, Onishi Y, Prieto-Martin P, Yamaguchi T, Miyara M, Fehervari Z, et al. CTLA-4 control over $Foxp3^{+}$ regulatory T cell function. *Science* 2008;**322**:271–5.
142. Azuma M, Yssel H, Phillips JH, Spits H, Lanier LL. Functional expression of B7/BB1 on activated T lymphocytes. *J Exp Med* 1993;**177**:845–50.
143. Sansom DM, Hall ND. B7/BB1, the ligand for CD28, is expressed on repeatedly activated human T cells in vitro. *Eur J Immunol* 1993;**23**:295–8.
144. Taylor PA, Lees CJ, Fournier S, Allison JP, Sharpe AH, Blazar BR. B7 expression on T cells down-regulates immune responses through CTLA-4 ligation via T-T interactions [corrections]. *J Immunol* 2004;**172**:34–9.
145. Paust S, Lu L, McCarty N, Cantor H. Engagement of B7 on effector T cells by regulatory T cells prevents autoimmune disease. *Proc Natl Acad Sci USA* 2004;**101**:10398–403.
146. Marazzi S, Blum S, Hartmann R, Gundersen D, Schreyer M, Argraves S. Characterization of human fibroleukin, a fibrinogen-like protein secreted by T lymphocytes. *J Immunol* 1998;**161**:138–47.

147. Chen Z, Herman AE, Matos M, Mathis D, Benoist C. Where $CD4^{+}CD25^{+}$ T reg cells impinge on autoimmune diabetes. *J Exp Med* 2005;**202**:1387–97.
148. Chan CW, et al. Soluble fibrinogen-like protein 2/fibroleukin exhibits immunosuppressive properties: suppressing T cell proliferation and inhibiting maturation of bone marrow-derived dendritic cells. *J Immunol* 2003;**170**:4036–44.
149. Shalev I, Liu H, Koscik C, Bartczak A, Javadi M, Wong KM, et al. Targeted deletion of fgl2 leads to impaired regulatory T cell activity and development of autoimmune glomerulonephritis. *J Immunol* 2008;**180**:249–60.
150. Gondek DC, Lu LF, Quezada SA, Sakaguchi S, Noelle RJ. Cutting edge: contact-mediated suppression by $CD4^{+}CD25^{+}$ regulatory cells involves a granzyme B-dependent, perforin-independent mechanism. *J Immunol* 2005;**174**:1783–6.
151. Grossman WJ, Verbsky JW, Barchet W, Colonna M, Atkinson JP, Ley TJ. Human T regulatory cells can use the perforin pathway to cause autologous target cell death. *Immunity* 2004;**21**:589–601.
152. Browne KA, Blink E, Sutton VR, Froelich CJ, Jans DA, Trapani JA. Cytosolic delivery of granzyme B by bacterial toxins: evidence that endosomal disruption, in addition to transmembrane pore formation, is an important function of perforin. *Mol Cell Biol* 1999;**19**:8604–15.
153. Deaglio S, Dwyer KM, Gao W, Friedman D, Usheva A, Erat A, et al. Adenosine generation catalyzed by CD39 and CD73 expressed on regulatory T cells mediates immune suppression. *J Exp Med* 2007;**204**:1257–65.
154. Kobie JJ, Shah PR, Yang L, Rebhahn JA, Fowell DJ, Mosmann TR. T regulatory and primed uncommitted CD4 T cells express CD73, which suppresses effector CD4 T cells by converting 5′-adenosine monophosphate to adenosine. *J Immunol* 2006;**177**:6780–6.
155. Kambayashi T, Wallin RP, Ljunggren HG. cAMP-elevating agents suppress dendritic cell function. *J Leukoc Biol* 2001;**70**:903–10.
156. Panther E, Idzko M, Herouy Y, Rheinen H, Gebicke-Haerter PJ, Mrowietz U, et al. Expression and function of adenosine receptors in human dendritic cells. *FASEB J* 2001;**15**:1963–70.

Is Foxp3 the Master Regulator of Regulatory T Cells?

ADRIAN LISTON

VIB and University of Leuven, Leuven, Belgium

Comment on

Thymic Selection and Lineage Commitment of $CD4^+Foxp3^+$ Regulatory T Cells.

Paola Romagnoli and Joost P. M. van Meerwijk

and

Molecular Mechanisms of Regulatory T cell Development and Suppressive Function.

Jeong M. Kim

In the two preceding review chapters—on lineage commitment of $CD4^+Foxp3^+$ T cells by Romagnoli and Meerwijk and molecular mechanisms of regulatory T cell development by Kim—both groups have concluded that Foxp3 is important for regulatory T cell differentiation, but that it does not constitute a bona fide master switch. The rationale for this hypothesis is based on several recent publications that analysed Foxp3 functional null ($Foxp3^{FN}$) T cells—that is T cells that received the signal to upregulate Foxp3 *in vivo* under physiological conditions, but were unable to do so due to a genetic deletion.[1–3] In each of these publications, the $Foxp3^{FN}$ T cells were tracked by markers under the control of the Foxp3 promoter, allowing these cells to be identified and functionally analysed. Surprisingly, in each case the $Foxp3^{FN}$ T cells maintained some characteristics of $Foxp3^+$ regulatory T cells, despite their inability to express Foxp3. $Foxp3^{FN}$ T cells are anergic *in vitro*, are dependent on exogenous IL-2, and express a substantial proportion of the transcriptional profile of $Foxp3^+$ regulatory T cells.[1–3] On this basis, it is convincingly argued that Foxp3 is not the master switch of the $Foxp3^+$ lineage, as Foxp3 is not sufficient to induce the full transcriptome of $Foxp3^+$ regulatory T cells. In this scenario, another unknown factor would act upstream of Foxp3 to initiate lineage entry, with a key event being the induction of Foxp3.

Progress in Molecular Biology
and Translational Science, Vol. 92
DOI: 10.1016/S1877-1173(10)92017-6

1877-1173/10 $35.00

The basis of this model is the assumption that a transcriptional profile is the best measure of cell lineage. However, it could be argued that suppressive capacity is the sine qua non of regulatory T cells, rather than the transcriptional profile that accompanies this suppressive capacity. In this regard, the function of Foxp3 remains unchallenged—ectopic expression of Foxp3 in nonregulatory T cells imparts a suppressive capacity,[4,5] and excision of Foxp3 from Foxp3$^+$ regulatory T cells results in loss of suppressive capacity, allowing effector function to develop.[6] However, Foxp3^{FN} T cells do not have any suppressive function, despite partial expression of the transcriptional profile.[1–3] How then can these two scenarios be reconciled? Foxp3 is clearly necessary and sufficient for Foxp3$^+$ regulatory T cells to gain suppressive function. Yet, Foxp3 is not sufficient to induce the full transcriptional profile observed in Foxp3$^+$ cells.

An alternative model to integrate these data points is one in which Foxp3 is indeed the master regulator of the regulatory T cell lineage, but the conditions under which Foxp3 is induced also leave their imprint on the transcriptional profile (Fig. 1). In this scenario, it is important to also consider TCR stimulation. TCR activation is not monolithic, despite using a common antigen receptor. Thus, while TCR activation will impact common transcriptional programs, the transcriptional imprint of this activation will vary depending on the context, which could be either chronic tolerogenic stimulation or acute immunogenic stimulation.[7] Lineage-defining transcription factors, such as Foxp3, T-bet,

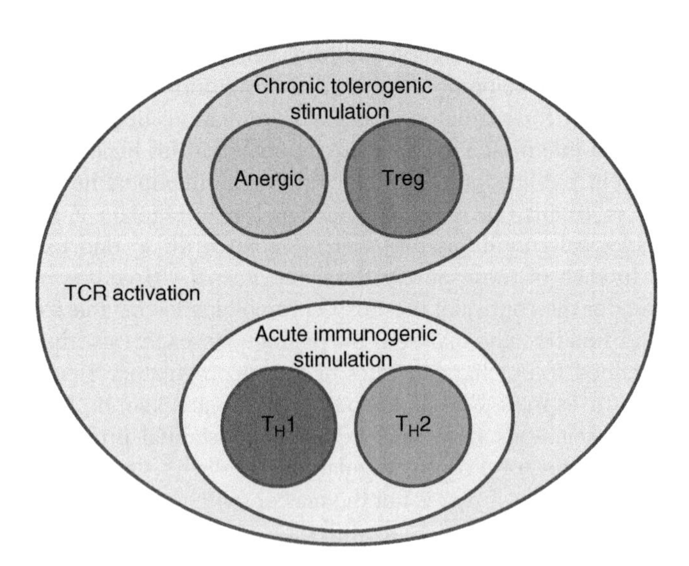

Fig. 1. Under a "nested signature" model, distinct T cell lineages such as anergic T cells, regulatory T cells, TH1 cells and TH2 cells contain a transcriptional imprint not only from the master switch (Foxp3, T-bet, etc) but also from the signalling context.

GATA-3 or RORγt, would then drive a lineage-specific transcriptional profile nested within the contextual TCR activation profile. Thus TH1 and TH2 cells will share a transcriptional imprint that is distinct from regulatory T cells, based purely upon the acute immunogenic nature of TCR activation. Here, the Foxp3-dependent transcriptional profile is the only component necessary for the regulatory T cell lineage, with the Foxp3-independent transcriptional profile simply reflecting the TCR activation history of the cell. There are several advantages to such a "nested signature" model: (1) it is consistent with Foxp3 being necessary and sufficient for suppressive activity; (2) it is consistent with $Foxp3^{FN}$ T cells having a portion of the $Foxp3^{+}$ T cell transcriptional signature, based on the common context of TCR activation (indeed, it would predict $GATA3^{FN}$ T cells to have a portion of the T_H2 transcriptional signature, etc.); and (3) it does not invoke an unknown transcription factor that controls the facets of the $Foxp3^{+}$ T cell transcriptional profile.

REFERENCES

1. Rasmussen JP, Gavin MA, Vasta V, Fontenot JD, Beavo JA, Manganiello VC, et al. Foxp3-dependent programme of regulatory T-cell differentiation. *Nature* 2007;**445**:771–5.
2. Hill JA, Feuerer M, Tash K, Haxhinasto S, Perez J, Melamed R, et al. Foxp3 transcription-factor-dependent and -independent regulation of the regulatory T-cell transcriptional signature. *Immunity* 2007;**27**:786–800.
3. Lin W, Haribhai D, Relland LM, Truong N, Carlson MR, Williams CB, et al. Regulatory T cell development in the absence of functional Foxp3. *Nat Immunol* 2007;**8**:359–68.
4. Hori S, Nomura T, Sakaguchi S. Control of regulatory T cell development by the transcription factor Foxp3. *Science* 2003;**299**:1057–61.
5. Fontenot JD, Gavin MA, Rudensky AY. Foxp3 programs the development and function of $CD4^{+}CD25^{+}$ regulatory T cells. *Nat Immunol* 2003;**4**:330–6.
6. Williams LM, Rudensky AY. Maintenance of the Foxp3-dependent developmental program in mature regulatory T cells requires continued expression of Foxp3. *Nat Immunol* 2007;**8**:277–84.
7. Jun JE, Goodnow CC. Scaffolding of antigen receptors for immunogenic versus tolerogenic signaling. *Nat Immunol* 2003;**4**:1057–64.

Index

D

E

F

G

H

I

J

L

M

N

O

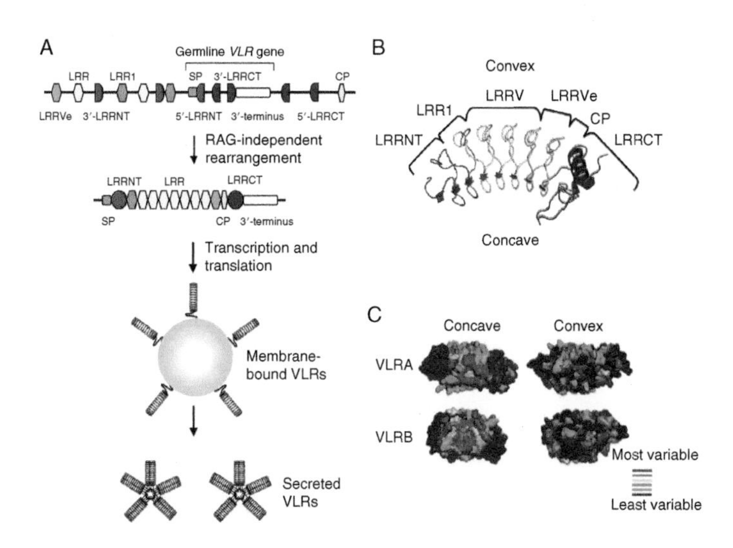

Masanori Kasahara, Fig. 5. Organization of the *VLR* gene and the structure of VLR proteins. (A) The germ line *VLR* gene has a structure incapable of encoding proteins. Modules coding for N-terminal caps (LRRNT), leucine-rich repeats (LRR), connecting peptides (CP) and C-terminal caps (LRRCT) occur in multiple copies adjacent to the germ line *VLR* gene. During the development of lymphoid cells, these modules are incorporated into the *VLR* gene. The rearranged *VLR* gene encodes a membrane-bound protein. The product of the *VLRB*, but not *VLRA*, gene is secreted and functions as antibodies. Whether membrane-bound VLRs occur as a monomer or multimer is not known. LRR1 and LRRVe denote LRR modules located at the N- and C-termini, respectively. The organization of the *VLR* locus shows considerable variation depending on loci and species. This figure, modified from Flajnik and Kasahara,[7] is intended to show salient features of *VLR* genes and does not accurately reproduce the organization of a specific *VLR* locus. (B) Crystal structure of hagfish VLRB molecules (PDB ID: 2o6S). The figure was generated using the PyMOL graphics tool (http://pymol.sourceforge.net/). (C) Sequence variability of hagfish VLR proteins. Variability is indicated by the color gradation from red to blue, where the most variable and the least variable patches are indicated in red and blue, respectively. Concave view (left); convex view (right). This figure was reproduced from Kim *et al.*[133]

A

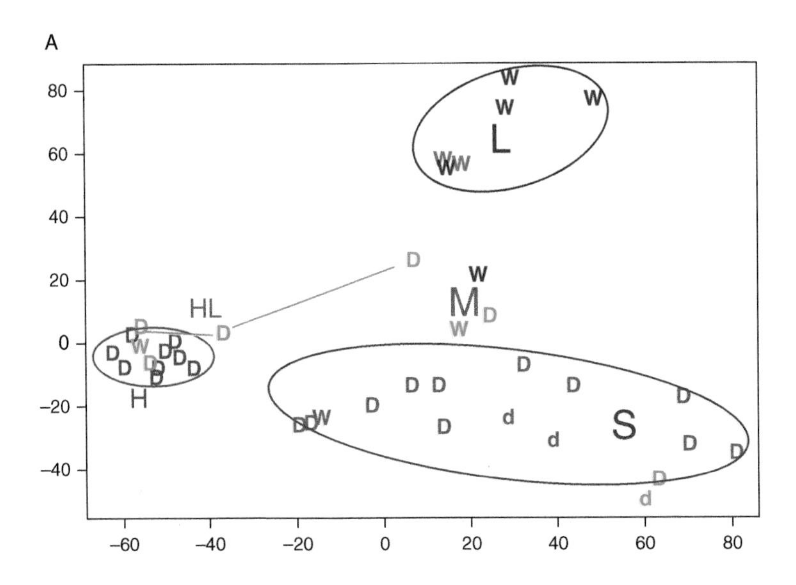

B

	Top 5 Kegg pathways		
	No. of genes	*p*-value	FWER. adjusted
Hematopoietic cell lineage	119	3.394e−06	0.00051
Cytokine–cytokine receptor interaction	243	4.510e−06	0.00068
Focal adhesion	234	2.710e−05	0.00409
T cell receptor signaling pathway	91	3.093e−05	0.00463
Phosphatidylinositol signaling system	77	6.003e−05	0.00894

C

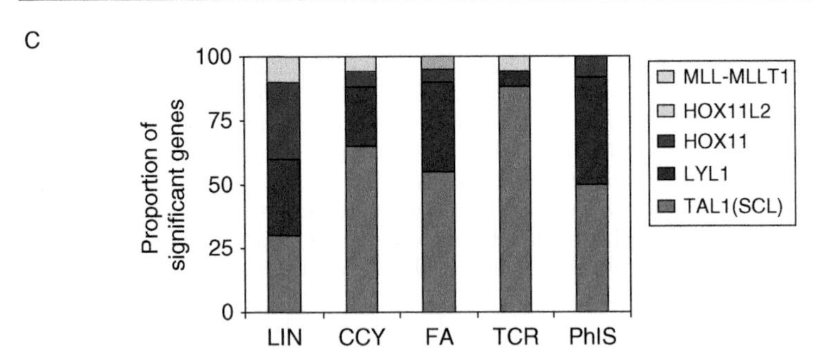

TREMBLAY *ET AL.*, FIG. 5. Analysis of gene expression signatures identifies molecular pathways correlating with patient survival. This analysis is based on microarray results published by Ferrando *et al.*[16] (A) Five prognostic groups in T-ALL based on global gene expression and patient survival. We performed a survival analysis of pediatric T-ALL based on gene expression data available from Ferrando *et al.*[16] using the Survival Forest (SF) method. This method predicted for each patient a survival curve that in turn can serve to define prognostic groups. Multidimensional scaling provided the graphical representation of these individual survival curves in a geometric space. In this two-dimensional survival dot plot, each point corresponds to a patient for whom a survival curve was predicted using SF. High similarities between survival curves correspond to small distances and clustered points on the plot and conversely. The plot shows three major prognostic groups T, H, and L (for *TAL1+* and the like, *HOX11+* and the like, and *LYL1+* and the like, respectively) at the edge of a virtual triangle and two minor groups, HL and M (for *HOX11L1+* and *MLL-ENL+*). Among the 10 samples (illustrated in green) that could not be classified previously,[16] two were found here to belong to the T prognostic group, and two to the H prognostic group, two to the M group, and three to the HL group according to survival times predicted from gene expression profiles. Furthermore, one *SCL+* case was previously clustered in the H group[16] and was found to carry the 10q24 translocation commonly associated with *HOX11* rearrangement. While the *LYL1* group was previously classified within the SCL bHLH family,[17,25] this graphical representation clearly demarcates the *LYL1+* group from the others. The status of the *CDKN2A* gene is shown for each patient sample: D, homozygous deletion; d, hemizygote deletion; W, wild type. Note that *LYL1* turns out to be the only group in T-ALL in which the *CDKN2A* locus is wild type in all samples. (B) Top five Kegg pathways derived from Goeman's global test that best discriminate the five prognostic groups. Global test gives one p-value for a pathway and adjusted p-values in multiple testing of many pathways, as well as z-scores for measuring the impact of individual genes on the test results. (C) The proportions of influent genes are represented by Kegg pathways and prognostic groups. LIN, hematopoietic cell lineage; CCY, cytokine–cytokine receptor interaction; FA, focal adhesion; TCR, T cell receptor signaling pathway; PhIS, phosphatidylinositol signaling pathway.

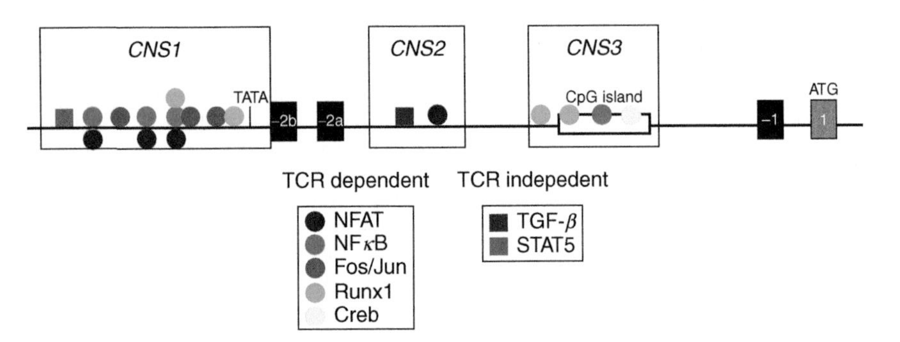

JEONG M. KIM, FIG. 1. Schematic map of *Foxp3* CNS 1–3 and its associated transcription factors. The relative location the TCR dependent and cytokine dependent transcription factors are represented as circles and squares, respectively. The CpG methylation island located in CNS3 is designated as a white rectangle.

Printed and bound by CPI Group (UK) Ltd, Croydon, CR0 4YY
08/05/2025
01864953-0002